THE
SILENT
BOMB

THE
SILENT
BOMB

A Guide to the Nuclear Energy Controversy

Edited by Peter Faulkner

Random House · New York

Friends of the Earth International
San Francisco · New York · London · Paris

Library of Congress Cataloging in Publication Data
Main entry under title:
The Silent bomb.
Bibliography: p.   Includes index.
1. Atomic power-plants.   2. Atomic energy industries—United States.
I. Faulkner, Peter T., 1933–
TK1078.S55 1977   621.48′35   76–53687
ISBN 0–394–41323–7

Manufactured in the United States of America
2 4 6 8 9 7 5 3
First Edition

Cover photo courtesy:
U.S. Energy Research and Development Administation.

*Grateful acknowledgment is made to the following for permission to reprint previously pub-
lished material:*

Ballinger Publishing Company: Excerpts from pages xviii–xix and 13 of *Non-Nuclear Futures:
The Case For An Ethical Energy Strategy* by Amory B. Lovins and John Price. Copyright
© 1975 by Friends of the Earth, Inc.

The Boxwood Press, Pacific Grove, CA.: Excerpt from page 20 of *Basic Ecology* by Ralph
and Mildred Buchsbaum. (1957)

Dr. L. Douglas DeNike: Selected portions of Dr. DeNike's testimony before the California
Assembly Committee on Resources, Land Use and Energy, November 19, 1975.

Environmental Action Foundation and the Honorable Mike Gravel: Excerpts from "Insur-
ance and Nuclear Power Risks by Senator Mike Gravel. Reprinted from *Countdown to a
Nuclear Moratorium,* edited by Richard Munson. Copyright © 1976 by the Environmental
Action Foundation.

Friends of the Earth, Inc.: Introduction; excerpts from a letter by Amory Lovins to David
Brower and excerpts from "The Incident at Brown's Ferry: Alabama's Nightmare in Candle-
light" by David Comey; a table and excerpts from an article by John Berger which appeared
in the mid-September issue of *Not Man Apart;* and additional excerpts from the March 1976
and the November 1976 issues of *Not Man Apart.*

Harper's Magazine Co., Inc.: "The Scarcity of Society" by William Ophuls. Reprinted from
Harper's Magazine, April 1974 issue. Copyright © 1974 by Harper's Magazine.

Scott Meredith Literary Agency on behalf of Richard Curtis and John Schaffner, Literary
Agent on behalf of Elizabeth Hogan: A condensation of the section entitled "What You
Don't Know Will Kill You," pages 40–54 from *Perils of the Peaceful Atom* by Elizabeth
Hogan and Richard Curtis. Copyright © 1969 by Richard Curtis and Elizabeth Hogan.

Michigan Law Review: Excerpt from "Nuclear Power: Risk, Liability and Indemnity" by
Harold Green. Reprinted from pages 509–510 of the *Michigan Law Review,* Volume 71,
January 1973.

The MIT Press, Cambridge, Mass.: Summarization of material contained in *The Nuclear Fuel
Cycle* by Thomas Hollocher.

Terry R. Lash and Richard Cotton (Natural Resources Defense Council, Inc.): Material
paraphrased and quoted directly from an unpublished paper entitled "Radioactive Wastes"
by Terry R. Lash and Richard Cotton, and also prepared testimony by Terry R. Lash before
the Assembly Committee on Energy and Diminishing Material, California Legislature.

To John, Richard and Thomas,
who will build a better earth from
the one we leave them

Editor's Note

Friends of the Earth in the United States, and sister organizations of the same name in other countries, are working for the preservation, restoration, and more rational use of the earth. To achieve these objectives, we urge people to make more intensive use of the branches of government that society has set up for itself. Within the limits of support given us, we try to represent the public's interest in the environment before administrative and legislative bodies and in court. We add to the useful diversity of the conservation front in its vital effort to build greater respect for the earth and for its living resources, including man.

We lobby for this idea. We work closely with our sister organizations abroad and with other environmental groups everywhere to strengthen ecological conscience. From this commitment, we seek to advance the sensitivity of all citizens to the fragile and fascinating systems that sustain our lives and the creatures with whom we share the earth.

We publish books and an environmental newspaper, "Not Man Apart," because of our belief in the power of words and also to help support ourselves.

If the public press is the fourth estate, perhaps we are the fifth. We speak out for you; we invite your support.

Foreword

Suppose that nuclear plants had been operating during Christ's lifetime. Assume that their operators had stored the radioactive wastes in giant shielded canisters, as the Energy Research and Development Administration has proposed to do with the wastes produced by modern plants. We would have been guarding these wastes for less than one percent of the time that they would have to be isolated from the environment.

Nuclear power represents the greatest single threat to the health and safely of humanity. These dangers are immediate, awesome, and unprecedented: risk of catastrophic accident, release of poisonous wastes, sabotage, the building of atomic bombs by terrorists—these are only a few of the ominous possibilities. Any one of these can have devastating short-run consequences for the thousands of people who will die slowly, and for the tens of thousands who will suffer.

The long-run effect promises to be worse than anything our species has ever known. In fact, it may have virtually no end, for the damage could persist for fifty times the length of recorded history.

Can we afford to take that chance? Does anyone have the *right* to do so? Why has debate on these issues been postponed until now? These questions are addressed in *The Silent Bomb*. How you, the reader, answer them may go a long way toward assuring the persistence of society as we know it.

A substantial portion of the scientific community opposes further construction of nuclear power plants. There are three main reasons for their concern. The first is that after decades of futile attempts, no satisfactory method has been found for the permanent, fail-safe storage of the extremely dangerous wastes produced during the nuclear fuel cycle. By building temporary storage facilities throughout the world, humanity grabs an immortal tiger by the tail on the assumption that someone will eventually discover a safe way to let go.

The second basis for concern arises from the massive devastation which could result from a reactor meltdown, followed by containment rupture* and the release of radioactive materials into the environment. Every engineering and scientific study of the consequences of this type of nuclear accident acknowledges that property damage could be in the *billions,* contamination of an enormous region could result, and illness and premature death could afflict tens of thousands of people over numerous generations.

To soften the impact of these horrible statistics, industry and government assure us that the probabilities of such an accident occurring are infinitesimal. They offer fundamentally dishonest studies like the recently released Rasmussen report (which, among other flaws and deceptions, ignores altogether the possibility of sabotage, and which is based on an extremely questionable analytic technique). The nuclear establishment has used these studies at least three times to hoodwink a dull and complacent Congress into supplying nuclear plant insurance that no sane insurance company will write. In fact, nuclear power is so dangerous that no utility would go ahead with it until Congress, through the Price-Anderson Act, limited the amount that injured American citizens could recover in damages following a major accident. This limit of $560 million covers less than 4 percent of the estimated $17 billion property damage that could follow a nuclear accident.

The third reason for concern turns on predictions that, sooner or later, terrorist groups will construct atomic bombs clandestinely for their own purposes, or criminals may divert nuclear materials for

---

* Fissioning nuclear fuel is enclosed in a steel pressure vessel surrounded by a bell-shaped containment. If the flow of coolant to the hot fuel stops, the fuel and the vessel may melt. The resulting pressure may rupture the surrounding containment shell and poisonous radioactive material will be released into the air.

radioactive blackmail. Associated with this problem is the prospect of living in a police state designed to minimize the annual number of cities destroyed by terrorists' A-bombs. Furthermore, by selling nuclear technology to all comers, we assure that dozens of countries will possess enough nuclear material to manufacture bombs by the end of the century. At least 20 percent of these countries will actually develop weapons from the materials *we* have supplied. If the ultimate threat to public health is nuclear war, we may be writing a death warrant for society by boosting nuclear reactor sales worldwide.

In spite of these imminent dangers, the nuclear industry argues that we must have hundreds of nuclear plants throughout the country to maintain a high living standard. This is nonsense, for how can the quality of our lives be high when we are facing the dangers described above? Recently, Dr. John W. Gofman, emeritus professor of medical physics at the University of California, pointed out that patriotism is no longer the scoundrel's last refuge. Instead, he spotlighted today's industrial and government scoundrels who threaten people by telling them that if they don't go along with what the nuclear establishment advises, they will lose their jobs, their lights will go out, their food supply will dwindle, and all will return to the Stone Age.

Instead of falling for this obvious trap, you might examine Schipper and Lichtenberg's report[1]* that the Swedes manage to achieve a standard of living as good as that enjoyed by Americans while using only 65 percent as much energy. Of course, you won't see this on the front pages of our newspapers or in electric utility company advertisements. The reason is obvious: the utility companies won't make as much money if the American people take energy conservation seriously. Budget policies that deliberately slight conservation and that, instead, promote unlimited energy consumption in both public and private sectors reveal the insulting and dehumanizing assumption, cherished by far too many industry and government leaders, that society exists merely to serve the economy, rather than the other way around.

Somehow, people in Sweden and several other countries have managed to do what we've never tried on a national scale: to live modestly and simply, to penalize waste instead of promote it, and to develop a steady-state economy that consumes only as much

* Notes begin on page 342.

energy as is needed for a comfortable life. If we apply some American ingenuity, we can follow their lead and forgo the enormous costs and even greater danger of nuclear power. We may even be surprised to see the quality of our lives improve as we become less extravagant.

But it will take more than a little ingenuity to shut down the nuclear industry, rearrange our priorities, and redesign our economic system. Unless each person starts now by studying the important issues and by giving at least 10 percent of his or her time to helping society achieve these objectives, our children have no hope of living a life even remotely as good as ours. Concerned citizens must rededicate themselves immediately to the task of building a sane, compassionate society and to preserving the environment for future generations.

This book is a joint effort, blending the skills and knowledge of engineers, scientists, and ecologists. While it was being prepared, three leading nuclear engineers resigned their jobs with General Electric to bring their warning message to the American public. Their testimony before the Congressional Joint Committee on Atomic Energy appears in Appendix A. It presents the most detailed and persuasive case yet for closing down the nuclear industry. The men who gave this testimony helped design many plants operating today, and it is based on extensive experience.

These men have taken the first step in organizing their lives and those of their families according to a new conservation ethic. Their sacrifice sets an example for all of us. The next step is for *us* to realize that our entire society is locked into a way of life that will take decades to change, just as it took decades to develop. *You* can initiate that process today, for your own choice to abide by this ethic will affect your family, friends, and neighbors. Live it first, then talk about it. Show those around you that their lives will be healthier and saner if they cut energy consumption and promote energy sources other than nuclear power.

PAUL R. EHRLICH
Stanford, California
September 22, 1976

Acknowledgments

I would like to thank the contributing authors whose excerpts, articles, and testimony are republished here: Dr. John Gofman, Arthur Tamplin, John Berger, David Comey, Howard Kohn, Robert Pollard, Ralph Nader, Richard Curtis, Elizabeth Hogan, John Fuller, David Pesonen, Philip Herrera, L. Douglas DeNike, Theodore Taylor, David Brower, Dale Bridenbaugh, Gregory Minor, Amory Lovins, Senator Mike Gravel, Richard Hubbard, and William Ophuls. Daniel Ford and Henry Kendall of the Union of Concerned Scientists offered suggestions for adapting their testimony, which, beginning with the 1972 AEC Rulemaking Hearings, contributed substantially to public awareness of nuclear industry problems. Several other people generously contributed their time and thoughts. I am indebted especially to Paul Ehrlich, David Brower, and Dow Woodward, whose encouragement and advice led to the initial draft, and to Joseph Cotchett, who earlier suggested that I write a book on nuclear power for the general public. I am also grateful to Susan Bolotin, editor of Vintage Books, for her patience and timely suggestions, and to copyeditor Cordelia Jason; to Gary Pierazzi, for preparing the splendid illustrations; and to Pat Sharp, who typed several drafts of the manuscript in record time. Jacqueline Mathes helped compile the Notes and Bibliography from the Project Survival library. Many others offered comments and other assistance: Robert Pollard, Terry Lash, Marion Lewenstein, Jacqueline Benhamou, Donna Lynn, Larry Klein, Ellen Rabinowitz, Amnon Goldworth, Thomas Cooney, Alison Finch, Jim M'Guinness, Roger Adams, Nancy Badger, Carol Bain, Leslie Grimm, Kathy Pering, Dan Dippery, Bill and Dorothy Dworsky, Denis Hayes, Dick Jaqua, David MacCuish, Malcolm McWhorter, Gary Williams, John Badger, Jim Harding, Sherry Paine, Anthony Roisman, Fred Slautterback, Alfred Stahler, Paul Valentine, Dan Wilbur, Dick Jorgensen, John Leshy, Doug Daetz, Betsy Amster, Charles Komanoff, Richard Glass, Susan Bacon, Kathleen and Bill Epperly, Philip and Susan Schneider, and Ed and Barbara Thomas. Important support from Lynn Bacon, Robert Walker, Dora Downs, and Margery Brisacher, and encouragement and affection from my parents, Louis and Katharine Faulkner, sustained my commitment to prepare this book. The engineers and executives with whom I worked on six nuclear projects as a systems applications engineer until 1974 were the first to point out many of the problems reviewed here; they contributed substantially but remain anonymous because reprisals would certainly follow if I identified them by name.

I owe special thanks to the entire staff of Friends of the Earth and *Not Man Apart* for continuous support and advice while *The Silent Bomb* was being prepared. I am grateful especially to Tom Turner, David Gancher, Hugh Nash, Anne Chamberlain, Bruce Colman, Jeffrey Knight, Connie Parrish, Lorna Salzman, Natalie Roberts, and David Brower, FOE president.

PETER FAULKNER
Stanford, California
October 7, 1976

Contents

Introduction

Nuclear power is rapidly becoming a major international controversy because millions of citizens now reject claims that the risks are minimal and that the benefits justify them. One of the most serious dangers is that other nations will follow India's lead by using imported nuclear technology to produce atomic weapons-grade material. After obtaining a reactor "for peaceful purposes" from Canada, India exploded a nuclear bomb in May of 1974 which contained nuclear material extracted from this reactor. Ten days later, India and Argentina agreed to exchange nuclear information for five years. These participants also described their agreement as promoting "peaceful nuclear research."

It makes little difference whether Iran or Egypt, Argentina or Brazil, is the next member of the international nuclear club. More than thirty countries will have large reactors operating by 1978. Altogether, 360 nuclear reactors are expected to provide a generating capacity of over 170,000 MWe (megawatts of electricity). Many of these countries already have military jet aircraft capable of delivering atomic weapons accurately and with devastating effect.

Since 1970, the countries have apparently realized that the nuclear option bestows electrical, political, and military power, and they are wasting no time in constructing nuclear plants over which the United States, Russia, and the United Nations can exercise only limited control.

In a few years, when these countries begin exploding their first test devices, the international balance of power may become dangerously uncertain. The effect on the stability of economic and social institutions of several antagonists consecutively exploding similar devices "for peaceful purposes" may be even more dramatic and irreversible. Unquestionably, international nuclear war will then seem less remote than today; and future debates over civilian reactor safety, waste disposal, and terrorism may be upstaged by the consideration that any nation wealthy enough to buy a nuclear reactor may flex its military muscle with derivative nuclear weapons.

U.S. corporate and government leaders long ago decided that these considerations were not serious enough to justify curtailing the development of the nuclear industry. As a result, U.S. reactor manufacturers and architect-engineers are aggressively marketing nuclear plants to foreign nations in hopes that either the international scientific establishment, the International Atomic Energy Agency (IAEA), or the United Nations will step in to ensure that nuclear materials will be used only for electric power production. To charges that they are irresponsibly making nuclear weapons available to any nation able to pay the price, U.S. corporations reply that if this country unilaterally refuses to sell nuclear technology, other nations, such as France and West Germany, will sell these systems instead. Amory Lovins points out that

> many nuclear advocates deny that the U.S. can stop nuclear proliferation . . . [and argue instead that] we might as well get the business and try to use it as a lever to slow the inevitable spread of nuclear weapons to nations and subnational groups in other regions. This approach is all too like Paul Ylvasaker's definition: "A region is an area safely larger than the one whose problems we last failed to solve."[1]

The scope of *The Silent Bomb* is limited to the human, environmental, economic, and engineering aspects of the present nuclear power controversy and to the contributing roles played by government and industry. Some time ago, these institutions persuaded the

American public to suspend judgment on nuclear fission until demonstration reactors provided enough data on which to base a reasonable decision. After more than 300 reactor-years of operation (number of reactors × number of years of operation) the industry can point to a perfect safety record, but it may have made a serious mistake in trying to build hundreds of commercial reactors while the demonstration program was under way, for

> a significant fraction of the public is deeply concerned by this [commercial] commitment. Sufficient concern has been voiced so that it is not an exaggeration to state that a national debate is now taking place over the acceptability of nuclear fission. This debate should have taken place prior to the inception of the "Peaceful Atom" program; it did not. Hence we now find ourselves debating the acceptability of a technology that is already being deployed and behind which has gathered a formidable array of political, economic and personal commitments.[2]

A 1976 study by the California Assembly Committee on Resources, Land Use and Energy initiated a reassessment of nuclear power and pointed to problems that "are widely acknowledged within the nuclear industry and the federal government," including the fact that "no method for the disposal of nuclear wastes has been adopted."[3] Paul Ehrlich also emphasizes this point in the Foreword to this book. Other problems acknowledged by industry and government are that "fuel reprocessing capacity is inadequate to handle the spent fuel from reactors soon to be in operation, fuel enrichment capacity will soon be exceeded, [and] fuel cycle and construction costs have been increasing very rapidly for nuclear plants."[4] The more controversial issues of accident risk, uranium shortage, plant reliability, vulnerability of the fuel cycle to transportation accidents and theft, and the limits on liability following a nuclear accident are still unresolved. The California study points out that

> the reason for the dispute among highly qualified and knowledgeable men is that the issues are not . . . resolvable [exclusively] through the application of scientific expertise. The debate is more the result of differing views of human abilities, human fallibility and human behavior than anything else. To have confidence in the safety of a reactor, we must have confidence in the degree of perfection man can attain in building and operating

complex devices. To have confidence in the perpetual isolation of nuclear wastes, we must have confidence in the longevity of our social institutions and the rationality of future generations. To have confidence in the security of the bomb-grade fuels which are present in the cycle which results in electricity, we must have confidence in the abilities of society's deterrents to prevent actions by fanatics or organized crime.[5]

In spite of concerns expressed by nuclear opponents, the nuclear industry urges the American public to accept a "plutonium economy" in exchange for electrical power they would *not* even need if they were to adopt rigorous and practical conservation measures.[6] If nuclear promoters have their way, hundreds of plutonium breeder reactors will replace light water nuclear plants in a few years. Instead of imported uranium, the price of which has increased 400 percent since 1972, industry plans to use homemade plutonium for nuclear fuel. If you inhale ten micrograms of plutonium and if this particle lodges in your lung, your chances of dying of lung cancer are excellent.[7] Consider that a softball-sized sphere of plutonium could, if properly dispersed, infect every human being on earth today.[8] Will the nuclear industry manage plutonium as reliably as it has stored radioactive wastes?

The Price-Anderson Act assures that if an accident occurs the public will end up holding the bag for all damages which exceed $560 million, and which government studies (such as the 1965 Brookhaven report, suppressed by the Atomic Energy Commission until 1972) predict may be thirty times as great. This legislation sets a precedent—government underwriting of insurance for private industry—that leads us to ask whether private insurance companies have any confidence in the nuclear industry's safety assurances. If nuclear power is as safe as its proponents claim, why don't utilities and reactor manufacturers accept full liability for the consequences of an accident?

The nuclear power controversy focuses especially on questions like these. You can begin participating today in this controversy by learning as much as you can about nuclear power, by studying carefully some of the publications suggested at the end of this book, and by reading *The Silent Bomb* critically, knowing that it presents as forcefully as possible the viewpoint of a number of thoughtful people, many of whom have had firsthand experience with the Atomic Industrial Complex (AIC).*

This book examines the strategies of the AIC and suggests how it attained unprecedented political power in our society. You will learn that:

- An international uranium price-fixing scheme has been operating for several years. This cartel may have already cost U.S. citizens hundreds of millions of dollars due to artificially high nuclear-plant fuel costs and may involve over a dozen U.S. corporations. Its existence was revealed in July of 1976 through documents obtained by Friends of the Earth. The U.S. Department of Justice is investigating the relationship between this cartel and the cost of uranium, which began skyrocketing shortly after uranium executives started meeting covertly. Meanwhile, Westinghouse Corporation has sued twenty-nine uranium suppliers, charging them with operating a cartel that fixed international uranium prices and attempted to punish Westinghouse for undercutting these prices.
- General Electric Corporation (GE) experienced certain problems in designing and developing its most recent reactor and containment package. The technical problems, described here in layman's language, pertain to the Mark III reactor containment, a concept that GE successfully marketed before several potentially dangerous design defects had been resolved.
- Gulf Oil Corporation, with interests in uranium mining and reactor manufacturing, channeled millions of dollars in the form of "political contributions" to key U.S. senators and congressmen, several of whom sat on the congressional Joint Committee on Atomic Energy. These men were in powerful positions to make decisions that secured industry interests, especially those of General Atomic Company, a Gulf subsidiary.
- A study released by Common Cause in June of 1976 reveals that over half the executive positions in the federal Energy Research and Development Administration (ERDA) and Nuclear Regulatory Commission (NRC) are filled by former employees of private corporations involved in energy activities. Although it is common practice in Washington to appoint experienced people from the private sector, the large proportion of energy industry

* Peter Faulkner is the author of all articles for which no other author is designated.

personnel in executive positions raises questions of conflict of interest (in ERDA's case) and of whether NRC can regulate effectively. Overrepresentation of industry in government may have reached a point in this case where the public will insist that this practice be ended.

- Published research papers by Dr. John W. Gofman, Dr. Arthur Tamplin, and others assert that there is no "safe" level of radiation exposure to the public. Several of these scientists and other critics have accused the U.S. government of rubber-stamping arbitrary standards for low-level radiation emission so that private industry may proceed unimpeded to develop nuclear technology.

- Karen Silkwood, a lab analyst for Kerr-McGee Corporation, may have been murdered while she was en route to a meeting with David Burnham of *The New York Times* with potentially explosive documents concerning conditions at Kerr-McGee. The analysis published here adds credence to suspicions that her death was not an accident.

- A uranium shortage may be likely in a few years unless huge deposits of ore are discovered and unless government and industry develop a system for reprocessing spent fuel that does not injure the environment. Meanwhile, thousands of tons of spent fuel are accumulating at reactor sites. In addition, uranium prices are 400 percent higher than they were in 1972 and climbing, with the result that Westinghouse Corporation may lose as much as *$2.5 billion* if it abides by earlier contracts which guarantee fuel to its customers at obsolete prices.

The Silent Bomb offers evidence that the U.S. nuclear program has been shaped and directed according to the fiscal interests of the nuclear industry and with wholly inadequate attention to the rights of present and future generations to live free from unacceptable risk. The federal government has, from the beginning, been an enthusiastic partner even to the extent of occasionally stacking the deck in favor of the nuclear industry. Recent evidence has emerged that the AEC concealed studies by its own scientists of serious nuclear plant risks,[9] that it may have manipulated plant licensing proceedings to avoid airing basic safety issues,[10] and that it may have denied concerned citizens substantive due process during plant licensing activities.[11]

As this book was being prepared, the U.S. District Court of Appeals in Washington, D.C., delivered a landmark decision in which much of this evidence was explicitly reviewed. On July 21, 1976, the court ordered the Nuclear Regulatory Commission (NRC) to consider the full implications of environmental hazards posed by fuel reprocessing and radioactive wastes before licensing new nuclear plants. Three weeks after the court order, the NRC announced that no more full-power operating licenses or construction permits would be issued until further environmental-impact studies were completed.

The July, 1976, decision thus led to an unprecedented national moratorium on the licensing of new plants. However temporary or limited in scope, it represents a significant milestone in citizen efforts to stop nuclear power and may lead to further industry cutbacks, for plants that are idle because they lack licenses to operate accrue enormous costs. According to Bernard Rusche, director of NRC's Office of Reactor Regulation, a single day's delay suffered by a 1,000-MWe plant in gearing up to full operation will cost up to $400,000 for generating the replacement power with oil and thousands of additional dollars in interest charges.[12]

Of the three branches of the federal government, the judiciary seems to be the one where the most lively and rational debate over nuclear power is taking place. Organizations including the Natural Resources Defense Council stand ready to continue legal battles, and these efforts may succeed where state initiatives have failed in restricting nuclear plant development. The U. S. Court of Appeals for the District of Columbia had no difficulty in grasping the safety implications of nuclear power and will not automatically acknowledge that the short-run effects of a nuclear moratorium on employment, taxes, and the cost of living—issues stressed in industry media campaigns to defeat state initiatives—will justify overriding federal environmental laws and regulations.

This book is addressed to the layman who would like to view the controversy from a broad perspective and who is concerned about industry claims that, for every problem raised by the nuclear controversy, there is a technical or scientific fix. In their concise and carefully documented book, *Non-Nuclear Futures: The Case for an Ethical Energy Strategy*, Amory Lovins and John Price note that

As one becomes aware of the real gravity and difficulty of the problems, of the disturbing but promptly forgotten failures of many engineering measures taken to try to solve these problems in the past, and of the limits of engineering solutions to problems of human nature, one starts to realize that the problems might be too difficult to solve. . . . Many people (though fewer each week) neither know nor care about these problems. Some other people are aware of at least some of the problems but they religiously share the engineer's faith that he can solve them even though they are quite different from the problems of keeping an airplane or a building or a bridge from falling down. And some other people suspect or believe . . . that the engineer's faith in nuclear safety is both unfounded in principle and unattainable in practice: indeed, that no matter what low level of risk society may somehow deem "acceptable," the nuclear engineer cannot be certain of attaining it without having utterly infallible people working within a society of unprecedented homogeneity and stability through countless millennia. . . . The most important parts of the debate about nuclear power are not technical disputes and are the legitimate province of every citizen, whether technically trained or not. In the past, nuclear experts often said that we must trust them because these matters are too complex for laymen to understand. Such statements worried laymen who thought that wise democracies do not make decisions that laymen cannot understand. But the laymen are now discovering, to their relief, that though responsible participation in the nuclear debate may be easier with a little technical knowledge, too much expertise tends to obscure rather than to illuminate the basic questions at issue, for they are not questions of nuclear engineering but of ethics, history, psychology, politics and plain common sense . . .[13]

MAXIMUM INCREDIBLE ACCIDENT

1
The Incident
at Browns Ferry
David Dinsmore Comey

At noon on March 22, 1975, both Units 1 and 2 of the Browns Ferry plant in Alabama were operating at full power, delivering over 2,100 megawatts of electricity to the Tennessee Valley Authority (TVA).

Just below the plant's control room, an electrician and an inspector were trying to seal air leaks in the cable-spreader room, where the electrical cables that control the two reactors are separated and routed through tunnels to the reactor buildings. They were using strips of spongy foam rubber to seal the leaks. They were also using candles to determine whether or not the leaks had been successfully plugged, by observing how the flame was affected by the escaping air.

The electrical inspector put the candle too close to the foam rubber, and it burst into flame. The resulting fire disabled a large number of engineered safety features at the plant, including the entire emergency core cooling system (ECCS) on Unit 1, and almost resulted in a meltdown accident. During the weeks that followed, nuclear industry critics pointed out that the fire demonstrated the vulnerability of nuclear plants to "single-failure" events and to human fallibility.

The official Nuclear Regulatory Commission (NRC) report on the incident affirmed that the fire was started by the inspector (referred to in the report as "C"), working with the electrician, "D," who said,

> Because the wall is about 30 inches thick and the opening deep, I could not reach in far enough, so C (the inspector) asked me for the foam and he stuffed it into the hole. The foam is in sheet form; it is a "plastic," about 2 inches thick, that we use as a backing material.

The inspector, "C," described what happened next:

> We found a 2- by -4-inch opening in a penetration window in a tray with three or four cables going through it. The candle flame was pulled out horizontally showing a strong draft. D (the electrician) tore off two pieces of foam sheet for packing into the hole. I rechecked the hole with the candle. The draft sucked the flame into the hole and ignited the foam, which started to smolder and glow. D handed me his flashlight with which I tried to knock out the fire. This did not work and then I tried to smother the fire with rags stuffed in the hole. This also did not work and we removed the rags. Someone passed me a CO_2 extinguisher with a horn which blew right through the hole without putting out the fire which had gotten back into the wall. I then used a dry chemical extinguisher, and then another, neither of which put out the fire.

In its report on the cause of the fire, the TVA stated:

> The material ignited by the candle flame was resilient polyurethane foam. Once the foam was ignited, the flame spread very rapidly. After the first application of the CO_2, the fire had spread through to the reactor building side of the penetration. Once ignited, the resilient polyurethane foam splattered as it burned. After the second extinguisher was applied, there was a roaring sound from the fire and a blowtorch effect due to the airflow through the penetration.

The airflow through the penetration pulled the material from discharging fire extinguishers through the penetration into the reactor building. Dry chemicals would extinguish the flames, but the flame would start back up.

Approximately fifteen minutes passed between the time the fire started (12:20 P.M.) and the time at which a fire alarm was turned in. It was not until one of the electricians told a plant guard inside the turbine building that a fire had broken out that an alarm was sounded. However, confusion over the correct telephone number for the fire alarm delayed its being sounded. As the NRC report on the incident noted,

> The Browns Ferry Nuclear Plant Emergency Procedure lists two different telephone numbers to be used in reporting a fire, one in a table of emergency numbers and the second in the text of the procedure. The appropriate number (299) is the one in the text. Dialing the number automatically sounds the fire alarm and rings the Unit 1 operator's telephone.
>
> The Emergency Procedure was not followed by those involved when reporting the fire. The construction workers first attempted to extinguish the fire, whereas the procedure specifies that the fire alarm be sounded first. The guard reporting the fire telephoned the shift engineer's office rather than calling either of the numbers listed in the procedure.

The fire alarm was sounded only after the shift engineer called the control room on extension 299 to get the reactor operator. It was fortunate that the shift engineer was in an office with a PAX phone (the plant's internal telephone system), which allowed him to call the 299 number. Had he been at a Construction Department extension, he could not have placed the call, as the TVA's investigative report later revealed:

> "BFNP Standard Practice BFS3, Fire Protection and Prevention," instructs DPP (Department of Power Production) personnel discovering a fire, whether in a construction area or an area for which DPP is responsible, to report the fire to the Construction Fire Department, telephone 235. "BFNP Fire, Explosion and Natural Disaster Plan" instructs personnel discovering a fire to call 299 (PAX). The construction extension cannot be dialed from the PAX system, and the plant extension cannot be dialed from the construction phone system.

Despite the fire alarm, the reactor operators in the plant control room did not shut down the two reactors, but continued to let them run. At 12:40 P.M., five minutes after the fire alarm sounded, the Unit 1 reactor operator noticed that all the pumps in the emergency core cooling system had started. In addition, according to the official TVA report,

> Control board indicating lights were randomly glowing brightly, dimming, and going out; numerous alarms occurring; and smoke coming from beneath panel 9–3, which is the control panel for the emergency core cooling system (ECCS). The operator shut down equipment that he determined was not needed, only to have them restart again.

The flashing lights, alarms, smoke, and continual restarting of ECCS pumps went on for a full ten minutes before the reactor operators began to wonder whether it might be prudent to shut down the reactors.

After the power level on the Unit 1 reactor began to drop inexplicably, the operator started to reduce the flow of the reactor's recirculating pumps. When the pumps suddenly quit at 12:51 P.M., he finally shut the reactor down by inserting the control rods.

An electrical supply network both controls and powers the ECCS and other reactor shutdown equipment on Unit 1. Beginning at 12:55 P.M., this key electrical network was lost. The normal feedwater system was lost, the high-pressure ECCS was lost, the reactor core spray system was lost, the low-pressure ECCS was lost, the reactor core isolation cooling system was lost, and most of the instrumentation which tells the control room what is going on in the reactor was lost. The Unit 1 operator said,

> I checked and found that the only water supply to the reactor at this time was the control rod drive pump, so I increased its output to maximum.

Meanwhile, a few feet away, on the Unit 2 side of the control room, warning lights had also been behaving in a peculiar way for some time. A shift engineer stated,

> Panel lights were changing color, going on and off. I noticed the annunciators on all four diesel generator control circuits

showed ground alarms. I notified the shift engineer of this condition and said I didn't think they would start.

According to the official TVA report,

At 1:00 P.M., the Unit 2 operator observed decreasing reactor power, many scram alarms, and the loss of some indicating lights. The operator put the reactor in shutdown mode.

Some of the shutdown equipment began failing on Unit 2, and the high-pressure ECCS was lost at 1:45 P.M. Control over the reactor relief valves was lost at 1:20 P.M. and not restored until 2:15 P.M., at which time the reactor was depressurized by using the relief valves and brought under control.

On the Unit 1 side of the control room, things were not going so well. According to the Unit 1 operator,

At about 1:15, I lost my nuclear instrumentation. I only had control of four relief valves. . . . At about 1:30, I knew that the reactor water level could not be maintained, and I was concerned about uncovering the core.

Had the core become uncovered, a meltdown of the reactor fuel would have begun because of the radioactive decay heat in the fuel.

To prevent the Unit 1 reactor water from boiling off, it was necessary to get more water into the core than the single high-pressure control rod drive pump could provide. It was decided that by opening the reactor relief valves, the reactor would be depressurized from 1,020 to below 350 pounds per square inch, where a low-pressure pump would be capable of forcing water in to keep the core covered. None of the normal or emergency low-pressure pumps were working, however, so a makeshift arrangement was made, using a condensate booster pump. This provided a temporarily adequate supply of water to the reactor, although the level dropped from its normal 200 inches above the core down to only 48 inches. Using the makeshift system, the Unit 1 reactor was under control for the time being.

Unit 2 was also under control, but by a rather thin margin. The "A" and "C" subsystems of the low-pressure ECCS and the core spray system had been lost early in the incident, and the "B" system failed intermittently between 1:35 and 4:35 P.M. With only one subsystem of the low-pressure ECCS available, the Unit 2 operator resorted to using the condensate booster pump arrangement similar to the one that had been rigged up for Unit 1.

Many instrumentation and warning lights in the control room were inoperative. The reactor protection system and nuclear instrumentation on both reactors had been lost shortly after they were shut down. Most of the reactor water level indicators were not working. The control rod position indicator system was not operative. The process computer on Unit 1 was lost at 1:21 P.M. (The computer on Unit 2 was inoperative because it was down for reprogramming.) Other systems were failing; at 2:43 P.M. one of the plant's four diesel generators failed, leaving the plant with a bare minimum of emergency on-site power.

To add to the confusion, the PAX telephone system failed at 1:57 P.M., making it impossible to place outgoing calls from the control room for several hours. This represented a considerable hardship, because the control room had lost control over most of the plant's valves, and the plant telephone system was being used to instruct equipment operators to manually adjust certain key valves in the condensate booster system pumping water into the reactor core.

Moreover, the Unit 1 operator did not know the level or the temperature of the water in the torus (the reactor containment suppression chamber) because the monitors were not working. Yet, as a General Electric supervisor's log showed,

> With the relief valves in operation, the need for torus cooling became vital. The RHR (residual heat removal) system was unavailable for torus cooling.

Unless the RHR system could be put into operation, there was the danger that the water in the torus would begin to boil, and this would eventually overpressurize the containment and rupture it. As the NRC report remarks, in its restrained way,

> After condensate flow to the reactor was established, the major concern was to establish torus cooling and shutdown cooling using the RHR as quickly as possible.
> Operator M stated that he made a list of RHR valves needed to obtain torus cooling. He further stated that at approximately 3:15 P.M. all torus temperature and level instrumentation was inoperable. . . .
> From about 2:00 P.M. until the fire was extinguished, several

attempts were made to enter the reactor building and manually align the RHR for torus cooling and shutdown cooling modes. . . .

None of these attempts resulted in establishing torus or reactor shutdown cooling. The attempts were severely limited by dense smoke and inadequate breathing apparatus.

The fire-fighting effort was not going well. Soon after the electricians had fled the cable-spreader room, a shift engineer had tried to turn on the built-in Cardox system in order to flood the room with carbon dioxide (CO_2) and put out the fire. He discovered that electricians had purposely disabled the electrical system that initiated the Cardox. He stated:

> I tried to use the manual crank system and discovered that it had a metal construction plate on under the glass and I tried to remove it. This was difficult without a screwdriver. . . . The next day, I checked other manual Cardox initiators and found that almost all of them had these construction plates attached.

He finally turned on the power, but the Cardox system drove smoke up into the control room above the cable-spreader room. One person present described the scene in the control room as follows:

> The control room was filling with thick smoke and fumes. The shift engineer and others were choking and coughing on the smoke. It was obvious the control room would have to be evacuated in a very short time unless ventilation was provided.

After the carbon dioxide system was turned off, the smoke stopped pouring into the control room. It had not put out the fire in the spreader room, however. A safety officer fighting the fire pointed out,

> The CO_2 in the spreader room may have slowed down the fire but did not put it out. We opened the doors for air, as the smoke in the whole area had become dense and sickening. Another employee and I each donned a breathing apparatus and went into the spreader room. We used hand lamps for illumination, but they penetrated the smoke only a few inches. The neoprene covers on the cables were burning, giving off dense black smoke and sickening fumes. . . . It was impossible to not swallow some smoke. I got sick several times.

Because of the close quarters in the spreader room, fighting the fire was difficult. One safety officer said,

I went into the spreader room wearing a Scott air pack and mask and carrying a fire extinguisher. I had to crawl under the cable trays. The air pack cylinder was too cumbersome to wear on my back so I took it off and slid it and the fire extinguisher under the trays toward the fire about 30 feet.

Inoperative equipment also hampered the fire-fighting effort. For example, one assistant shift engineer said,

I returned to the spreader room to direct the fire-fighting effort. A wheeled-dry-chemical extinguisher had been brought to the spreader room, but its nozzle was broken off at the bottle and I told some of the men to get it out of there and find another unit.

The official NRC report noted other deficiencies:

Breathing apparatus was in short supply and not all of the Scott air packs were serviceable. Some did not have face masks and others were not fully charged at the time of the start of the fire. The breathing apparatus was recharged from precharged bulk cylinders by pressure equalization. As the pressure in the bulk cylinders decreased, the resulting pressure decrease in the Scott packs limited the length of the time that the personnel could remain at the scene of the fire.

One of the assistant unit operators who was sent into the reactor building to manually open the RHR cooling valves reported,

We made three tries, but could not get to the valves. Our breathing equipment could only supply 18 minutes of air per tank, which was not sufficient to enable us to get to the valves and back out of the area. The air tanks were being recharged, but the pressure in the main tanks was not strong enough to fill the tanks [completely]. After the third attempt we went back to the control room and told the assistant shift engineer of the problem and that we needed different equipment or fully charged tanks to succeed.

The electrical cables continued to burn for another six hours because the fire fighting was carried out by plant employees, despite the fact that professional firemen from the Athens, Alabama, fire department had been on the scene since 1:30 P.M. As the Athens fire chief pointed out,

> I was aware that my effort was in support of, and under the direction of, Browns Ferry plant personnel, but I did recommend, after I saw the fire in the cable-spreader room, to put water on it. The Plant Superintendent was not receptive to my ideas.
>
> I informed him this was not an electrical fire and that water could and should be used because the CO_2 and dry chemical were not proper for this type of fire. The problem was to cool the hot wires to prevent recurring combustion. CO_2 and dry chemical were not capable of providing the required cooling. Throughout the afternoon, I continued to recommend the use of water to the Plant Superintendent. He consulted with people over the phone, but apparently was told to continue to use CO_2 and dry chemical. Around 6:00 P.M., I again suggested the use of water. . . . The Plant Superintendent finally agreed and his men put out the fire in about 20 minutes. . . .
>
> They were using type B and C extinguishers on a type A fire; the use of water would have immediately put the fire out.

Even when the decision to put the fire out with water had been made, further difficulties developed. The fire hose had not been completely removed from the hose rack, so that full water pressure did not reach the nozzle. The fire fighters did not know this, however, and decided that the nozzle was defective. They borrowed a nozzle from the Athens fire department, but the threads were mismatched and the nozzle would not stay on the hose.

Once the fire was put out, it was possible for plant employees to go into the reactor building and manually open valves to get the RHR system operating.

On Unit 1, however, a new emergency developed. About 6:00 P.M., control of the last four relief valves was lost and the reactor pressure increased to above 350 pounds per square inch, making it impossible for the makeshift condensate booster pump system to inject water into the reactor. As in the early stage of the accident, the only source of water for the Unit 1 reactor was now the control rod drive pump, and this probably would not prevent a boil-off accident that would turn into a core meltdown in just a few hours.

The spare control rod drive pump was inoperative. Although it was later determined that a series of valves could have been turned to allow the Unit 2 control rod drive pump to supply water for the Unit 1 reactor, the reactor operators did not know this at the time.

With the reactor pressure mounting higher and higher, the relief valves were finally brought back into operation at 9:50 P.M. About 10:20 P.M., the reactor was depressurized to the point that the condensate booster pump could again get water into the reactor.

Normal shutdown was established on the Unit 1 reactor at 4:10 A.M. the next morning. The nightmare at Browns Ferry was over.

Had the reactor boiloff continued to the point where a core meltdown took place, however, it is doubtful that the endangered surrounding population could have been evacuated in time. Evacuation of the county's residents was the responsibility of the civil defense coordinator for Limestone County, but, as he admitted to NRC inspectors,

> I heard about the fire at Browns Ferry on the morning of Monday, March 24, 1975 [two days later]. No one in the Civil Defense System notified me or attempted to do so. . . . I feel that our county should have been notified since the plant is located in our county.

The sheriff of Limestone County said:

> I heard about the fire at the Browns Ferry plant after it was over. . . . I have not had any updating of procedures proposed to me since the initial plan was outlined in 1972. I do not have a copy of the emergency plan.

The sheriff of neighboring Morgan County did hear about the fire four hours after it started, but said, "I was asked to keep quiet about the incident to avoid any panic."

The NRC noted in its investigative report:

> No official notification was made to the State of Alabama Highway Patrol by the State of Alabama Department of Public Health or by TVA. . . .
>
> An attempt was made to notify the Lawrence County Sheriff at 4:08 P.M., but no answer was received. Only one attempt was made to locate the sheriff.

In fact, this try-once-and-fail procedure was more or less the norm. The NRC investigation stated:

> The State of Alabama Emergency Plan for the Browns Ferry Nuclear Plant was implemented at 3:30 P.M. to the extent that notifications were made to designated state personnel and principal support agencies. . . . Only one attempt was made to contact principal support agencies that were located in counties surrounding the site regardless of whether the agency was contacted or not. The notification process was discontinued at 4:40 P.M.

The fire knocked out the radiation monitors on the Unit 1 reactor building vent almost immediately, and the Unit 2 vent monitor was inoperable from about 2:00 P.M. until 9:00 P.M. Both the NRC and TVA state unequivocally that no significant radiation release occurred, but there were continuing difficulties in obtaining air samples both at the plant site and in the surrounding area. For example, as noted in the NRC report,

> At 5:05 P.M., the Director, Environs Emergency Center, directed that additional environmental air samples be obtained. . . . At this time, individuals in the Site Emergency Center observed smoke emanating from the reactor building and the decision was made to evacuate the meteorological tower.

Radiation sampling of the air was started at 4:45 P.M. at Athens, 10 miles northeast of the plant; at Hillsboro, 10 miles southwest; at Rogersville, 35 miles northwest; but not at Decatur, 20 miles southeast and directly downwind of the plant. Investigation revealed that no air sampler was available at Decatur. At approximately 10:00 P.M., air sampling was initiated at Decatur after the state Air Pollution Control Commission loaned them one of its samplers.

Other equipment failures continued to plague the plant. Shortly after nightfall, the aircraft warning lights on the plant's radioactive gas release stack went out. Since the stack is 600 feet tall, loss of the lights could have resulted in an aircraft colliding with the stack. The NRC report describes what was done next:

> At 8:37 P.M., a member of the environmental staff made an attempt to telephone the gatehouse by using a public telephone to inform the security guards that the warning lights on the

plant stack were not operating. Since the gatehouse could not be reached, the environmental representative telephoned the EEC (Environs Emergency Center) and explained the condition. The EEC Director directed the information to the plant because of the need to contact FAA [Federal Aviation Administration] authorities immediately.

It is unclear why no one thought to phone the FAA directly instead of giving the information to the plant, which had enough problems on its hands at the time.

Information on the time and sequence of restoration of control circuits was lost after 4:30 P.M. on the afternoon of the fire because the plant's electric sequence printer ran out of tape. No one replaced the tape until 2:00 P.M. the following day. Plant telephone calls to TVA management were intermittently recorded due to mechanical problems with the tape recorder. Some of what was recorded, however, is indicative of what was going through the minds of TVA and NRC personnel. The following is an excerpt from a conversation at 7:47 P.M. between J. R. Calhoun, chief of TVA's Nuclear Generation Branch, and H. J. Green at the Browns Ferry Plant:

> *Green:* I got a call that Sullivan, Little and some other NRC inspector are on their way and will get here some time tonight and so all our problems will be over.
> *Calhoun:* (laughs) They will square you away, I am sure.
> *Green:* We probably have a violation. We've kept very poor logs.
> *Calhoun:* (laughs) No doubt!

At about 9:00 P.M., Calhoun phoned Frank Long, in the U.S. Nuclear Regulatory Commission's Region II office in Atlanta:

> *Long:* The doggone public news media types will probably drive you out of your mind. Okay, your people did put out a news press release?
> *Calhoun:* Yeh, we put one out about 4:30. Somewhere close to 4:30. . . . Only thing we can say right now is that it could have been a hell of a lot worse.
> *Long:* Oh, yeh.
> *Calhoun:* You know, when you talk about a fire in the spreader room, you've really got problems.
> *Long:* It would affect just about everything.

Calhoun: Yeh, you know everything for those two units comes through that one room. It's common to both units, just like the control room is common to both units.

Long: That sorta shoots your redundancy.

Emergency procedures inside the Browns Ferry plant were also deficient. Many employees did not know what the sound of the fire alarm meant, and few had been trained in emergency procedures despite earlier fires at the plant.

Large numbers of TVA employees went into the plant control room, adding to the chaotic situation there. One assistant shift engineer reported that, instead of the six persons normally there, "the maximum number of people in the control room at any one time I guessed to be between fifty and seventy-five." Indicative of the tension felt in the control room is the later comment of the Unit 2 shift engineer that "the Plant Superintendent asked me if I had control of Unit 2 and if everything was O.K. *almost continuously."*

What is the significance of the Browns Ferry incident?

One question is: What the devil were the electricians doing using a candle to test for air leaks? Although it is perfectly possible to design an inexpensive anemometer to test for air leaks, or even to use smoke from a cigarette, these methods were rejected two years ago by the Browns Ferry plant personnel in favor of using candles.

Some senior personnel at the plant thought that the urethane sheet foam used to seal the cable penetrations was fireproof. The leader of the electrical conduit division at the plant said:

The practice of using RTV-102 and sheet foam to seal air leaks has been the practice for two or three years. We believed that the urethane would not sustain a fire. Urethane samples had been tested several years ago and it needed a flame for 20 minutes to sustain a fire.

They had only tested two of the polyurethane samples, however, using an American Society for Testing Materials (ASTM) test that the Marshall Space Flight Center later found to be of marginal value. No test had been made of the foam polyurethane, however, and the NRC's consultants, from the Marshall Space Flight Center, found that:

a cursory match test on a piece of the foam rubber disclosed almost instantaneous ignition, very rapid burning, and release of molten flaming drippings.

Even though some people at the plant thought the ASTM tests showed the penetration sealant material to be nonflammable, senior management knew it was highly flammable. The plant instrument engineer told NRC inspectors:

> During the test and startup period of Unit #1 [in 1973], I demonstrated the flammability of the sealing material to the Plant Superintendent. I burned the material in the Plant Superintendent's office. He immediately called someone with Construction and they discussed the situation. . . . I feel the Plant Superintendent did all that was immediately possible to investigate the situation as it appeared that Construction was not going to change the material.

The Plant Superintendent admitted to the NRC inspectors, "I was aware that polyurethane was flammable, but it never occurred to me that these penetrations were being tested using candles."

Many senior management personnel at the plant denied knowing of the practice of using candles to test cable penetrations. The rest indicated that they knew candles were being used, but thought the sealant materials were not flammable.

The electricians seemed to be the only group that knew both that the foam rubber was flammable and that candles were being used as the testing method. As one electrician recounted,

> The electrical engineer called the group [of electricians] together and warned us how hazardous this method was. "Why, just the other day," the electrical engineer said [in effect], "I caught some of that foam on fire and put it out with my bare hands, burning them in the process."

One of the electricians who started the fire said that candles had been used for more than two years, but said, "I thought that everybody knew that the material we were using to seal our leaks in penetrations would burn. . . . I never did like it."

The real irony of the Browns Ferry fire was that two days before, a similar fire had started but had been put out successfully. After the fire on Thursday night, the shift engineers and three assistant shift engineers met. According to one of them,

We discussed among the group the procedure of using lighted candles to check for air leaks. Our conclusion was that the procedure should be stopped.

Yet nothing was done. The fire was noted in the plant log and briefly discussed the next day at the plant management meeting. No one on the management level seemed to consider it a safety problem worth following up. This was the standard operating procedure; as the NRC investigative report notes,

> Previous fires in the polyurethane foam material had not always been reported to the appropriate levels of management, and, on the occasions when reported, no action was taken to prevent recurrence.

Considering these practices, it was probably not a question of *whether* the Browns Ferry plant would have a major fire, but *when*.

What does the fire mean for other nuclear plants? That depends on whether the NRC carries out the recommendations made by the Factory Mutual Engineering Association of Norwood, Massachusetts, the fire underwriters the NRC engaged as consultants after the Browns Ferry accident:

> Conclusions and Recommendations:
>
> The original plant design did not adequately evaluate the fire hazards of grouped electrical cables in trays, grouped cable trays and materials of construction [wall sealants] in accordance with recognized industrial "highly protected risk" criteria. . . .
>
> It is obvious that vital electrical circuitry controlling critical safe shutdown functions and control of more than one production unit were located in an area where normal and redundant controls were susceptible to a single localized accident. . . . A re-evaluation should be made of the arrangement of important electrical circuitry and control systems, to establish that safe shutdown controls in the normal and redundant systems are routed in separated and adequately protected areas.

There is a cable-spreader room below every U.S. nuclear plant control room. Despite requirements for separation and redundancy of reactor protection and control systems, every reactor has been permitted to go into operation with this sort of configuration, which lends itself to a single failure's wiping out all redundant systems.

If every plant currently operating and under construction were required to rewire so as to achieve true redundancy and eliminate cable trays bunched together, the cost would range between $7.680 and $12.343 billion, according to my calculations. It will be interesting to see whether the new commissioners of the NRC will require such changes.

Except for one news release, dated March 27, 1975, NRC headquarters in Washington has remained silent about Browns Ferry. That news release, quoted below, does not make one optimistic that any meaningful lesson has been learned from the Browns Ferry incident.

> The functioning of some in-plant operating and safety systems, including emergency core cooling systems, was impaired due to damage to the cables.
>
> The two reactors were safely shut down and cooled during the fire. NRC inspectors report that there was redundant cooling equipment available during the reactor cooldown. . . .
>
> Although some instrumentation was lost, certain critical instrumentation such as reactor water level, temperature and pressure indicators continued to function and both plants were safely shut down. . . .
>
> On Unit 1, although a loss-of-coolant accident had not occurred, the emergency core cooling system was activated and supplied additional water to the reactor. It was manually shut down to prevent over-filling. Later, during cooldown, when ECCS was called for manually as one of the several alternate means of supplying cooling water, it did not activate; the alternate methods had more than sufficient capability to cool the core.

Whether the NRC has sufficient capability to cool the public's reaction, once the facts about Browns Ferry are known, will be interesting to observe.

Issues and Comments

According to Comey, both the NRC and the TVA clamped a tight lid on specific details of the fire immediately after it occurred. These details are evidently absent from the news releases issued later by these agencies;[1] omission of these details may affect a reader's interpretation of the fire's severity. Until TVA employees fed details of the Browns Ferry fire to Comey on March 24, 1975, the public had no way of knowing that it had resulted in significant damage. Without delay, Comey released these details to CBS, *The New York Times,* and the *Wall Street Journal.* At this point the nuclear industry lost control of what subsequently became a national news story.

On March 24, Comey also made a formal legal request to the NRC and was given a TVA preliminary report to NRC[2] and an inspection report from the Atlanta office of the NRC.[3] The details and quotations in Comey's article are drawn from these reports.

The way that NRC and TVA handled details of the Browns Ferry fire indicates how industry and government in the future may deal with the news of a nuclear accident. It may be that the public's right to be told the truth in understandable terms may be coming into conflict with vested interests which tend to withhold derogatory information. News suppression[4] often occurs when the agency that controls information happens to be the one that would suffer if the information was made public.

One's position on these issues may depend on one's role in society and commitment to an organization. For example, the position taken by an industry public-relations executive will differ in some important respects from that of an environmental spokesperson. The following excerpt is from an imaginary argument over the Browns Ferry fire between two such people:

Industry Position: This was an incident, not an accident. No release of radioactivity occurred. There was no explosion. The fire was stopped in time and both reactors were shut down. This incident shows that we can prevent system failures from getting out of control.[5]

Environmentalist Position: If this was an "incident," why did it cost the TVA $100 million?[6] The *total* cost includes damage repair, rewiring the control room to protect redundant systems, income loss due to power outages of Units 1 and 2, and buying power from other utilities while Units 1 and 2 were out of commission. This was a multimillion-dollar fire that would have had multi-*billion*-dollar consequences if there had been a meltdown.

In the Browns Ferry case, the nuclear industry may have rationalized its low-key news releases on the grounds that the fire was a "minor" one. How industry construed citizens' rights to information in this case depended very much on its initial assessments of the fire's severity. But to what extent did the nuclear industry's concern with its public image affect its ability to accept and report the facts without distortion?

Besides the legal issues associated with citizens' rights to information, the Browns Ferry fire raises questions of safety and reliability that have been the subject of debate since the first commercial nuclear plants were completed over ten years ago. During the last decade, the space industry was able to launch a number of successful moon missions by rigorously enforcing a comprehensive system of quality controls throughout the design, manufacturing, and mission stages. Both nuclear and space programs have suffered disastrous fires, and both have undoubtedly learned a great deal from them. The important difference here is that the health and environmental consequences of a single nuclear plant accident will be substantially greater than if every one of the Saturn rockets had exploded on the launching pad.

Among the engineers and executives who build and operate nuclear plants, there may be a few infallible people. Their personal and professional infallibility is mostly intrinsic and unique, and may not carry over to include the thousands of people who work for them: the manufacturer who supplies the reactor systems, the shippers who transport the fuel rods, the inspectors who check for quality assurance, the NRC representatives who monitor the inspectors, can't *all* be equally infallible. Human error is impossible to escape and safeguards are as fallible as the people who design, install, and operate them.

As before, our hypothetical antagonists would view the Browns Ferry fire and its implications for safety from entirely different positions:

Industry View: Focus on the facts. The fire provides ample evidence that we've built in so many safeguards that no meltdown or major accident will ever occur. It's true that plenty of things went wrong at Browns Ferry, but enough systems operated properly and enough correct decisions were made so that the public was protected.

Environmentalist View: Focus on what *almost* happened. *How close* did we come to suffering a major accident? How many critical systems *should* have worked but didn't? *But for* a few safeguard systems, did we not come very close to catastrophe?[7]

What is the minimum standard of care the public should expect of the nuclear industry? In answering this, consider (1) short- and long-run costs of a major accident: fiscal, human, and environmental;

(2) the number of people affected after an accident; (3) the mutation burden imposed on all future generations and all forms of life as the result of radiation-induced genetic damage.

Once you establish the minimum standard of care, address the question of whether the nuclear industry is able to maintain it. If the industry can't, should we, and can we, restrain it from constructing new plants and operating existing ones? Are there other alternatives?

Does the Browns Ferry fire suggest that the industry is having trouble maintaining the minimum standard of care? If you answer "partly yes and partly no," should the issue be resolved in favor of public safety or industry profits? Who should enjoy the benefit of the doubt, if one exists: the present and future voting public or the nuclear industry?

Partly because the nuclear power industry has already committed an enormous amount of resources and money to electric generation, it has always been given the benefit of the doubt during the past several years, despite hundreds of "incidents." There is quite a large backlog of these incidents, but until recently, these and a great many other facts about nuclear power were somehow kept, as Thomas Whiteside described the process, "by the workings of the prevailing system of power in our society from reaching a popularly perceived threshold that triggers irresistible public pressure for reform."[8]

In this age of cost overruns and coverups, a certain degree of voter apathy and cynicism is understandable, especially in light of evidence that the Atomic Energy Commission (AEC) and its successor, the Nuclear Regulatory Commission, have suppressed the warnings of their own engineers and scientists[9] so that technically documented criticisms of the industry are seldom publicized. Conclusive evidence of this policy became national news in February, 1976, when Robert D. Pollard,* an NRC reactor engineer and project manager involved in reactor licensing, resigned from his job and presented his criticisms in a series of press conferences. A six-year veteran of AEC and NRC politics, Pollard forwarded a letter with an attached report to William A. Anders, NRC chairman. These two documents detailed several ways that dissent within both agencies was suppressed and alleged that NRC staff reviews were inadequate; that NRC allowed reactors to continue to operate without resolving major safety problems; and that safety considerations were subordinated to a

---

* See Appendix B.

pervasive attitude in the NRC that our most important job [was] to get the licenses out as quickly as possible and to keep the plants running as long as possible. . . . As a result, safety problems whose resolution would [interfere with plant development] were not resolved or were resolved in a way that created the minimum interference with licensing.[10]

On June 10, 1976, the Union of Concerned Scientists, a leading intervenor group in Cambridge, Massachusetts, that has successfully pinpointed federal regulatory inadequacies since 1971, published a carefully documented report entitled *Browns Ferry: The Regulatory Failure,* which confirmed the gist of Pollard's charges. Despite this concise critical report and the nationwide publicity surrounding the Browns Ferry fire, as of October, 1976, neither the NRC nor the nuclear industry had instituted substantial reforms.[11]

2
Malignant Giant:
The Nuclear Industry's Terrible Power and How It Silenced Karen Silkwood
Howard Kohn

She was 28, a slight woman, dark hair pushing past slender shoulders, haunting beauty nurtured in a small-child look. She was alone that chilly autumn night, driving her tiny Honda across the prairie. The Oklahoma fields lay flattened under the crude brushmarks of the wind, the grass unable to snap back to attention. Every few miles a big-boned rabbit, mangled and broken, littered the roadside. A few years earlier she had fired off angry letters when sheep ranchers staged rabbit roundups, clubbing to death the furry army that had threatened their grazing land. She was like that, thrusting her opinions where they weren't welcome.

In the early evening darkness of Wednesday, November 13th, 1974, Karen Silkwood was on an environmental mission of another sort. On the seat beside her lay a manila folder with [what she said were frightening accounts of conditions] at the plutonium plant where she worked. Waiting at a Holiday Inn thirty miles away were a union official and a *New York Times* reporter who had just flown from Washington, D. C., to meet with her.

Hours later, Karen Silkwood's body was found in a small rivulet along Highway 74 where rabbits often came to drink. Her car had swerved left across the highway, skittered about 270 feet along an embankment, smashed head-on into a culvert wingwall, lurched through the air and caromed off another culvert wall, coming to rest in the muddy stream.

Her death was ruled an accident; the police decided she was asleep at the wheel. But the union official was not satisfied. The manila folder was missing. And a private investigator discovered two fresh dents in the rear of her car, telltale marks of a hit-and-run.

Karen had been an intense, serious girl who shunned the local teenage spots for library reading and volunteer work at a hospital. Her acquaintances remember only one irritating characteristic: She talked back to her teachers, firmly correcting them when they slipped up, say, on the atomic weight of tritium. "She was," says an old friend, "a very nice person who always wanted to be right about everything."

She graduated in 1964 with a college scholarship and best wishes from everyone. At nearby Lamar College, Karen pursued her science interests, hoping for a career as a laboratory analyst.

Before her junior year, she dropped out to marry a pipeline supervisor with Mobil Oil. Seven years, three children, one bankruptcy and a divorce later, she resumed her earlier ambition and accepted a job as a laboratory technician with one of the nuclear elite, Kerr-McGee Corporation of Oklahoma City.

Uranium, an unpretentious metal buried mostly in isolated pockets under western deserts, would fuel tomorrow's nuclear plants, and the oil companies were on the ground floor. Kerr-McGee, which flies its K-M trademark topmast at hundreds of service stations in the Southwest, grabbed up all the uranium fields it could sink a shaft in. On a Navajo reservation near Shiprock, New Mexico, Kerr-McGee discovered a cache of uranium under the parched turf. Navajos were paid as little as $1.60 an hour to exhume the metal, hauling it out in wheelbarrows from the stifling, scratchy air below. By 1969, the Navajo mines were exhausted. Only then did the miners learn that uranium dust had infected many of them with a rare lung cancer that resists early diagnosis.

By June of 1973, the cancer had killed 18 of the 100 Navajo miners, and 21 more were feared dying. But Kerr-McGee refused to take responsibility or pay medical expenses. "I couldn't possibly tell you what happened at some small mines on an Indian reservation," Kerr-McGee spokesman Bill Phillips told a Washington reporter. "We have uranium interests all over the world."

By the seventies, Kerr-McGee had mined and milled tons of yellow-cake uranium and had acquired 800,000 acres of uranium leases and a substantial share of the market. With assets approaching a billion dollars, it was the nation's largest uranium producer. But even greater profits lay in another nuclear fuel: plutonium.

Uranium, and fossil fuels such as coal and petroleum, are in limited supply. In fact, by the mid-1980s, we may be faced with a uranium shortage. But plutonium is the alchemist's dream: It can reproduce itself. An industry brochure puts it like this: "Question— How many pounds of plutonium will you have left after you use three pounds in a nuclear reactor? Answer—four pounds!"

Plutonium barely exists in nature; our present supply is entirely man-made. It was first discovered in 1940 among the waste products of fissioned uranium. Plutonium can take several forms, but it is usually a gray, soft metal, a slushy liquid nitrate or a fluffy yellow-green oxide powder fine enough to be inhaled. In any form it is "fiendishly toxic," according to one of its discoverers, Dr. Glenn Seaborg.

Plutonium is much more dangerous than uranium. It is incredibly combustible, readily convertible into nuclear weapons and, once let loose in the atmosphere, it stays deadly for 250,000 years and cannot be recaptured. Swallowing it in a quantity that can be seen can sear the digestive tract. It is also carcinogenic but, because only a few hundred people have ever handled it, scientists disagree as to what amount can cause cancer. A millionth of a gram has induced cancer in lab animals and some experts say that a softball-sized bag of plutonium, if properly dispersed, could visit cancer on every home on earth.

Kerr-McGee's plutonium plant, built next to one of its uranium plants, opened in 1970 shortly before 8,583 fish turned belly-up in the nearby Cimarron River following a big ammonia spill. Raised against the flat harshness of rural Oklahoma, the barnlike plant is unimposing; only a chain-link fence and armed guards hint at the devil's brew within.

Kerr-McGee had assured the Atomic Energy Commission (AEC) that it could handle plutonium safely. But the AEC, a government agency in the curious role of both promoting and policing the nuclear industry, soon received numerous reports of irregularities and accidents at the Kerr-McGee plant. Though the commercial processing of plutonium left no margin for error, things kept getting bungled.

In October, 1970, soon after the plant opened, two workers were contaminated when a radioactive storage container was left in the open for three days. Twenty-two more workers were exposed to plutonium in January, 1971, when defective equipment allowed plutonium oxide to escape into the air. Less serious incidents were common. The protective "glove boxes" used by the workers often had holes. Sometimes the "Super Tiger" and "Polly Panther" drums, specially designed to store the volatile liquid, unaccountably leaked. An improperly designed piping system once sent plutonium sloshing to wrong parts of the plant.

One day a worker bent to adjust a compressor unit; it exploded, ripping through his hand and tearing off the top of his face, scattering tissue over the ceiling. He died instantly. "When I got there," remembers a former lab technician, "they were washing the goo down the drain."

In April, 1972, two maintenance men repairing a pump at the plant were splashed with a rain of plutonium particles which settled on their hands, faces, hair and clothes. At noon they left the plant for lunch in a nearby town, not discovering their contamination until they returned. They were scrubbed clean, along with their car. But Kerr-McGee neglected to inspect the restaurant where the men had eaten.

When Karen Silkwood arrived at the Kerr-McGee plant in late summer, 1972, she was eager to begin a career as a nuclear laboratory technician. But after spending only three months testing the plutonium fuel rods, Karen was outside the chain-link fence marching with a strike placard.

The Oil, Chemical and Atomic Workers International Union (OCAW), representing the plutonium workers, was at loggerheads with Kerr-McGee. The company had managed to keep the unions out until 1966. Now the OCAW was demanding a new contract with higher wages, safer conditions and better training. Kerr-McGee had replied with an offer worse than the old contract. Then, as soon as workers went on strike, the company rushed scabs onto the job, barely missing a beat in fuel rod production.

Even Kerr-McGee officials later conceded, in a letter to the Sierra Club, that thrusting untrained strikebreakers into the plant led to more plutonium spills and leaks. ("Some scabs got only four hours of training when they should have gotten five days," fumed one striker.) Among the inexperienced substitutes hired during the strike was the plant's safety officer.

Outside on the picket lines, meanwhile, 26-year-old Karen started spending time with 22-year-old Drew Stephens, a short-haired, brainy lab analyst with an easy smile. Stephens had grown disenchanted after the rash of accidents. Now he and Karen shared in the outrage building among workers at the plant. As they became closer, Stephens began living with Silkwood, though they were not deeply in love.

The strike lasted ten weeks. Those pickets whose jobs had not been lost to scabs returned to work in January, 1973, reluctantly signing a new contract that stripped away many of their previous rights, including certain protections against arbitrary firings and reassignments. A few weeks later, a plant employee was emptying a bag of plutonium wastes when a fire spontaneously erupted, shooting radioactive dust into the air. Seven workers sucked in the poisonous particles. But Kerr-McGee supervisors waited a day before calling in a physician. Four days later, the seven workers still had not been tested for lung contamination.

After several months with Drew Stephens, Karen moved out of their shared apartment, unwilling to risk another marriage. She wanted her own place, and though she and Stephens remained friends, her career came first. She began working overtime and became involved in the union, OCAW Local 5–283.

Silkwood looked to the union as the only outlet for her growing frustration with management. When suddenly exposed to a swirl of airborne plutonium in July, 1974, she was not wearing a respirator. For over a year she had been asking the company to buy a special respirator mask to fit over her tiny, narrow face. It hadn't arrived.

When union elections came up the next month, Silkwood ran for and won one of the three seats on the steering committee. Fellow workers knew her as the spunky chick who talked back to her bosses. "Goddammit, I am right and *you* are wrong," she once raged at a supervisor. "If you want to tell *me* what to do, *you* oughta learn how to do the job right."

Despite growing anticompany jabber at the plant, most workers did not want a fight. Many simply quit; the annual turnover rate among the 115 hourly workers, according to the union, averaged 60 percent. Some complained of being harassed out of their jobs; three workers who griped to AEC officials about safety conditions early in 1974 were reportedly transferred to menial jobs in the chilly warehouse.

In the fall of 1973, Ilene Younghein, an Oklahoma City housewife, began a one-woman campaign to shut down the plant. Angry workers simply wanted the company to improve training procedures and apply safety precautions rather than lock its doors. But they supplied inside scuttlebut to Younghein, the Sierra Club and Friends of the Earth, hoping that outside pressure would prod Kerr-McGee to clean up its operations. Younghein did her best, collecting 500 signatures on a petition for stricter federal controls and penciling two lengthy doomsday articles for the Oklahoma *Observer*, a maverick semiweekly unintimidated by Kerr-McGee.

By August, 1974, Karen Silkwood and two other union steering committee members had started to prepare a declaration of war against the company. Interviews of workers revealed a grim tale of corporate callousness: new employees often were sent directly into production without safety training (one such worker had been badly contaminated and had quit the next day before receiving medical attention); production schedules sometimes forced workers to stay on the job even when the air wasn't safe to breathe—supervisors ordering them to wear respirators rather than hunting the source of contamination; and plutonium was sometimes stored in such casual containers as desk drawers.

Karen and fellow steering committee members Gerald Brewer and Jack Tice prepared their case. A few weeks later, they left for Washington, D. C., to meet with OCAW officers at international headquarters. They arrived on September 26th and met Steve Wodka, an OCAW legislative assistant. Though only 25, he was among the union's best troubleshooters.

Wodka and his boss, Tony Mazzochi, pumped the Kerr-McGee trio for details, then marched them over to the AEC. Officials there listened to their complaints and promised an investigation.

"Both Tony Mazzochi and I felt this was a very serious situation [besides the reported incidents]," Wodka says. "But we felt it was premature to bring it to the attention of the AEC. We had to have proof before we could make any accusations. So we asked Karen to go back to the plant, [. . .] to document everything in specific detail."

Silkwood agreed to go undercover.

Back in Oklahoma she revealed her new role to Stephens. She stood in his living room, crouching over the radiator vent to shake off the autumn chill, and jabbed a delicate brown finger at the air: "We're really gonna get those mothers this time."

Six days before Silkwood's Washington trip, Stephens abruptly quit, riled by a sudden transfer. "When I first went to work there I wanted to be the world's greatest laboratory technician. Now I never wanted to see the place again.

"But Karen felt differently. She wanted to reform the place. She had tried to go through channels and she'd become very frustrated. But when she returned from Washington she was really excited. This was her chance to do something. She figured things were really going to change."

On October 10th, two of the nation's leading plutonium experts arrived in Oklahoma City from the University of Minnesota, summoned by the OCAW International to conduct crash courses for Kerr-McGee's plutonium workers. Their credentials were impressive: Dr. Donald Geesaman, a top AEC scientist for 13 years, had crusaded for stiffer plutonium standards until he was fired; Dr. Dean Abrahamson was both a physicist and a physician.

The two professors were told that 73 workers had been internally contaminated by plutonium during the previous four years. (Dozens more had accidentally brushed plutonium or been sprinkled with it, but had washed it off their skin.) The seventy-three had been exposed to airborne plutonium; any inhaled into their lungs could not be washed out. The probability of cancer in such cases, Dr. Abrahamson warned, "is disturbingly high." Because it takes 10 or 15 years after exposure to detect cancer, no cases have yet been reported to Kerr-McGee. But those workers with internal contamination must live with the threat of cancer for years to come.

Karen Silkwood was one of those seventy-three, and she was shocked by Abrahamson's news. She had assumed that she would stay clear of cancer if she did not breathe in more plutonium than allowed under AEC guidelines. But Abrahamson was saying, "If you can measure plutonium in the air at all, it's too high." The AEC guidelines, he said, were meaningless.

Silkwood grew moody and restless, working nights and unable

to sleep during the days. She got a prescription for some sleeping pills. And she began to hunt for another job.

But first, she vowed to Stephens, she was going to get proof [of other problems at the plant] . . . She had already collected some evidence, she said, and was certain that she could get more.

Silkwood spent the weeks of October staying after-hours, pouring over files, recording every questionable procedure, building a dossier in a dog-eared manila folder. She did not know then that other employees had noticed her spying, and that the plant rumor mill was abuzz with suspicions about what she was up to.

"I have guilt feelings about those weeks," Stephens says. "I should have talked to her more, been with her more, helped her out. . . . But I just wanted to forget about the place."

On Tuesday, November 5, 1974, Silkwood discovered that she had been contaminated with plutonium again. Frightened, she later called OCAW International headquarters in Washington from a pay phone. The office number rang and the union operator connected her. Steve Wodka.

"Hello." An uncertain trickle started down her face. Her voice faltered. "Please come to Oklahoma," she said. "Something very weird is happening here."

Three times in the past three days Karen Silkwood had been contaminated with plutonium, and no one knew where it was coming from. A monitoring device had first discovered flecks of plutonium on her skin and clothing shortly after she reported for work on November 5th. She had quickly stepped under a brisk shower. But the next day the monitor flashed on again. More plutonium on her skin. Another shower. On the third day the mystery repeated itself—and a nasal smear indicated she also was contaminated internally.

How much plutonium, she wanted to know, could she ingest before it burned out her insides?

Wodka tried to reassure her and promised to fly in. Silkwood hung up and sought out Drew Stephens. "She was damn near incoherent," says Stephens. "She was crying and shaking like a leaf; she kept saying she was going to die."

Again she picked up the phone and called long distance. Minneapolis. Dr. Dean Abrahamson. She wanted medical advice from a physician. She told him that somehow, somewhere, she had gotten plutonium all over her, inside and out. "She knew what the medical implications were," recalls Dr. Abrahamson, "and she was worried."

A team of Kerr-McGee inspectors, armed with alpha counters, full-face respirators, special galoshes, taped-up gloves and white coveralls, were called in to hunt down the source of the plutonium. There had been no recent accident at the plant to account for her contamination. So, at Silkwood's request, they trekked to her apartment. There the alpha counters commenced eerie gibberings. Plutonium, in small quantities, was everywhere. Outside on the lawn, the inspectors filled a 55-gallon drum with alarm clocks, cosmetics, record albums, drapes, pots and pans, and bedsheets. Alongside they stacked chairs, a bed, stove, refrigerator, television, and other items to be trucked to the Kerr-McGee plant for later burial in an AEC-approved site.

The plutonium trail turned hottest in the kitchen, inside the refrigerator. Packages of cheese and bologna were the two most contaminated items. Apparently, the plutonium had been tracked around the apartment from the refrigerator. But no one could explain how the foods had become the source of contamination. The apartment was sealed off and the AEC notified.

Silkwood, however, was more worried about the plutonium inside her. She kept popping the Quaaludes that had been prescribed a few weeks before. "The Quaaludes were supposed to be taken for sleeping at nights," Stephens says. "But she was using them during the day, just to calm down. I'd never seen her so scared."

Wodka had jetted in from Washington, and, after talking to Kerr-McGee and AEC officials, helped make arrangements for Silkwood to fly to an AEC laboratory in New Mexico to be checked out for poisoning. On Sunday, November 10th, five days after her first contamination, she boarded a Braniff airliner.

That same morning, a front-page *New York Times* story reported that, according to the AEC's own internal documents, the AEC had "repeatedly sought to suppress studies by its own scientists alleging that nuclear reactors were more dangerous than officially acknowledged or that raised questions about reactor safety devices." One AEC study, kept confidential for seven years, predicted that a major nuclear accident could kill up to 45,000 persons and pollute an area the size of Pennsylvania. *Times* reporter David Burnham, who in 1970 interviewed Frank Serpico and broke open the New York police corruption scandal, had sifted through hundreds of memos and letters and learned the AEC had a ten-year record of blue-penciling data, soft-soaping test failures and glad-handing an industry that increasingly appeared not to know what it was doing.

The report gave scant comfort to Silkwood as she flew to Los Alamos, New Mexico, site of the world's first plutonium explosion during the A-bomb tests of World War II. With her were Stephens and Sherri Ellis, her roommate of the past few months.

The three shared the same fears; all had been contaminated in the apartment.

For two days they underwent a "whole body count," a meticulous probing of skin, orifices, intestines and lungs, urinating at intervals into plastic bottles and defecating into Freezette box containers.

After the first day, the three had cause for relief. Dr. George Voelz, the health division leader, assured them they had suffered no immediate damage. Even Silkwood, by far the most contaminated, was told she was in no danger of dying from plutonium poisoning.

On Tuesday, November 12th, Silkwood called her mother to announce the good news revealed by the tests, but added, "I'm still a little scared. I still don't know how I got contaminated. I feel like someone's using me for a guinea pig."

"I told her to come home," her mother recalled. "And she said she would. She said she was ready for a vacation . . . she just had to do a couple of things first."

After more laboratory tests at Los Alamos, the three travelers flew back to Oklahoma City, landing about 10:30 Tuesday night. Because the women's apartment had been gutted of furniture, they checked in at Stephens's bungalow. Silkwood wandered over to her favorite radiator vent, rubbing to warm up, then went to bed early. She had a busy day ahead. She told Wodka she would give him the evidence she was collecting as soon as she returned from Los Alamos, and Wodka had set up a meeting with her and David Burnham, the *Times* reporter, who was winging in from the east coast. The meeting was scheduled for Wednesday night at the Holiday Inn Northwest in Oklahoma City.

On Wednesday morning, Silkwood drove to work. Contract negotiations between Local 5–283 and Kerr-McGee had begun the week before and, as a committeewoman, she was supposed to take part in the bargaining. She spent the morning in negotiations, arguing the union demands for better safety training and higher injury benefits. In the afternoon she met for several hours with AEC inspectors who were trying to unravel the mystery of her contamination.

At 5:15 P.M. she drove to Crescent, about five miles from the plant, and stopped at the Hub Cafe for a supper meeting, sans supper, to discuss negotiations strategy with Local 5–283. Jack Tice, who headed the negotiating team, told the assembled union members that, as expected, Kerr-McGee was not budging off its hard line.

Silkwood excused herself about 6 P.M. to telephone Stephens, reminding him to pick up Wodka and Burnham at the airport and to expect her at the motor hotel about 8 P.M. She sounded normal, Stephens remembers, perhaps a bit excited about having an audience with the *New York Times*. At 7:15 P.M. Silkwood left the Hub Cafe and headed for Highway 74 and the Holiday Inn Northwest. A fellow union member would later swear in an affidavit that Silkwood, minutes before she left the restaurant, was carrying a manila folder an inch-thick with papers. [. . .]

Thirty miles away, Wodka, Burnham and Stephens waited for the papers until 8:45. Then they picked up the phone; but for some reason the Holiday Inn lines were out of order, and another hour passed before the three could get through.

Meanwhile, at 8:05 P.M. a truck driver, sitting high up in his cab rolling along the two-lane highway, spotted the white Honda, almost hidden in the muddy culvert. Silkwood had traveled about seven miles from the Hub Cafe, a ten-minute drive.

By the time Stephens, Wodka and Burnham learned the news from a local union member, her car had been towed to Ted Sebring's garage in Crescent. And Silkwood had been pronounced dead on arrival at the Guthrie Hospital, the victim of multiple and compound fractures.

The three men raced to the culvert, only a mile from the plutonium plant, and prowled about, stepping gingerly through the mud. All they could find were shards of aluminum trim, the orange roadside reflectors that had been trampled by the bouncing car, and Silkwood's uncashed paycheck.

Later they found the wreck locked up in Sebring's garage and peered at it through the window. They stopped at the home of union committeeman Jack Tice, one of the last to see Silkwood alive. Stephens called Silkwood's parents. Then they returned to the culvert, searching for clues in the tire tracks and the scraps of metal.

The explanation the State Highway Patrol offered was that Karen Silkwood, exhausted after driving 600 miles from Los Alamos to Oklahoma City, had fallen asleep and drifted off the road to an accidental death. Almost immediately the police had to alter their official version when they were told Silkwood had flown from Los Alamos and had gained a full night's sleep only 12 hours before the crash.

The second official version was somewhat more convincing. Sometime during the afternoon of November 13th, Silkwood had gulped down at least one of the pasty white Quaaludes from the vial in her coat pocket. Oklahoma City's chief forensic toxicologist, Richard W. Prouty, discovered 0.35 milligrams of methaqualone in her bloodstream, conceivably enough to lull her to sleep on the highway.

But that was not sufficient for Steve Wodka.

Silkwood had swallowed several Quaaludes in the past week without nodding out. Why would she fall into a trance on her way to an extremely crucial meeting? And the proof of fraud she was supposedly carrying had disappeared. Her present effects, listed by the medical examiner, included an ID badge, an electronic security key (for the plant), two marijuana cigarettes, a small notebook, her clothes, $7 in bills and $1.69 in change. But there was no manila folder heavy with Kerr-McGee documents.

Trooper Rick Fagan, however, had mentioned finding dozens of loose papers blowing about the accident scene when he first arrived. Fagan had plucked up the papers, he told his superiors, and shoved them into the Honda. According to the highway patrol's information officer, Lieutenant Kenneth Vanhoy, the papers were in the Honda when Ted Sebring hauled the car away.

Presumably they were still there at 12:30 A.M.—five hours after the accident—when Sebring unlocked his garage for a group of Kerr-McGee and AEC representatives who said they wanted to check out Silkwood's car for plutonium contamination.

But by the next afternoon when Stephens, Wodka and Burnham claimed Silkwood's car from Sebring, no papers were inside.

Three days after Silkwood's death, an auto-crash expert hired by OCAW International arrived in Oklahoma City from the Accident Reconstruction Lab of Dallas. A. O. Pipkin, an ex-cop, is a veteran of 2,000 accidents and 300 court trials, a no-nonsense pro considered the best man around for piecing together an accident scenario.

Dressed in a Day-Glo orange jumpsuit, Pipkin examined the Honda and found two curious dents, one in the rear bumper, another in the rear fender. They were fresh; there was no road dirt in them.

And they appeared to have been made by a car bumper.

At the scene, Pipkin noted that the Honda had crossed over the yellow lines and hit the culvert on the *left* side of the highway. If Silkwood had nodded into a stupor, he reasoned, she would have drifted to the *right.* In the red clay, Pipkin found something else the police apparently disregarded: tire tracks indicating the car had been out of control before it left the highway.

Pipkin's disconcerting conclusion: Karen Silkwood's Honda had been hit from the rear by another vehicle.

On December 20th, five weeks after Karen Silkwood's death, Kerr-McGee temporarily closed its plutonium plant. These were trying days for the company. Supporters of Kerr-McGee found it necessary to print ads reminding Oklahomans that Dun & Bradstreet had recently named it among the five best-managed corporations in the country. But headlines kept popping up all over, thanks to the *New York Times* wire service, telling of a mysterious death [. . .] and ill-trained workers sent in to handle one of the world's most dangerous poisons.

Nuclear proponents were worried, especially those of the nuclear elite like Dean McGee, who had been helping babysit plans for a multibillion-dollar "nuclear park" near Muskogee in northeastern Oklahoma. The initial rendering included over 20 facilities, the boldest assortment of nuclear props ever assembled. Even Muskogee's citizens were beginning to flinch. "The bad publicity," complained Senator Henry Bellmon, a big McGee booster, "is making it more difficult to get what we want in the Muskogee area."

But the controversy around Kerr-McGee would not quit. Hints of strange goings-on salted the news. Robert G. Bathe, a plutonium worker, reported to police that a motorist had "harassed" him as he drove home from the plant a few nights after Silkwood's death. When Bathe's statement leaked to the press, he and the police suddenly refused to discuss the incident. Shortly afterwards, however, David Burnham of the *Times* reported that security at the plant was so atrophied that 60 pounds of plutonium—enough for five Nagasaki bombs—were unaccounted for and possibly missing, an allegation Kerr-McGee heatedly denied.

The most serious development, though, was the AEC investigation, which promised a full report on Kerr-McGee.

On December 17th, at the height of the AEC investigation,

Kerr-McGee announced that five more employees had been contaminated at its plutonium plant. The company claimed it had evidence that the accidents were contrived, a modest slander suggesting that workers sniffed poison to embarrass their bosses. Though Kerr-McGee said it had given its evidence to the FBI, the FBI denied receiving it. Nonetheless, three days later, Kerr-McGee handed out lay-off slips, announcing the plant would not reopen until the payroll was checked for security.

Closing the plant five days before Christmas reminded the workers how close they were, in hard times, to standing in line for unemployment checks; some feared that talking to AEC investigators might further jeopardize their jobs. (Earlier in December, rather than risk losing their jobs to strikebreakers, the workers of Local 5–283 had ratified a new contract that again fell far short of their demands.)

Predictably, the plant shutdown ruptured the tentative alliance between the plutonium workers and local environmentalists. To Ilene Younghein, the shutdown was a first step to victory; to Frank Murch, a middle-aged man with seven years invested in Kerr-McGee, it was a slap in the pocketbook: "You're damn right I'm bitter about this. I'm bitter at the environmentalists. It's a hell of a thing, putting this many people out of work." Some began blaming the dead. One worker who earlier had talked about honoring Karen Silkwood with a special grave marker now spat at the mention of her name.

"Attitudes changed," says Gerald Brewer. "People started to blame Karen for getting thrown out of work right before the holidays." Brewer was one of the two union committee members who accompanied Silkwood to Washington in September. He had worked at the plant three years.

In early January, after plutonium production resumed, Brewer was demoted from his job and transferred to an isolated warehouse. Two weeks later he was fired. There was no official explanation; a company spokesman was still denying the firing five days later.

Brewer's apparent sin, besides his role in compiling grievances, was his refusal to submit to a polygraph test that asked questions such as: "Have you ever talked to the media?" Although of questionable legality, the polygraphs were required of most plutonium workers as a "security precaution" before they could return to their jobs. A Kerr-McGee official described company strategy in a conversation with Jack Taylor, ace reporter for the *Daily Oklahoman:* "We're going to tool back up slowly and hire people who are trustworthy and not involved [in the union]." As for undesirables—"You don't have to tell them [anything]. You can just say, 'You didn't clear security.' "

Along with Brewer, five other workers who snubbed or failed the polygraph tests were handed pink slips. Jack Tice, the third union committeeman who made the trip to Washington with Karen, was transferred to the most isolated part of the plant. Among the six employees fired was Sherri Ellis, the girl who shared the contaminated apartment with Silkwood. After her roommate's death, Ellis initially cooperated with Kerr-McGee, refusing to talk to either the OCAW or the media. At one point she was seen, red-eyed and distraught, being escorted by two company detectives away from the Edmond Broadway Motor Inn where she had been staying, compliments of Kerr-McGee. Then Ellis—without explanation—aired a suggestion that Silkwood may have been pilfering plutonium from the plant. Shortly thereafter, Kerr-McGee reportedly offered Ellis $1,000 as payment for any claims she might have outstanding against the company.

But Ellis turned down the offer. She began to worry that she had been more seriously contaminated than she had been told; her gums bothered her and she had trouble sleeping. In late December she hired a lawyer and threatened to sue the company for copies of her health records. Three weeks later she was fired. Two weeks after that, in early February, Ellis told friends that twice someone had tried, and failed, to break into her apartment.

During the month between the plant shutdown and the firings, the AEC had published the results of its investigation. (According to a *Daily Oklahoman* story, Kerr-McGee officials received a copy of the report well ahead of its official release apparently in violation of AEC rules.) Company officials, who had been refusing comment since Karen Silkwood's death except to say, "We will let the AEC speak for us," pronounced themselves pleased with the findings.

On the question of falsified records the AEC did locate one former worker who admitted using a felt-tip pen to touch up photo negatives that measured the welding on plutonium fuel rods. The worker, however, said he acted only to make his job easier and not under orders from Kerr-McGee. Without Silkwood's documents, the AEC reported, it could find no other proof. But the OCAW questioned whether the AEC was really looking. According to the OCAW, the AEC lied when it claimed to have interviewed a worker who disputed Silkwood's allegations of fraud. This worker, the OCAW said, gave the union a sworn affidavit that the AEC never interviewed him—and that he believed quality controls are not adequate.

Whether Kerr-McGee's plutonium fuel rods are safe is still unknown; they have yet to be tested at the AEC facility in Richland.

On the question of plant safety, the AEC reported that 20 of the 39 grievances it examined were true or partially true: Plutonium had been stored in a desk drawer instead of a prescribed vault; in various incidents, employees had been forced to work in areas not tested for contamination or where leaks remained; in another, the company failed to report a serious leak that had forced it to close the plant in May, 1974; generally, respirators had not been checked regularly for deficiencies; few workers had been properly trained.

Such disregard for safety, the AEC decided, merited no censure beyond adding these new citations to those already in the Kerr-McGee files. Kerr-McGee was free to resume its role in the AEC's fast-breeder program.

The Silkwoods were still trying to sort out what had happened, to resolve the many unanswered questions. They kept pondering over the central mystery—how had Karen been killed? Had the Quaalude rendered her in an unconscious or hypnotic state, or had someone stolen out of the darkness, a mugger armed with a powerdrive, to dead-end her into the culvert?

"I know that working for that company is what killed her," her father said. "But I would still like to know if it was an accident or if it was a murder."

On November 1, 1976, Karen's parents filed a civil suit against Kerr-McGee. It may be one way, they have been advised, to flush out some answers. Kerr-McGee, according to one source, is already preparing for a suit.

So far the investigation is a stalemate.

In January, the Oklahoma Highway Patrol reopened the case for six days, reexamining the evidence and reaching the same conclusion. Actually, it didn't reexamine all the evidence. It couldn't.

Along Highway 74, the Honda's tire tracks had been obliterated by a tractor-grader—reportedly less than 24 hours after A. O. Pipkin had inspected the scene. The stretch of highway had then been repaved on one side, making it difficult to determine in which direction a sleepy driver might have drifted.

The Honda was still available. But the State Highway Patrol regarded it as unreliable evidence since it had been out of its possession.

At the request of OCAW International, however, three other auto-crash experts scrutinized the car. All three agreed with Pipkin that the dents could not have been caused by the concrete culvert. Dr. E. L. Martin of Albuquerque, New Mexico, who scrutinized the bumper with a microscope, said that the bumper dent resulted from "contact between two metal surfaces." It is highly probable, according to these experts, that another car slammed into the Honda as Silkwood drove toward the Holiday Inn Northwest.

Was the other driver a hired killer? Or a loose drunk? No investigator knows.

The three men who waited for Silkwood that night think someone was trying a scare tactic that got out of hand.

If Silkwood did have proof of fraud, then several people conceivably had motives for intimidating her: a plant supervisor afraid of going to prison for falsifying records; a company higher-up who feared a scandal would mean multimillion-dollar losses; a plant worker who felt that Silkwood was threatening his livelihood; or an AEC official who worried that she would jeopardize the entire fast-breeder program.

At OCAW headquarters, Steve Wodka has found it difficult to return to other chores. The Silkwood case keeps nagging him. There are too many unanswered questions. For instance, how did Silkwood become contaminated a week before her death? For weeks afterwards, Wodka kept the results of her Los Alamos tests scribbled on an OCAW blackboard, trying to puzzle out the mystery. The most logical explanation, he decided, was that Silkwood had been contaminated at the plant and unknowingly carried the plutonium home with her. However, the AEC reported that this would have been virtually impossible, given her duties at the plant during the time immediately preceding her contamination.

So now Wodka has come reluctantly to believe she was poisoned. "Someone must have entered her apartment and placed the plutonium in her refrigerator. That's the only way it could have gotten on the cheese and bologna. We've heard from several sources, including the AEC, that Karen had been seen going through the files, looking for records. Someone apparently figured out what she was up to. One sure way of preventing her from gathering any more evidence would have been to poison her, maybe scare her into leaving."

Wodka also cites another AEC finding: Extra plutonium apparently had been added to four of the urine samples Silkwood gave to Kerr-McGee for analysis in late October and early November. "I think someone tampered with these samples, hoping to get her out of the plant or at least to confuse the issue."

Kerr-McGee officials have advanced a different conspiracy theory, passed along in off-the-record conversations with local reporters. They suggested that Silkwood contaminated herself to embarrass the company. But even if one assumes that Silkwood had become a frenzied zealot, Kerr-McGee's theory does not explain why she thought getting contaminated in her apartment would embarrass the company, or how one more contamination would have any impact on the company after 73 cases in four years.

The OCAW International has pledged not to give up until the case is solved.

"Karen was a very unusual person," Wodka says. "She stood up to the company. She was outspoken. She was very brave, now that we look back on it; in many ways she was a lone voice. She was willing to go ahead when other people were afraid."

"She died for a cause," agrees Ilene Younghein. "She will be remembered as a martyr."

At NRC, the regulatory division of the new AEC, she will be remembered, too. The commission has begun a file on her. It reads: "Silkwood, Karen, . . . Former Employee, Kerr-McGee."

Issues and Comments

Kohn's article implies more than it provides in the form of concrete evidence. Once Karen Silkwood's personality quirks and other irrelevancies are set aside, several important questions remain: What do the Kerr-McGee and Browns Ferry cases have in common? Does there seem to be a breakdown in leadership, personal ethics, and technical care in either or both instances? Did the AEC recognize any of the problems or attempt to remedy them by more rigorous regulation?

Is it possible that neither private industry nor the federal government can be trusted with nuclear technology? Both the Kerr-McGee and Browns Ferry cases reveal a tendency for one or the other to suppress derogatory information.[1] To what extent were coverup tactics employed in each case?

Kohn implies that the documents Karen was carrying the night she died were removed from her car at Sebring's garage by the anonymous group of AEC and Kerr-McGee officials. How did these men learn of her accident so quickly? Also, checking Karen's car for plutonium contamination could have been done the following morning; yet the group used this as an excuse for gaining access to her car immediately after the accident.

If Karen was murdered, her assassin achieved several things by silencing her besides preventing *The New York Times* from obtaining evidence of Kerr-McGee records falsification. Especially important is the intimidating effect her death had on other nuclear critics. To the extent that reprisals, employer pressure, and other forms of manipulation are common, do the darker possibilities suggested by Kohn become believable?

One might ask why the intimidators are going to such extremes. Is it possible that the fiscal investment, both private and federal, in fission power is now so large that unusual steps are being taken to protect it? Thomas Scortia and Frank Robinson point out that "a retrenchment from fission power could well bankrupt the major power companies of the country."[2]

In Chapter 16, evidence is presented that corporations engage daily in a variety of underhanded and, in some cases, blatantly dishonest practices. Can we trust nuclear industry executives and engineers with the responsibility of developing safe, reliable nuclear power systems in the best engineering tradition?

To what extent is the federal government backing away from its responsibility for enforcing the high engineering standards required for safe electrical generation by nuclear fission? To what extent is private industry shortcutting its responsibility for implementing these standards? If shortcuts cannot be prevented, is our society approaching an upper limit to the degree of engineering precision attainable in *any* undertaking? This threshold may be viewed in terms of human fallibility—the dishonest response wherein those who cannot or do not want to attain high engineering standards take steps to suppress evidence of the consequences. Evidence of the threshold may also appear in the form of plant performance data—frequent outages, equipment and systems failures, and inadequate electric generation.

If we are approaching this limit, should we close down the nuclear program before an accident reveals that this limit has been exceeded? What are the possible consequences if we don't? If we do impose a nuclear moratorium, we will suffer certain costs attributable to reduced electric power supply and cancelled nuclear projects. Are these costs acceptable in light of the alternatives?

Is there a difference between safety designed into a mechanical system and safety practiced by people? Are the differences important, and do there seem to be deficiencies in *both,* so far as the nuclear industry is concerned?

In Kohn's account of Karen's personality, is her insistence on perfection, on holding others to a high standard, believable? Are her standards unrealistically high? If they are both reasonable *and* high, why did she have so much trouble getting others to accept them?

3
We Almost
Lost Detroit
John G. Fuller

EDITOR'S PREFACE: On August 4, 1956, the Atomic Energy Commission (AEC) issued a construction permit to the Power Reactor Development Company (PRDC) of Michigan for the Enrico Fermi fast breeder test reactor near Detroit. The Supreme Court dismissed a suit challenging the validity of the permit in 1961.[1] By deferring a definitive safety finding until construction was finished, the AEC assured that only narrow questions pertaining to plant safety would be reviewed while the project was vulnerable to cancellation. This procedural tactic enabled the AEC to sidestep the broader, far more important issues of waste management, public safety, and the rights of future generations to live free from unreasonable risk.

These issues should have been raised and openly debated before construction began on the first nuclear plants. They were not, and one reason may have been that both government and industry wanted to avoid alarming the public to a degree that might have interfered with plans for generating most electricity in the United States by nuclear fission.

The 1961 Supreme Court decision is notable for the dissenting, appended opinions of Justices Douglas and Black. They refer to the AEC safety committee report on the Fermi reactor as follows:

> Plainly these are not findings that the safety standards have been met. They presuppose . . . that safety findings can be made *after the construction is finished.* But when that point is reached, when millions have been invested, the momentum is on the side of the applicant and not on the side of the public. The momentum is not only generated by the desire to salvage an investment. No agency wants to be the architect of a white elephant. . . .[2]

Five years after the U.S. Supreme Court decision, the Fermi test reactor was ready to begin its final stages of power ascension. By that time, well over $100 million, and millions of man-hours, had been invested. In eight carefully plotted steps the Fermi staff brought the reactor's power toward the first goal of 100,000 kilowatts.

The following excerpt from John G. Fuller's book, published in 1975, details the last few hours of the Fermi reactor's operation, its partial meltdown, and its decommissioning. Despite meticulous care in the design, planning, and construction of the reactor, the team in charge of its operation faced a great many unknowns. Would a nuclear explosion occur if the breeder exceeded the safe fission level? Theoretical breeder-accident analyses predict that if fuel damage and melting take place, and if the coolant is not able to reach the tiny channels between the fuel rods and assemblies, the fuel could be jammed and compacted together.[3] Then a small explosion might occur that would push the fuel rods even closer. The closer the fuel rods are packed, the greater the unwanted power surge. After that, a larger explosion might occur. In other words, a small, unplanned power surge in the core could lead to a small explosion causing a large power surge in the reactor, followed by a large explosion.

Any one of these unknowns could turn a carefully planned test into a holocaust. Luckily for Detroit, the damage at Fermi was limited and the control room staff terminated reactor operation in time. The accident nevertheless underlined Justice Douglas's earlier concern. His criticism of the majority opinion which gave a green light to Fermi five years before the accident applies equally well to President Gerald Ford's approval of $500 million for a new breeder in Tennessee: "a lighthearted approach to the most awesome, the most deadly, the most dangerous process that man has ever conceived."[4]

By October 4, 1966, the Fermi engineers had things in good enough shape to make another try at reaching their first-stage high-power goal. They planned to run the reactor for a while at idling speed, slowly raising the temperature of the viscous sodium fluid to a little over 500° F., a temperature high enough to make pressurized water boil, but not sodium. There were three routine tests to make, mainly checking pressures and temperatures. At eight o'clock at night on October 4, the system was ready to make the approach to criticality, that point at which the reactor would stand by for the next step.

Here, the process was stopped for the control room operators to check everything out. The readings on the instruments were exactly as predicted for the amount of distance the control rods had been withdrawn. This is an important check, because the higher the long, thin control rods are pulled out of the core, the greater the power that should be coming through. If the rods are out some distance and the power is less than predicted, it is an immediate sign of trouble.

There was no trouble apparent at this time, however. Things looked good for the big push which was to begin at 8 A.M., October 5. The Fermi crew was naturally itchy to get on with the job after so many delays.

The first problem that was discovered on the morning of the fifth was a malfunction in one of the steam generator valves. It took until nearly 2 P.M. to clear it up. Then another power push was made. But this was barely started before there was trouble with the boiler feed water pump.

Again the control rods were pushed in to reduce the power while this was taken care of. After another start-up, there was a brief hold to put the reactor on automatic control. Then the power began rising again. By 3 P.M., the power was up to about twenty percent of its 100,000-kilowatt limit in the current series of tests.

It was at this point that Mike Wilber, the assistant nuclear engineer in the control room, noticed some erratic changes in the neutron activity of the reactor, [possibly] . . . a pickup of electronic noise in the control system. To be on the safe side, the reactor was put on manual control again, and the reactor behavior watched carefully.

In a few moments the apparent noise disappeared from the instrument readings. Again, the reactor was put on automatic control. Any decision at the control board now would be critical.

Just a few minutes after the first ominous signs at the control panel of the Fermi reactor, at 3:05 P.M. to be exact, Mike Wilber noticed another problem. For the amount of heat and power that was coming out of the reactor, the control rods should have been raised only six inches out of the core. Instead, they were a full nine inches out. This was not a comfortable situation. Further, the reactivity signal was again moving crazily and Wilber's first thought was that the core temperature was too high.

The instruments that showed the temperatures of the individual subassemblies were rather awkwardly installed, about twenty or thirty feet away from the main control board, behind the relay panel —a wide bank of instruments stretching along the width of the control room.

The operator stopped the power increase immediately, and Wilber went behind the control board to check the core outlet temperature instruments. He scanned them quickly. It was immediately obvious that two subassemblies were showing high outlet temperatures: M-140 and M-098. Each tall, slim can that wrapped a bundle of slender fuel pins had its own designation in the core, just as a crossword puzzle has its squares identified. M-140 had been acting up before. It still wasn't clear whether it was the instrument that was off, or whether the subassembly itself was actually overheating. The instrument had been reshuffled to a new position to check this, because of the previous misbehavior. But M-098 had never been a problem child. And it had never been moved from its original position.

It was hard to get a complete picture of the blistering hot core, because only one out of every four subassemblies was equipped with a thermocouple. A most disturbing thing was that M-140 should be reading about 580°. It was now showing over 700°.

Suddenly, as Wilber was standing in front of the temperature instruments behind the control panel, radiation alarms went off. It was exactly 3:09 P.M. The air horn began blasting—two blasts every three seconds. Then over the intercom, a laconic announcement: "Now hear this. Now hear this. The containment building and the fission product detector building have been secured. There are high radiation readings, and they are sealed off. Do not attempt to enter. Stay out. Both buildings are isolated. This is a Class I emergency. Stand by for further instructions. Stand by for further instructions."

The crew began scrambling about on its assigned emergency procedures. All doors and windows were closed. The fresh air intakes in all buildings were shut down. Plant guards closed off the entrances to the site. The health physics team rushed to the control room.

A Class I emergency was in effect at Lagoona Beach, and no one could say what would happen next.

Regardless of how well-trained and prepared the Fermi crew was to meet emergencies, the situation in the control room was tense and dangerous. A sudden appearance of radiation, of poisonous fission products leaking out of the reactor vessel and into the containment building, needed immediate expert attention and extremely cautious action.

It was a delicate situation. The alarms had gone off. Radiation was leaking. Some core temperatures were inexplicably high. Direct inspection of the core of the reactor vessel was impossible. Even if the containment building were not sealed off, there was no way to see if the fuel rods were melting, how much damage to the core had been sustained, what direction the accident was taking, or the shape of any melted fuel.

The maintenance engineer, Ken Johnson, was at his desk in his office, not far from the control room, when the alarms went off. He ran down the short corridor to the relay room, where there was a panel of gauges monitoring the radiation levels. They were reading high, especially in the containment building. His first thought was whether there was anyone in it. No one could enter without clearance. The only entrance and exit to the containment shell was through an enormous double door that formed an air lock. Anyone entering would have to step into a chamber and wait for the outer door to close. Then the inner door—as thick and enormous as that of a huge bank safe—would open. The process was timed for thirty seconds. Johnson picked up a phone and called Bob Carter, his maintenance foreman. He asked for an immediate count of the crew. They were all present and accounted for.

At the control room console, the operator had begun to pull down the power as soon as the radiation alarm sounded, dropping the rods slowly to see if the reactor could be brought under control. No one knew yet what had happened, or why.

A natural impulse, of course, would be to scram the reactor immediately. But thermal shock, due to sudden changes in the sodium temperature, had to be guarded against, in both the blazing hot core and the channels that carried the coolant.

This sort of problem left the operating crew between the devil and a runaway meltdown. Could an engineer or reactor operator be cool enough to handle the complexities in a crisis situation? Even if a technician memorized every factor, every golden rule in the industry's bible, how could they all be considered in the seconds— or minutes, if they were lucky—that were allowed in a nuclear accident crisis?

Mike Wilber was still trying to put his finger on what was happening inside the reactor core. So far, at least, the radiation was not threatening the control room, and it was within reasonable limits where it was coming out of the tall stack. It had not yet reached intolerable limits outside the containment shell. Already, a team of health physicists under John Feldes was circling the outside of the containment building with Geiger counters.

In this type of emergency situation, time is crucial; confusions and complications create frustrating delays. One complication was that a Fermi instrument engineer had been working on the fission product monitor, checking the calibration on the panel. When he saw the steep climb in radiation at the time of the alarm, he thought immediately that he had merely triggered a false alarm while working with the instruments. But the temperature readings on the subassemblies and other indicators showed that something was happening in the core that was very real. And so, at 3:20 P.M., eleven minutes after the radiation alarm had gone off, the decision was made to manually scram the reactor. The question: Was this too soon or too late?

All the rods went down into the core normally, except one. It stopped six inches from the full "down" position. This was no time to take a chance. A second manual scram signal was activated. The reluctant rod finally closed down fully.

Ken Johnson made his way to the control room. The red light in the corridor, which had read REACTOR ON, was no longer on, so he knew now that the reactor had been scrammed. The control room was quiet; operators for the new shift were coming in; several staff members were checking instruments and charts, trying to find out what the trouble was. All the signs seemed to be pointing to a fuel melting situation, and every nuclear engineer in the business knew what that could mean.

The prime questions were: Was the reactor secure? Would it

stay secure? What could be done to explore the accident that wouldn't trigger a secondary accident more serious than the first? The urgent priority was to make sure that no hazardous condition existed in the core. The potential hazard was, of course, enormous, and the lack of experience in handling fast-breeder accidents made the situation fraught with danger. Further, no provision had been made in the design for investigating and recovering damaged fuel elements.

To say that the Fermi team was sitting on top of a powder keg would be a major understatement. The threat of a secondary accident was, as McCarthy [PRDC assistant general manager] was to say later, "a terrifying thought . . ."

However terrifying the situation, it was staring the Fermi crew in the face. The keynote was *uncertainty*. There were few road maps to go by. No one at the hastily called meeting knew exactly what had happened within the reactor core. No one knew what would happen if they tried to look inside it—or *how* to look inside it. The most probable cause of fuel melting was the blocking of the sodium coolant. (See Figs. 1 and 2.)

McCarthy took command by saying: *"We will go at this very, very slowly."* Before any kind of exploration of the condition of the reactor, a procedure would have to be written. It would have to be checked and double-checked before any attempt to put it into action would be permitted. Again, there could be no margin for error.

There were a couple of public laws in Michigan, dating as far back as 1953, which provided for the attempt to cope with nuclear catastrophe. The department of public health was named the official radiation control agency. But how could a cloud of radiation that could fan out to cover an area the size of a state be controlled, even by the most expert public health officer? The department of state police was designated as the coordinator of civil defense activities if and when the governor proclaimed an emergency. But how could a handful of state police handle a gigantic exodus from the city of Detroit, or even from Monroe County?

The scene was almost unimaginable. Trucks, cars, buses stalled in massive traffic jams along the superhighways. Long streams of people carrying blankets, pots, pans, children, moving out of Detroit toward Ann Arbor, Lansing, Grand Rapids, Ontario—themselves places of dubious safety under the silent plume of radiation. And yet,

Artist's Sketch of Fermi Reactor

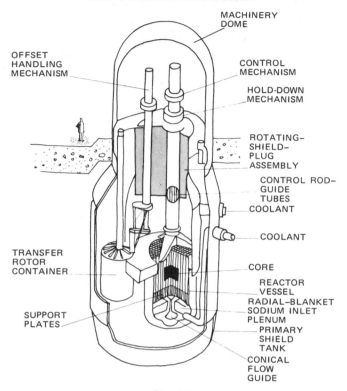

MACHINERY
DOME

OFFSET
HANDLING
MECHANISM

CONTROL
MECHANISM

HOLD-DOWN
MECHANISM

ROTATING–
SHIELD–
PLUG
ASSEMBLY

CONTROL ROD–
GUIDE
TUBES

COOLANT

COOLANT

TRANSFER
ROTOR
CONTAINER

CORE

REACTOR
VESSEL

RADIAL–BLANKET

SODIUM INLET
PLENUM

SUPPORT
PLATES

PRIMARY
SHIELD
TANK

CONICAL
FLOW
GUIDE

Figure 1

in the AEC meeting at Brookhaven, the only answer that had come up in the discussions was evacuation.

The state of Michigan plan reads with simple eloquence:

> In the event that an incident occurs which releases radioactive materials in concentrations that may be a public health hazard, this plan will be implemented. Implementation will commence by proclamation of an emergency by the Governor or by the order of the Director of the Department of Public Health . . . This department will: perform monitoring, evaluate data, and establish emergency response actions.

But what would these actions be?

The Michigan radiation emergency plan had many provisions. One of them was that the state police would notify the bordering states and counties of the approaching danger. But what, in turn,

Schematic Diagram of Lower Portion of Reactor and Vessel

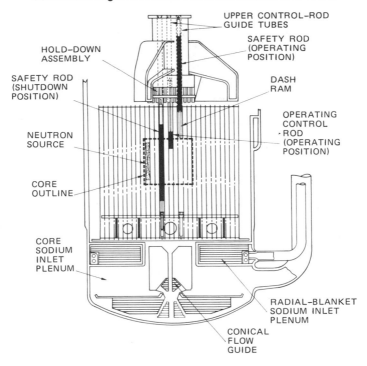

Figure 2

could these states or counties do, aside from the vagaries of "establishing emergency response actions"? What is the "response action" for a cone of radiation that will settle an invisible mantle of contamination only God knows where?

The plan included neat and tidy classifications of exposure conditions. There were three main classes: "Whole body exposure, including eyes, gonads, and blood forming organs," exposure to "the thyroid of a one-year-old child," and "liquid discharges." The only real answer was to vacate the area.

The plan itself seemed an exercise in futility. But so was the meeting in the conference room of the Fermi plant. The few confirmed facts that could be accurately determined at the time were that, because of the combined readings of several meters, there had been both "fuel melting" as well as "fuel redistribution," meaning that the fuel had shifted as well as melted. This would automatically leave the way open for further and more serious accidents to happen. And there was still the question as to whether the reactor had been scrammed soon enough.

Many explanations of what might have happened were brought up at the meeting: broken fuel pins, strainers, foreign material on the pins, fuel swelling, and other possibilities that might have blocked the coolant from coming through the subassemblies. Somehow, somewhere, the melted subassemblies must have been starved from their protective sodium coolant, either by some foreign matter that blocked the nozzle, or by the flow behavior of the sodium itself.

McCarthy laid down two programs. One was to work out a detailed analysis and experimental program to find out just what were the chances of a secondary accident. The other was to try to find out what the cause was, and to try to get the reactor back into service.

But the first problem would be the one hanging over not only the heads of the crew, but the entire state of Michigan as well.

The slightest disturbance of a partially melted core could easily set off a more powerful secondary accident. Worse, the primary accident itself had gone beyond the confident predictions for a "maximum credible accident" of both McCarthy and Hans Bethe. McCarthy had stated flatly that only one subassembly could melt in the Fermi reactor. The instruments already showed that two were affected, and there were probably more. He had also stated that the reactor would shut down automatically in such a situation. It had not; it had required a manual shutdown. Hans Bethe had testified that a core meltdown accident was "incredible and impossible." Both were experts, and both were wrong.

Now, with the reactor shut down, and no one knowing what possible shape the core and the melted uranium were in, what could be believed about the other predictions?

With the belated scramming of the reactor, the radiation leakage had begun to drop off. It was some comfort, but the concern was what might now happen in the core. The only way to get at the core was through the fuel-loading contraption, the awkward and clumsy Lazy Susan mechanism that provided no vision of what was going on, and that could easily jar a partly melted core into a secondary accident. It was like trying to look inside a gasoline tank with a lighted match. How could they explore a reactor drenched in radioactive poisons without the risk of wiping out Detroit and a big chunk of Michigan with it? Ironically, hardly anyone in Detroit, or the state of Michigan, had any idea of the potential danger.

Even though the control rods were shoved down into the core, there was no assurance that a secondary accident could not take place. This was the problem the Fermi engineers faced—and what Detroit and its surroundings would face if the worst did happen as a result of probing into the causes.

Under these conditions, there is little wonder that the Fermi engineers talked of "hair-raising decisions" and "terrifying thoughts." In effect, they were sitting on top of a volcano, which if left alone might be all right; but, if they tried to take a peek inside, it might erupt. Yet they were forced to take some action. They could not walk away from the accident, even if they wanted to. Aside from the hundred-plus million investment at stake, it would be impossible to leave a hot reactor sitting there, loaded with deadly fission products, soaked in radioactive sodium, and choked in an unknown configuration of melted uranium. The reactor could not be sealed in a tomb of cement and forgotten about. Sooner or later, its poisons would eat down through the base, no matter how thick, and contaminate the water table and soil below it. A cement tombstone, even if poured lavishly over the core, would eventually sweat out the buried poisons and continue the contamination for thousands of years. Not only that, but if the concrete became wet when it was irradiated, there could be what is called a "radiation dissociation" of the water to hydrogen and oxygen which would actually create an explosive mixture, turning the entombed reactor into a bomb. It was too late to go back to the drawing boards.

For several days, the only thing the Fermi crew could do was gather ideas, and write up procedures as to how to carefully carry them out. But the time bomb was still ticking, quietly and relentlessly. During those days, the weather grew less and less cooperative, with the wind shifting so that any escape of radiation would cover the maximum population of Detroit and its sprawling suburbs. The day of the accident marked the beginning of a warm spell, so any escaping radiation would tend to hang lazily under the nocturnal inversion conditions that existed through each night—the warm air trapped by the cool lid that boxed it in. It was, in fact, perfect Indian Summer weather—the worst possible kind for any ground release—a time when the haze from burning leaves hangs heavy and stagnant in the air, creating a smoky veil that lingers pungently for days.

Finally they decided to run some cautious tests. In one tense and timid exploration, the control rods were withdrawn one at a time to test the reactivity level, and then shoved back into the core. On reading the instruments, it was confirmed that fuel melting had definitely taken place.

A decision had to be made. The engineers could no longer wait and watch. Inside the bowels of the reactor vessel was a maze of unknown geometry. Exploring it could be disastrous if the wrong steps were taken. A slight jar or bump under the wrong conditions could catapult the reactor into a helpless runaway. But somehow that wrecked fuel had to be hauled out, and the unknown faced.

Finally, a consensus was reached. A surgical incision had to be made, and the scalpels prepared for an operation to remove the diseased fuel.

The situation changed from one of trepidation to one of crisis.

There was practically no experience to go on, and any attempt to remove melted fuel from the reactor was fraught with catastrophic potentialities. To get at the damaged subassemblies meant raising the giant hold-down device that sat on the top of the core like an enormous spider. If any of the fuel subassemblies were sticking to its huge fingers when it was lifted, there could be hell to pay.

In an atmosphere of controlled tension, the Fermi engineers began the job. Using every measuring instrument in the core that could apply, the hold-down device was raised very slowly. It worked. Apparently nothing was stuck to its claws. Then, very slowly and cautiously, a mechanical arm was swept over the top of the core to check on whether any of the subassemblies were poking up above the top level of the core. Again, there was success. The chances of a secondary accident appeared less with each step.

Then the big lobster claw that would sweep over the core to lift out the subassemblies was brought into action. A special weight gauge was installed on it. There was no way the reactor shield tank could be opened to look inside. It was blistering with radioactivity, and filled with the argon gas that kept the sodium away from the air.

The idea was that any subassembly that had melted would weigh less than the normal ones. In this way, the damaged fuel could be located, and with luck, removed and examined far away from the reactor in what is known as a "hot cell."

The detective work that began a month after the accident continued week after week, month after month, at a snail's pace. It was essential to learn exactly how much fuel had melted to eliminate the possibility of a secondary critical accident in case some of the fuel shifted during the exploration process. The investigators finally learned that not two, but four subassemblies had been damaged, with two of them stuck together.

It took from October, 1966, to January, 1967, to determine this, and from January to May, 1967, to remove the damaged subassemblies. Removing them was a precarious and overwhelmingly difficult five-months-long job. Special optical devices and cameras were devised. Part of the thick, opaque sodium syrup had to be drained from the reactor, although there was no provision for this in the reactor's design. A shielded viewing window had to be inserted in the plug at the top of the vessel. A borescope placed on the end of a flexible tube was pushed down into the reactor.

By August, 1967, more of the sodium was drained out to expose the meltdown pan at the very bottom of the reactor vessel. So far, even the warped and twisted subassemblies gave no clue as to the cause of the accident—an accident, they said, that could never happen. By September, nearly a year after the meltdown, they were able to lower a periscope through a stainless steel pipe that was shoved down through a hole in the plug that circled the top of the reactor vessel. A quartz light was rigged to slide down it. The device finally reached the meltdown pan, forty feet down, at the very bottom of the reactor. There, the inverted ice-cream cone, known as the conical flow guide, was located to spread out any melted uranium that had spilled down onto the meltdown pan.

As the periscope scanned the bottom of the vessel, it became apparent that there was no melted uranium there. But there was something else. Manipulating the forty-foot-long periscope and light, the engineers saw what looked for all the world like a crushed beer can lying innocently on the meltdown pan. Here, at last, could be the cause of blockage of the coolant nozzles of the subassemblies; a flattened piece of metal that could easily stave off the sodium and allow the uranium to melt, the cladding to rupture, the subassemblies to warp and twist, and the fission products to burst out.

But how did the beer can get there? Had some worker carelessly dropped it from his lunch pail and unwittingly nullified all the carefully planned safety devices that would protect against a meltdown? And *was* it a beer can? And if not, what was it? The investigation had only begun.

The first problem was to identify the crunched metal more clearly. There were fifteen optical relay lenses in the pipe that reached to the bottom of the reactor. The first pictures failed to give enough clarity. Somehow, they had to find a device that would shove the object nearer the camera, and flip it over so they could get a better view from all sides.

News of the Fermi meltdown was kept quiet, but it continued to attract the attention of critics. George Weil, one of the most competent and qualified of the critics, amplified his position this way: "Under current plans for the accelerated growth of nuclear fission to meet our energy requirements, we are committing ourselves to the nightmarish possibility of a radioactive-poisoned planet. Today's nuclear power plant projects are a dead-end street. There are too many, too large, too soon, too inefficient; in short, they offer too little in exchange for too many risks. The commitment of billions of dollars to the development of breeders . . . would almost certainly be an irreversible decision, foreclosing any serious consideration and adequate federal funding of alternative energy sources. . . . With determined efforts to harness fusion, solar, and other energy systems, we may well escape the threat of ecological radioactive disaster."

The tools for fishing out the mysterious "beer can" were created at enormous expense, were dubbed such names as the "hawk-bill cutter," and the "organ pipe tool," and were of unbelievable complexity. The shifts worked twelve hours a day, fishing down through the two pipes and manipulating the tools and light by remote control. Ken Johnson worked from the top of the reactor, forty feet up, while Phil Harrigan, his assistant, worked the snake-like tool from the side of the reactor, thirty-five feet away from the base. An intercom system kept them in touch. All the work was done in specially-built, locked-air chambers to avoid radiation.

Johnson worked the quartz light like a hockey stick, trying to maneuver the elusive hunk of metal into the jaws of Harrigan's retrieval tool, with Harrigan working blind and following Johnson's instructions. The lenses in the periscope would cloud up, and it would take a day to clean them.

Finally, at 6:10 P.M. on a Friday night, almost a year and a half after the meltdown, the crushed metal was firmly gripped by the tool Harrigan was operating. It was drawn slowly up the coolant pipe. It took an hour and a half to lift it up. With the temperature inside the reactor at 350°, the metal was given time to cool as it was drawn up to the surface.

It was examined carefully, and finally identified. Ironically, it was one of five triangular pieces of zirconium that had been installed as an added safety measure to protect the upside-down, cone-shaped flow guide. Somehow, one of the five pieces had worked loose and clogged the coolant nozzles. It had been installed back in 1959 and forgotten. It wasn't even on the blueprints.

By January, 1972, the Fermi plant had operated less than thirty days at its licensed capacity, for a total of 378 hours—without producing meaningful electricity or breeding any large amount of plutonium.

On August 27, 1972, when the AEC issued a "Denial of Application for License Extension and Order Suspending Operations," the announcement was made that the plant would be closed forever.

It was Eldon Alexanderson who would preside over the burial of Fermi No. 1. His job was to figure out how to take apart a core full of three and a half tons of radioactive uranium (speckled with enough plutonium to cause a decided uneasiness), thirty thousand gallons of radioactive sodium, and a vessel so bombarded with radiation that no one could enter it even in a protective suit. The bets were that the $4 million set aside to do the job wouldn't come anywhere near handling it. In Minnesota, after a decision was made to close it down, a far less complicated reactor had been given burial rites costing $7.5 million. One short-cut idea of Alexanderson's was to remove as much of the radioactive guts from the Fermi reactor vessel as possible, and seal it up. But it would still be hot. And there was no way to guarantee that corrosion wouldn't occur, or that water wouldn't leak out of the mausoleum in generations to come.

The nuances of the decommissioning were incredibly complex. Although the reactor was in the state known as "subcritical," there could be a reactivity accident with little or no warning. Even a loud noise could actuate some mechanisms that could threaten the workers. The sodium could always hit air or water to create a tinder-box situation. It constantly had to be bathed in argon gas or nitrogen to avoid this. Health physicists continually had to monitor the reactor with Geiger counters. All kinds of contaminated equipment had to be dismantled and stored in what were called "equipment decay tanks."

All unused penetrations into the reactor vessel had to be covered and seal-welded, including the heating and ventilating ducts. The uranium-packed subassemblies would have to be chopped up in three sections to ship out of the plant to a burial ground. But first they had to be submerged in "swimming pools" to cool off for months.

The casks used to ship the subassemblies were cylinders, nine feet in diameter, weighing eighteen tons each. They had to be sealed in a coolant, shielded with seven inches of lead, and mounted on flat-bed truck trailers to be shipped out. But since the casks were tested only for a thirty-foot fall and a thirty-minute fire, what might happen in the event of a shock impact or fire beyond those arbitrary limits was too frightening to contemplate. Dr. Marc Ross of the department of physics of the University of Michigan has concluded that, if fire or impact distorted the shipping cask of a typical fuel shipment, the leakage of cesium from it would be particularly lethal, both directly through inhaling it and indirectly through contamination of the food chain. Children, infants, and weakened adults would die if they were half a mile downwind from the accident.

Even the loading of the cut-up fuel assemblies was a precarious process. Each fuel unit was on the verge of becoming critical if accidentally placed too close together in the cooling water. Plutonium is so hazardous that no way has yet been found to permanently store it underground. It must be kept in recoverable containers and constantly monitored, while some solution is sought for the problem—a problem not just for the Fermi plant, but for all nuclear plants worldwide. There were six private burial grounds for contaminated materials in the United States. None of them would accept the irradiated sodium or the blanket material which had bred small amounts of plutonium. Grudgingly, the AEC agreed to consider handling the material itself, but no firm plan was made. The question dangled, and the price of dismantling soared above the $4 million mark.

The atmosphere inside the Fermi No. 1 plant became more and more dispirited. "The decommissioning effort continues," Alexanderson said ruefully to a reporter in March of 1974. "Much of it is accompanied by a sinking feeling by staff personnel, as disassembly of many components gradually and irrevocably reduces the plant from the largest operating breeder reactor in the world to fully decommissioned, partially dismantled status. It is very sad to see its demise . . ."

The hollow shell of what had once been the Fermi No. 1 reactor now sits by the gray waves of Lake Erie. In a shed next to the reactor building, triple decks of shiny black steel drums, all marked DANGER: RADIOACTIVE SODIUM, sit in a roped-off area—30,000 gallons of it that nobody wants or is willing to cart away. It is a problem no one has ever faced up to before, or really knows how to cope with.

Near the stacks of drums sits a box-like structure, about the size of an enlarged phone booth, marked cryptically by the work crew "Merlin's Box." It was here that the liquid sodium had been poured into the barrels through sealed pipes before it "froze" into the deadly chalky powder inside the drums. Some three hundred cubic feet of radioactive junk—cladding, rods, sawed-up metal from the spent guts of the reactor—are also scattered about the site waiting to be carted away if takers approved by the AEC can be found. The radioactive blanket assemblies, specked with plutonium, still rest in "swimming pool" storage vaults. As welders make the final seals on the "hot" reactor vessel, plans are being made for the guards to set up their vigil—monitoring the shell with Geiger counters for generation after generation.

There are no trash barrels to hold this poisonous legacy.

Issues and Comments
The Genie's Offer

You are about to discard a soft drink can for recycling when a genie appears and says, "You've got a lot to live and Pepsi's got something to give. As the one-billionth customer to buy a Pepsi, you are offered the following proposition: I can grant you one, limited wish. You may provide the human race with any boon you choose, except that I cannot lengthen or save anyone's life, *provided that,* in return, you will allow me to take the lives of 10,000 people chosen at my whim, that you act on this offer immediately without consulting anyone else, and that you will not discard this can for recycling."

Would you take this boon, retain the can, and allow 10,000 lives to be sacrificed? Would your answer differ if the genie agreed that, once you had determined the boon, every individual could choose whether he wanted to utilize it and *only* those who chose to do so would risk being sacrificed? Does your answer depend on whether those who choose to utilize the boon would be informed of the risk?

Richard Danzig, associate professor at the Stanford School of Law, who heard this problem in law school, points out that "all decisions have a moral content but all of us operate with an extraordinarily simple moral apparatus."[5] He reports that some people, when faced with this problem state that, whatever the proper choice, they cannot assume the responsibility for this decision. But if only they have the choice and only at this moment, then not deciding *is* deciding; it is the equivalent of a negative decision.

Professor Danzig also reports that other people react to the genie's offer by stating categorically that one life can never be sacrificed for the benefit of other lives, save by the full and free choice of those who would be sacrificed. To stimulate thought on this line of argument, Professor Danzig asks, "If you believe that life is so sacred, how can you tolerate the existence of the private auto—surely a relatively small boon—which kills thousands every year? As legislators would you ban non-emergency vehicles? As individuals, would you forswear the luxury of driving to eliminate the undeniable risk of killing some pedestrian?"[6] If you would, what else must you forswear to be consistent?

The nuclear controversy presents a number of complex moral questions. Especially important are those which ask whether we have the right to mortgage the safety and resources of future generations, or to impose on them our radioactive wastes. Even more important are the genetic consequences of radioactive releases to the environment. These problems raise even more fundamental issues: who should bear the burden of risk? What should be our present standard of care? Should future generations have standing in our courts?

Before you address these questions, consider the genie's offer until you have an idea of what is involved. The transition from analyzing the genie's offer to what the nuclear industry offers should pose no problem. How are the offers similar? How different? Are the similarities more important than the differences?

Finally, if you had sole responsibility for cancelling or continuing present nuclear commitments, how would *you* choose? Bear in mind that neither the deaths of thousands nor the advantages of the nuclear boon are as clear as they are in the case of the genie's offer.

Nuclear power may have created control imperatives to which our democratic machinery cannot adequately respond without compromising political imperatives of the same magnitude. To what extent is this view sustained by your reading thus far? Justice Brennan delivered an even-handed analysis of the procedural questions raised by the Fermi case and wrote the majority opinion[8] which held that the Atomic Energy Commission was within both the substance and the intent of the federal law by granting the construction permit. One wonders why the majority ruling failed to reach some of the awesome and complex moral issues posed by nuclear power. This omission on the judiciary's part, compounded by silence from the White House and unsuccessful attempts to spark debate in the legislature, suggests that our government was either unaware of these broader questions or avoided them.

Only in the dissenting opinions of Justices Douglas and Black are these issues recognized. They surfaced again fifteen years later in 1976 when Chief Justice David Bazelon of the U.S. District Court of Appeals delivered a majority opinion which rejected the government's plea that the substantive issues of fuel reprocessing and waste disposal—and the derivative questions of feasibility, public health and moral responsibility to future generations—had already been effectively considered by the NRC in generic proceedings and therefore did not have to be considered at the plant licensing stage.[9] Bazelon's decision turned on a strict interpretation of the National Environmental Policy Act (NEPA), the statute of contention during the landmark Calvert Cliffs case five years earlier.[10] The 1976 interpretation may force NRC to publicly confront a possibility raised in 1975 by Lash, Bryson and Cotton of the Natural Resources Defense Council that "no reasonably assured means of permanent storage [for radioactive wastes] can ever be found."[11] If the burden of proof that such means *can* be found is shifted to the government, subsequent suits may close down the U.S. nuclear industry permanently.

While litigation over waste reprocessing, conservation and licensing continues, the government's plan for building breeder reactors appears intact, although progressing at a slower rate. In the meantime, the economics of this technology may trouble the public as much as its safety. John J. Berger points out in *Nuclear Power: The Unviable Option:**

> Although in the late 1960's the AEC estimated the breeder's capital cost would be only $150 per kilowatt (10 percent more than a conventional reactor) the General Accounting Office recently estimated the cost of a 100-megawatt breeder reactor to be 275 percent as much as an ordinary nuclear plant.[12] The

* See Bibliography. Especially recommended for further reading. Berger summarizes all recent reactor and conservation research in a readable, comprehensive paperback. His chapters on solar and other alternative energy sources are definitive for lay readers.

prototype 350–400 megawatt demonstration breeder reactor now under construction on the Clinch River near Oak Ridge, Tennessee, has grown in cost from about $700 million to about $1.8 billion and may cost $2 billion before it is completed.[13] The other major breeder project in the U.S., the Fast Flux Test Reactor at Richland, Washington, once bore an $87 million price tag and is now pegged at $750 million, including the cost of component testing. Altogether, $2.2 billion has been spent on breeder research and the total cost of the breeder program that the AEC in 1971 said would be $3.9 billion now is likely to cost $10 billion by 1982 . . . Although breeder advocates point out that 40 percent of the $10 billion is attributable to inflation, breeder costs . . . are enormous.[14]

Nor are the economic arguments[15] the only grounds for challenging the breeder. Several of its engineering characteristics pose significant threats to safety: It features no emergency core cooling system (although several redundant safety systems are included to prevent a major accident); it uses a nuclear fuel of approximately the same grade designed for atomic bombs; it also concentrates fissionable material in a smaller volume than light water reactors, operates at a high temperature, and, instead of a water coolant, it requires molten sodium approaching 1,000 degrees Fahrenheit to transfer heat from the fuel rods to the steam generator.

Pure sodium must be handled with extreme care. When sodium contacts air or water, violent explosion and fire may result. Because a water moderator is not present in the breeder core, the uranium-235 is bombarded by and emits more neutrons moving at higher velocities than is the case with light water reactors. When neutrons exit the inner core of the breeder reactor, they enter a blanket of non-fissionable uranium-238. An uranium atom in this blanket may be changed, if it absorbs a neutron, into plutonium. This is the method by which the LMFBR "breeds" plutonium, itself a fissionable fuel, which may be removed from one reactor and, after reprocessing, burned in another.

Recent experience with the French Phenix breeder suggests that it takes much longer than predicted to breed plutonium fuel.[16] Also, breeder reactions are much faster than those characteristic of light water reactors, and as a result operators have less time to react to danger signals. The aggregate of economic and safety objections to the LMFBR may have been the reason William A. Anders, former NRC chairman, recently "expressed grave doubt . . . about the heavy reliance placed by the nation on the breeder reactor as an energy provider."[17] Inasmuch as the safety problems of light water reactors have not been resolved by any means, it makes little sense to push forward with the development of a nuclear fission technology even more complex and with much greater potential for accident.

In its 1976 study, *Reassessment of Nuclear Energy in California,* the California Assembly Committee on Resources, Land Use and Energy commented on the role of the breeder in solving anticipated uranium shortages.

Faced with the uranium situation, critics have asked: If the uranium supply is so short that we must commit ourselves to a potentially even more dangerous breeder technology, then why begin using conventional [light water] reactors at all? And if there is much more uranium available, why are we devoting so much government money to developing the breeder? Some traditional supporters of the nuclear industry have begun to ask this latter question in addition to the usual critics.[18]

4
What You Don't Know Will Kill You

**Richard Curtis
and Elizabeth Hogan**

If a major reactor accident occurs, enormous amounts of radioactive fission products will be released into the atmosphere. A large number of people will be exposed directly to massive doses of radiation. What will happen to them?

- No one would survive at whole body doses of greater than about 600 rem* unless they received good supportive therapy; even then their chance of surviving would not be large.
- A dose of 300–350 rem will be deadly in half the cases, although good medical treatment might extend this to 500 rem.
- A dose of 200 rem or more received in a single brief exposure will be fatal in some cases.[1]

When the whole body receives a dose of 600 rem all at once, it commences a downhill course that nearly always ends in death:

* rem = roentgen equivalent man. This is the standard unit of human radiation exposure. See Glossary for further definitions of rem and rad (radiation absorbed dose).

The patient experiences a sudden nausea which progresses to vomiting and occasionally to diarrhea for two or three days. The onset is sudden and may occur within 30 minutes of exposure; in other cases, several hours elapse before the symptoms appear. Except for a few cases which progressively decline through exhaustion, fever, occasional delirium and death after several days, most patients experience a temporary remission.

A fall in white-blood-cell count and malaise between 14 and 30 days after exposure foreshadow the next ailments. The first overt symptoms may be hair loss, small hemorrhages in the mouth membranes and beneath the skin, and the susceptibility to bruises. Bleeding gums, mouth, throat and bowel ulcerations are followed by appetite loss, renewal of diarrhea and high fever.

By the fifth week, the number of white blood cells will have fallen to the point that the patient is vulnerable to many infections. From that point onwards, the mortality rate is high. Those who survive require a long convalescence that may suddenly end in death at any time due to what might be a minor infection in a healthy person.[2]

The above pattern, known as "marrow death," is the result of the failure of the cells of the bone marrow to produce blood. It appears that, if a few healthy marrow cells survive, or if the victim receives injections of marrow from healthy donors, this form of death can be averted. But the victim actually is only buying time, for the chances are high that cancer, leukemia, acute anemia, or some other ailment will strike in due course.

What about higher doses than the minimum lethal dose of 600 rem? If the victim receives a dose of 1,000 rem or more, it would appear that nothing can help him, and he is doomed to live no longer than a week.[3] The symptoms leading to his demise are characterized as "intestinal death."

At considerably higher doses, say 3,000 rem and up, the nerve tissue functions break down and a "central nervous system death" follows quite rapidly. There is a case on record[4] of an AEC employee who succumbed to the latter a day and a half after taking a whole-body dose estimated at 12,000 rem. Plutonium was being recovered from the Los Alamos reactor, and, through an error, three batches in different stages of purification were put in a large tank together.

When an electric stirrer was activated, the plutonium reached critical dimensions and exploded.

The workman operating the stirring device was thrown from a low ladder, got to his feet, and ran outside crying that he was burning up. His skin was indeed glowing cherry red. Within minutes he began vomiting and discharging a profuse, watery diarrhea. By the time he got to the hospital, he was in shock. Thirty-six hours after the accident, he was, mercifully, dead. Autopsy revealed symptoms of all three kinds of radiation death—marrow, intestinal, and central nervous system—plus considerable damage to the heart muscle.

We have been talking about "whole-body" doses. In the event of a nuclear accident, however, it is likely that many people will suffer damage to specific organs only. What are some of the effects of high doses on these organs and tissues?[5]

Female Sex Organs

The National Research Council reports that single doses to the ovaries between 125–150 rem may produce a cessation of menstruation in 50 percent of women.[6] A single dose of 170 rem can produce temporary sterility for 1–3 years, and a dose of 500 rem will mean permanent sterility in most women.

It should be borne in mind that the female child is born with all the ova she will ever produce. Exposure of her ovaries will thus affect eggs which are to be fertilized when she matures, leading either to defective offspring or to hereditary defects which will manifest themselves in future generations.

Male Sex Organs

Very small doses of radiation, 50 rem or less, can lead to inhibition of sperm production. A 250 rem dose may produce sterility for a year or more, and 500–600 rem will bring about permanent sterility.[7] Even if sterility is temporary, the sperm that do return may bear defective genes that may eventually appear as mutations in future generations.

Blood and Blood-related Organs

No more than 50 rem applied to the whole body will cause the number of lymphocytes, a variety of white blood cell, to drop by 50 percent and it will take about a week before the number returns to normal. Higher doses, or repeated or prolonged exposure, will lead to precipitous drops in blood cell count from which recovery is much slower.[8]

Destruction or disturbance of red and white blood cells will injure the body's ability to protect against infection, repair damaged tissue, and promote clotting. Many organs in the body responsible for blood cell production, removal of dead cells, blood storage, and other functions can also be hurt. Among these organs are the lymph nodes, spleen, and bone marrow.

Skin

Thousands of cases of skin cancer have been produced by overdoses of radiation, usually high doses in the thousands of roentgens.[9] Smaller doses may not produce immediate damage, but latent periods of twenty years or more may pass before ulcerations or malignancies appear. In 1955, two British doctors reported[10] a case of skin cancer in a seventy-year-old woman. She had received about 1,500 rem from a fluoroscope trained on her abdomen to detect kidney stones. A burn appeared shortly after the exposure, then healed. In 1947, however, a skin cancer, which ultimately proved fatal, developed in precisely that spot. How long had it taken? 49 years. The original fluoroscope irradiation had occurred in 1898!

Exposure of the skin to 300–400 rem and up can also cause temporary or permanent loss of hair, graying of hair, destruction of sweat glands, and loss of the skin's natural suppleness and glossy texture.

Eyes

Cataracts, which cause the lens to become opaque, have been produced thousands of times as a result of deliberate irradiation of patients for cancer and other conditions. The latent period, observed in studies made by Dr. G. R. Merriam, Jr., of the Institute of Ophthalmology of New York, averaged almost four years, but one case took as long at thirteen years to appear.[11]

Bones and Teeth

Bone cancer has often appeared years after exposure to heavy doses of radiation (1,500 rem and more).[12] The bones are particularly susceptible to radiation from certain isotopes, such as strontium-90, absorbed as a result of ingestion of radioactively contaminated food, such as fallout-tainted milk.

Radiation of children's bones has caused retardation of growth: about 150 rem will do it, and heavier doses have resulted in limb shortening.[13] Local irradiation of the jaws has slowed tooth growth, and large doses may be followed by infection about the teeth, loss of teeth, and destruction of the jaw bone.

Brain

Relatively small doses of radiation to localized regions of the brain have produced a variety of symptoms, depending on what functions those regions control. In one experiment, a couple of volunteers who had been given 100 rem in the diencephalon, the middle brain, experienced a number of symptoms ranging from ringing in the ears to apathy to hyperactivity. Single doses on the order of 500 rem and more have produced brain damage in children.[14] X-radiation for scalp conditions has produced such effects as blindness, paraplegia, epilepsy, delirium, twitches, and ulcerations.

Lung

Heavy doses of radiation, such as those received for treatment of breast cancer (2,000–3,000 rem) have produced swelling and scarring of lung tissue, and possibly tumors. Breathing of radioactive material, such as radon (the gas found in uranium mines), has produced numerous lung cancers, with a latent period as long as seventeen years.[15]

Cancer

Cancer can begin when a single cell is altered in such a way that its normal self-reproductive powers are affected. Such a cell begins multiplying rapidly, undeterred by influences which customarily inhibit cell growth. Of critical importance here is that the nucleus of a cell, in which are contained the cell's reproductive mechanisms, can be damaged by a burst of ionizing radiation. The only question remaining is, how long does it take before a malignancy manifests itself? The answer is: it may take generations of the given cell type. The "turnover" in some skin cells, for instance, is about 4 months, but it may take years or even decades before the "descendants" of a radiation-damaged skin cell form cancerous ulcerations.

The direct relation between cancer and infinitesimal amounts of radiation has been particularly well-illustrated in leukemia. Studies of Japanese cases show a straight-line relation between dosage and leukemia-induction down to nearly zero. Even as early as 1948, Dr. N. P. Knowlton at Los Alamos showed[16] that no more than 0.2 rem (1/5 of one roentgen) per week of gamma radiation was sufficient to depress white cell numbers, and even smaller radiation doses produced detectable abnormalities in the lymphocytes as demonstrated by Dr. M. Ingram II in 1952. Though such abnormalities did not produce immediate ill effects, it was thought that they could well be the forerunners of anemia, leukemia and other serious or fatal blood diseases.

One of the most highly regarded experts on environmental cancer is Dr. W. C. Hueper of the National Cancer Institute. In an exhaustive study, Dr. Hueper listed the reports of carcinomas and sarcomas attributable to radiation of one kind or another, natural and man-made.[17] His concluding sentence is worth quoting:

> The sum total of the numerous observations on occupational, medicinal, and environmental radiation cancers cited indicates that civilized and industrial mankind has entered an artificial carcinogenic environment, in which exposure to ionizing radiation of various types and numerous sources will play an increasingly important role in the production of cancers.

Genetic

Ordinarily a gene, the fundamental unit of heredity located in the chromosomes of cells, is stable. It copies itself unerringly generation after generation.[18] Occasionally, however, it undergoes a spontaneous change, presumably chemical, called a mutation. Then a gene that hitherto produced, say, brown eyes suddenly starts producing blue eyes. From then on, all future generations drawn from that gene are blue-eyed. Such alterations of hereditary characteristics can also result from breakage of chromosomes (the bodies containing genes) and rearrangement of the broken chromosome parts.

Under natural conditions, mutations are exceedingly rare, on the order of 1 in 100,000 generations. But Muller, in the 1920s, while exposing fruit flies to radiation, managed to increase the number of hereditary abnormalities in their descendants.[19] Subsequent studies in a wide range of plants and animals have confirmed Muller's discoveries: in every organism examined, it has been observed that high-energy radiation reaching the chromosomes will produce mutations.

The number of mutations has been shown to be directly proportional to the amount of radiation, right down to low levels. The important question is, how low does the level go before we can declare unequivocally that no genetic damage will occur? In 1960, James F. Crow, professor of genetics at the University of Wisconsin School of Medicine and president of the Genetics Society of America, provided the chilling answer:

> Geneticists are convinced that there is no threshold for radiation-induced mutations: that is, there is no dose so low that it produces no mutations at all. Each dose, however small, that reaches the germ cells between conception and reproduction carries a risk to future generations proportional to the dose.[20]

What could this mean for the future of mankind? Professor Crow, calculating the possibilities in an article titled "Radiation and Future Generations," suggested that, for every roentgen of slow radiation, the kind of radiation we can expect to receive in increasing doses from peacetime nuclear activity, about five mutations per hundred million genes exposed will manifest themselves, meaning that "after a number of generations of exposure to one roentgen per generation, about 1 in 8,000 of the population in each generation would have severe genetic defects attributable to the radiation."[21]

Some harmful effects* include hemophilia, erythroblastosis fetalis (a blood disease of the fetus or newborn child), familial periodic paralysis, nervous and mental diseases, metabolic and allergic disorders, and certain congenital diseases.

A variety of anomalies, many verging on the monstrous, are possible as a result of radioactive damage to the genes. Gigantism,

---

* Editor's note: These are only a very few of the countless genetic consequences of radiation. Every physiological characteristic has a genetic basis, and every gene is vulnerable to radiation damage. Thus, every physical characteristic is vulnerable to mutation due to radiation, and every mutation may be carried through dozens of generations.

dwarfism, albinism, club foot, hair lip, cleft palate, Siamese twins, Janus monsters (two faces on a single head and body) phocomelia (rudimentary limbs), sirenomelus (legs fused with no separate feet), hydrocephalus (grotesquely distended head), and hermaphroditism (physical bisexuality) may be induced by radioactive bombardment of reproductive cells.

Of course, in the evolutionary process, harmful mutations—and most mutations are decidedly harmful—tend to eliminate the hereditary lines that carry them, because of sterility, disease, and feebleness. Even if this process of natural selection does erase harmful strains, the elimination process itself entails suffering of every imaginable sort.

Issues and Comments

Earlier in *Perils of the Peaceful Atom,* the book from which the preceding excerpt is drawn, the authors point out that nuclear plants scheduled for future operation must "unavoidably produce large quantities of radiation and radioactive wastes known as 'fission products' . . . which are a million to a billion times more toxic than any industrial pollutants known heretofore."[22] There are an infinite number of routes by which these toxic materials can reach living things. In many cases, once they are released—accidentally or deliberately —there is no way that humans, animals, plants, or insects can be protected from contaminants before they are dispersed to the point of being harmless.[23] During this lethal interval all living things are exposed to what Ralph Nader describes as

> a silent, cumulative form of violence you cannot see, touch, taste, or smell. These new forms of violence do not draw attention to themselves by attacking pain thresholds on contact or by causing bleeding at the outset. Their effects show up years after the initial exposure in the statistics on lung cancer, . . . and genetic damage.[24]

But the radiation violence causing these disorders cannot be traced unless there is a sudden catastrophic dose. According to Nader, it is precisely because of the invisible, silent, cumulative havoc that radioactivity can wreak that we must take a *preventive* strategy.[25]

In simple terms, the irreversible effects of radioactivity urgently compel us to protect ourselves, future generations, and the environment from this kind of damage. If nuclear industry spokespersons complain that this is an inordinately cautious approach to radiation, it is because radiation is an inordinately lethal threat to all forms of life.

Neither the short-term nor the long-term, delayed effects of radiation exposure are likely to show up on a corporation's profit-and-loss statement, so we can't expect the industrial sector to protect the public aggressively, especially if radiation protection measures depress corporate earnings. Our government addresses the problem by setting permissive radiation standards that, according to Dr. John Gofman and other respected scientists, may compromise safety in favor of economic considerations.[26] Legislators will not revise these radiation standards to protect the public until it can be proved that small doses of radiation have been harmful.[27]

At the same time that our conventional institutions seem unable or unwilling to restrain the burgeoning nuclear industry, the human race is becoming biologically obsolete, for our senses are not equipped to detect radiation, nor is there any chance that our species will mutate fast enough to refine our sensory apparatus in this way. When individual citizens become aware of this and realize as well that corporations worldwide may resist stringent controls on radioactive releases because these controls are uneconomic, they may be moved to seek as much absolute protection as possible for themselves and their loved ones.

Now is a good time to start. You must learn all you can about radioactivity and the way it can hurt you, your family, and your neighbors. At the same time, you must educate them so that they can protect themselves. Next, you must work to end the irresponsible manufacture, deployment, and use of nuclear materials. Because electric power by nuclear fission promises to yield enormous amounts of radioactive material, *The Silent Bomb* focuses on the nuclear industry. Finally, as this industry shows signs of closing down, public attention must be shifted to the enormous cache of military weapons[28] and the problems of international proliferation of nuclear material.[29]

A GUIDE TO
THE NUCLEAR
CONTROVERSY

5
How a Nuclear Reactor Works

In a few minutes you can learn how a nuclear reactor generates electricity. There is nothing mysterious or complex about this process. In fact, if you know how to boil water to make coffee, you already know the essentials: apply heat to water until it boils, then pass the boiling water over coffee grounds. Figure 1 shows what the recipe looks like in the form of a flow diagram.

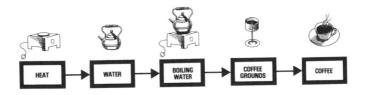

Figure 1

The diagram reads, from left to right: *WHEN ENOUGH HEAT IS APPLIED TO WATER YOU OBTAIN BOILING WATER. THIS BOILING WATER ACTS ON COFFEE GROUNDS TO CREATE THE BEVERAGE.*

Figure 2 shows the flow diagram changed slightly.

Figure 2

This diagram reads: *WHEN ENOUGH HEAT IS APPLIED TO WATER, THE RESULTING BOILING WATER YIELDS STEAM. THIS STEAM DRIVES (ACTS ON) A TURBINE, WHICH IS CONNECTED TO A GENERATOR. THE TURBINE AND GENERATOR TURN TOGETHER, AND ELECTRICITY IS GENERATED.*

This process applies to both types of generating plants studied here. Fossil-fuel-fired plants burn coal, petroleum, and natural gas, and generate about 82 percent of our electricity. *Nuclear plants* provide about 8 percent of our electricity and use uranium fuel to generate heat by nuclear fission. Both combustion (fossil-fueled) and fission (nuclear-fueled) plants produce great quantities of heat, which is used to convert water to steam.

The significant difference between nuclear and fossil plants is the fuel used by each. Most nuclear plants "burn" uranium-235 that has been mined, refined, and enriched (i.e. concentrated) and then molded into fuel pellets. When enough enriched uranium fuel pellets are brought together under the right conditions, uranium-235 atoms begin to split, or fission; this produces the heat needed to turn large quantities of water into steam.

In a fossil-fueled plant (see Figure 3), coal, petroleum, or natural gas is burned. The heat of combustion is used to create superheated, pressurized steam in a boiler. The steam is piped to a turbine-generator, where it impinges on the blades of the turbine, causing it to turn. The turbine is connected to the generator by a shaft; the turbine-generator turns at high speed as high-pressure steam pushes against the turbine blades. When the generator turns, electricity is generated.

The spent steam passes from the turbine to a condenser through which cooling water is pumped from an outside source. The condenser absorbs some of the steam's heat, converting it again to water, which is pumped back into the boiler. Notice that the water coolant

A Conventional Fossil Fueled Electric Power Plant

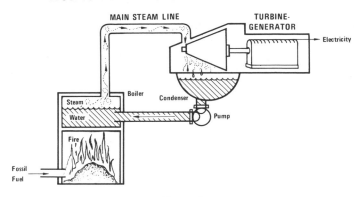

Figure 3

shown in Figure 3 circulates in a continuous "loop"—from boiler to turbine to condenser and back to the boiler.

A nuclear-fueled plant is diagrammed in Figure 4. When the nuclear fuel in the reactor core fissions, heat is produced, which boils water circulating through the reactor core. This design, one of three described in this article,* is called a boiling water reactor (BWR). Great quantities of steam† at approximately 1,000 pounds per square inch (psig) pressure are piped to the turbine-generator, causing it to

Boiling Water Reactor (BWR)

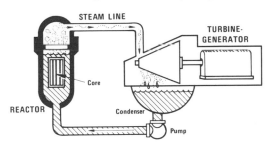

Figure 4

* Readers interested in learning more about other nuclear plant systems will find books referenced in the Notes and Bibliography sections helpful. Walter Patterson's *Nuclear Power* provides a more thorough treatment of reactor operation and design than appears in this article. Technically trained readers will enjoy *Nuclear Power Plant Systems and Equipment* by Kenneth C. Lish.

† As much as 16 million pounds per hour, depending on the output of the system.

turn. The spent steam is condensed and returned to the reactor core as "feedwater."

Although the steam is radioactive as it leaves the core, most of the radioactivity is short-lived and does not heavily contaminate the turbine or feedwater systems. The BWR is the simplest of operating reactor system designs. It features a single steam-water loop through which steam is piped directly from the reactor to the turbine. The BWR also features control rods to control the rate of nuclear fission; these rods are inserted from *beneath* the core by hydraulic drives electrically activated by operators in the control room. Nearly all commercial BWRs in the United States are designed and manufactured by General Electric–Nuclear Energy Division, San Jose, California.

The pressurized water reactor (PWR) diagrammed in Figure 5 is similar to the BWR design, but whereas the BWR circulates steam from the reactor to the turbine and back to the reactor through a direct "loop," the PWR features *two* loops. In the primary loop, pressurized water carries heat from the core to a steam generator, or boiler, where the heat is transferred through tube walls to water flowing through a second loop. Secondary-loop feedwater from the condenser is pumped into the steam generator, absorbs heat from the first loop, and turns to steam. The second loop carries the steam at about 1,000 pounds per square inch pressure to the turbine-generator and then to the condenser.

Note the similarity between the PWR and BWR. The PWR's secondary loop is designed to keep the turbine and associated pipes and pumps as free of radioactivity as possible. Thus the PWR's heat

Pressurized Water Reactor (PWR)

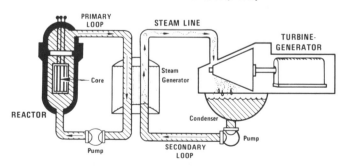

Figure 5

exchanger acts as a "buffer" between the two loops. If the tubes did not leak, the turbine, condenser, pumps and piping would not become radioactive and thus would be easier to replace and repair. Unfortunately, nearly all steam generator tubes do leak, and contamination of the secondary loop is inevitable in these cases. Pressurized water reactors have control rods inserted from *above* the reactor core. They are manufactured in the United States by Westinghouse, Combustion Engineering, and the Babcock & Wilcox Company.

The third reactor design examined here is shown in Figure 6. The liquid metal fast breeder reactor (LMFBR) uses molten sodium instead of water as the reactor core coolant. It is a more complex design than the BWR or PWR. The LMFBR is known as the "breeder" because it produces more plutonium fuel than it burns. The details of this process are discussed on page 62.

Note the three loops in Figure 6. Molten, radioactive sodium circulates in the primary loop through a heat exchanger which transfers primary-loop heat to molten sodium (nonradioactive) in the secondary loop. Secondary-loop heat converts water in the *third* loop to steam as it circulates through a steam generator. The nonradioactive steam is piped through the third loop to drive the turbine-generator.

Prototype or demonstration breeders are operating in France, England, and Russia, but in the United States several test reactors have developed serious problems.[1] A test breeder in the Soviet Union may have suffered an accident several years ago, although federal government reports have insisted that the massive environmental damage evident in the Ural Mountains followed a radioactive

Liquid Metal Fast Breeder Reactor (LMFBR)

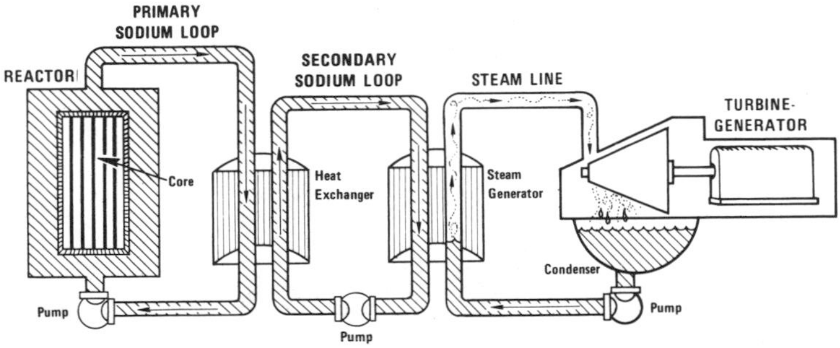

Figure 6

waste dump explosion.[2] Unresolved technical questions, problems with safeguarding dangerous plutonium, and the high cost* of building an LMFBR plant may persuade the U.S. government to cancel its plans to finish the Clinch River, Tennessee, breeder scheduled for operation during the 1980s.[3] A later step in the breeder development plan is to construct dozens of commercial breeders across the country, which would seriously aggravate all the problems associated with a "plutonium economy"—radioactive contamination of the environment and the risk of accident, theft, and sabotage.[4]

Figure 7 is a simplified diagram of the nuclear fission process. An atomic particle, a neutron in this case, strikes the nucleus of a uranium-235 atom, which fissions, or disintegrates, into fission fragments, heat, and one or more free neutrons. All this takes place in less than 1/1000 of a second. The fission fragments are usually nuclei of lighter elements, such as strontium or cesium. The fission heat is absorbed by the reactor coolant and contributes to the generation of steam. In light-water reactors, the coolant also acts as a "moderator" by slowing down fast neutrons emitted during fission. Given any volume of material, *slower* neutrons are more likely to split nearby fissionable nuclei than *fast* neutrons; thus more nuclei will split when *slow* neutrons are more abundant, and more energy will be released.

If the uranium nucleus emits free neutrons, these speed away and may similarly strike other nearby uranium nuclei. A fission chain

A Fission Reaction

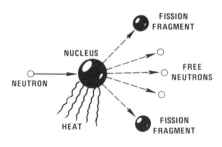

Figure 7

* The 1976 estimated cost was $1.95 billion and rising, as compared with the original estimate of $700 million. Fiscal and safety considerations may have persuaded President Carter to cut back sharply funds for the breeder shortly after taking office in 1977, a move that Washington observers predict will lead to shelving or eventual cancellation of the project.

reaction is self-sustaining when the number of neutrons released in a given time equals or exceeds the number of neutrons that are absorbed by nonfissioning material or that escape from the system. We say that a nuclear reactor is "critical" or has "reached criticality" when it is sustaining a chain reaction.

When an atomic bomb is exploded, this chain reaction occurs more rapidly. For example, one fissioned nucleus will emit two neutrons, which are absorbed by. two other uranium nuclei, which release *four* neutrons, and so on. At the same time, a gun or implosion mechanism contains or increases the density of the fissionable material to sustain a critical mass. The resulting explosion releases a tremendous amount of energy in a very short time and reaches temperatures in the millions of degrees. In a power reactor, the reaction is controlled to the desired energy release rate.

To control the multiplication of neutrons in a reactor core, control rods containing a neutron absorber such as boron are moved in and out. These control rods absorb some of the free neutrons emitted during fission. When *all* the control rods are inserted, the boron near the fissioning uranium absorbs enough of these free neutrons so that the rate of fission levels off. If enough control rods are withdrawn quickly, the fission reactions start multiplying very rapidly and may "run away." If the resulting energy release increases beyond the safe point, the fuel may be damaged, the steam pressure will increase, and possibly the steam system may suffer overpressurization. Such an event is called a "transient accident."

An accident may also occur because of insufficient cooling. The most feared event is a major "loss-of-coolant accident" (LOCA), where a large coolant pipe breaks and most of the coolant is lost. If this occurs, and if emergency core cooling systems do not function in a matter of seconds, the fuel will begin to melt. The molten fuel is expected to melt through the reactor vessel and the supporting concrete and into the earth. This "meltdown" accident could have catastrophic consequences if, as is expected, the buildup of steam and radioactive gas inside the containment overstresses the containment structures. If the containment fractures, and an uncontrolled release of the large quantities of radioactive material occurs, enormous damage and loss of life will occur. The risks and consequences of meltdown accidents are described in the chapter entitled "Reactor Safety."

For the moment, it is enough to know the basic facts about

nuclear fuel, reactor design, and how reactors generate and control heat. For the remainder of this article, the subject will be boiling water reactors.

The expensive and carefully refined uranium used in nuclear reactors is molded in the form of cylindrical pellets about one inch long. These pellets are sealed in rods between twelve and fourteen feet long made of an alloy called "zircalloy." Sixty-four of these rods are gathered together in a *fuel bundle* which is sheathed in a *fuel channel* to form a *fuel assembly,* shown in Figure 8. A typical 1,220-MWe BWR core is made up of 732 fuel assemblies and 177 control rods. A fully loaded core will contain about 46,000 fuel rods and more than 100 tons of uranium in a cylindrical space about

BWR Fuel Assembly

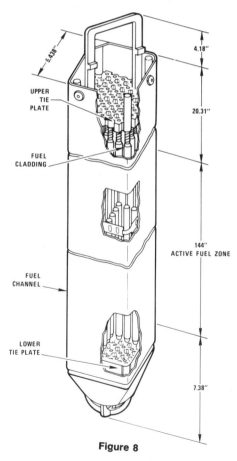

Figure 8

twenty feet in diameter and fourteen feet high.

It is extremely important that the fuel rods be slightly separated from one another so that the water coolant may flow freely among the rods, thereby ensuring good heat transfer. The fuel assemblies are spaced so that water coolant moves evenly throughout, and so that both partially fissioned and fresh fuel are positioned for optimum reactor heat generation.

During initial fuel loading, the work crew carefully adds one fuel assembly at a time into the reactor core. The fission process that could occur under these conditions is "poisoned"—or quieted—by the control rods inserted at equal intervals among the geometrically arranged fuel assemblies. (See Figure 2 in Appendix A.) These rods absorb free neutrons from spontaneous fissions so that there are not enough available to sustain a chain reaction.

When the plant is ready for "start-up," the control rods are slowly withdrawn in symmetrical order, one at a time. Gradually, as the density of free neutrons inside the core builds up, reactivity increases. When each free neutron absorbed by a uranium nucleus is replaced by a free neutron released by another fissioning atom, the reactor "goes critical" and the chain reaction becomes self-sustaining. As more free neutrons are released, the number of fissions and the amount of heat released also increases, the water coolant heats up and begins to boil, and steam is generated. As long as the rate of the chain reaction is stabilized and enormous quantities of coolant are pumped continuously into the reactor vessel, the reactor operators will maintain control of this new kind of fire. During normal operation, the control rods are raised and lowered to adjust the reactor power level. It may also be controlled by varying the rate of water through the recirculation loops (see Figure 9). In an emergency, the operators may "scram," or shut down, the reactor by quickly inserting all withdrawn control rods.

The term "control rod" conveys the idea that this is the chief means for assuring safe reactor operation. But control rods absorb neutrons, not heat. Even with all control rods inserted and reactor fissioning essentially stopped, if the coolant flow into the core stops after a sustained period of full-power reactor operation, the core could melt down in a matter of hours with catastrophic consequences because of the accumulating decay heat.* For this reason,

---

* As described earlier, fission fragments remain after a uranium-235 nucleus absorbs a free neutron and subsequently splits. These fragments, called "fission

there are a number of systems in a nuclear plant, many of them redundant, which are designed to assure a continuous water supply to the core under all circumstances.

Two major systems are provided in the plant to ensure against a catastrophic release of radioactive materials. These are the *cooling system* and the *containment system*. A brief review of the systems will provide a basis for understanding the technical material presented in later chapters. The reader is cautioned that the systems are described here in terms of how they are *designed* to work, rather than how they actually *do* work.

The reactor cooling system ensures that a substantial, continuous flow of water is maintained within and around the fuel core. This is necessary so that the fuel elements do not overheat and melt, and also to properly carry heat away from the core in the process of creating steam. Because these activities are the most critical from the standpoints of economics and safety, the reactor cooling system is composed of pipes, pumps, instruments, and controls of the highest quality. In addition, the system includes built-in backups and redundancies, only a few of which will be mentioned here. Among the subsystems that make up the reactor cooling system are:

- Feedwater pipes and pumps, which bring water into the reactor vessel in large quantities from the condenser.
- Recirculation pumps, which drive the jet pumps, which, in turn, force the feedwater toward the bottom of the core and provide the pressure to drive the coolant upward through the fuel assemblies (see Figure 9).
- The emergency core cooling system (ECCS), a combination of coolant injection systems automatically activated by conditions in the reactor vessel. One of these, a loss-of-coolant condition, could be created by failure of the feedwater pumps, a major leak due to a pipe break, or a combination of events. If the ECCS is activated, various combinations of pumps, driven by steam or electrical power, pump or spray large quantities of water into the reactor core. Presumably, this rapid injection of water will be

---

products," are isotopes which undergo spontaneous decay over time. As fission products decay into more stable elements, they emit radiation and heat. In the reactor core, every fissioned uranium-235 atom yields fission products whose accumulation results in considerable "decay heat." When a reactor undergoes a controlled shutdown, a residual heat removal system (RHR) is activated to remove the considerable amounts of decay heat.

Schematic Arrangement of BWR Reactor Pressure Vessel

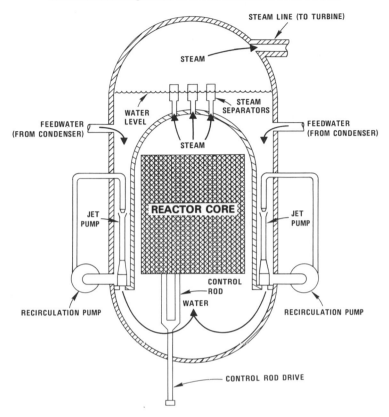

Figure 9

cnough to keep the fuel from melting. (See Figure 1 in Appendix A.)

A backup system, the automatic depressurization system (ADS), is provided in case both feedwater pumps and the high-pressure emergency cooling systems fail. The ADS will automatically depressurize the reactor by opening the relief valves, which discharge steam into the pressure suppression pool of water around and below the reactor vessel. Once reactor pressure is reduced satisfactorily, a *low-*pressure emergency cooling system may be initiated. The low-pressure core spray system is composed of two independent loops for core cooling after the reactor is depressurized. (See Figure 1 in Appendix A.)

Containment Concept MARK I

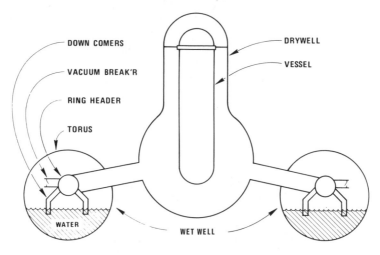

Figure 10

Another critical safety system examined in this book is the containment, an inclusive term for the protective shells around and including the reactor vessel. The containment is designed to ensure that no radioactive material escapes into the environment. The drywell, suppression pool, and reactor building of the General Electric Mark I design are shown in the cutaway drawings in Figures 10 and 11. The Mark I *primary* containment is inside the secondary containment—the reactor building. It consists of two steel chambers, one in the shape of an inverted light-bulb, and the second shaped like a large torus, or doughnut. The light-bulb-shaped chamber contains the reactor vessel and the major coolant piping systems. This chamber, called the drywell, encloses the reactor vessel and provides extra volume as well as a channel for steam that would escape from a break in the primary cooling system. The steam would be routed through vents into the "suppression pool," below and outside the drywell, where the steam will condense.

The secondary containment is the reinforced concrete building enclosing the primary containment. A cross-sectional diagram of both primary and secondary containments showing the drywell and suppression pool in a more recent design, the Mark III, may be found on page 175. Readers are encouraged to compare the Mark I and Mark III designs so that they will have a firmer basis for

Reactor Building

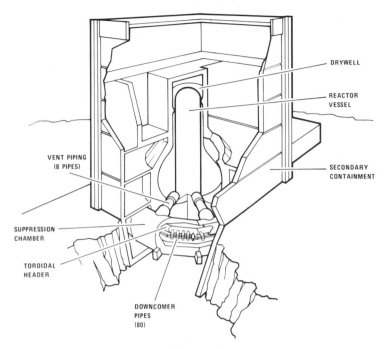

Figure 11

understanding the technical material presented in "future chapters." Detailed discussion of the Mark I, II, and III designs is included in Appendix A, pages 292–303.

Issues and Comments

Beginning with the Industrial Revolution, the engineering profession abided by the truism that safe operation depends on careful design verification and testing as well as on proper training of operators. Evidence is presented in later articles that careful nuclear plant design may have been neglected in the rush for lucrative contracts, and that the practice of selling systems before they were designed may have gained credence and momentum in the defense and aerospace industries. Recently, a number of aerospace and defense systems have been marketed this way and more than a few have turned out to have distinct limitations.[5] In retrospect, resources consumed in developing these systems might have been devoted to other purposes if their designs had been carefully prototyped and subjected to thorough testing before being marketed. Instead, sales imperatives often dictated that the sequence of prototyping, test and verification would take place *after,* rather than before, the systems were marketed.[6] The current view is that there is nothing intrinsically wrong with this approach, and its advocates point out that it is common practice in several industries today. Others have expressed the opinion that this practice represents a significant compromise of engineering ethics that makes sense only when short-run profits are more important than long-run considerations.

For lack of a better term, we will refer to this practice as "post-testing." Examples of this may be found in General Electric's efforts to market the Mark III containment design concept before several major design problems were resolved (described in this book in "Shakeup at General Electric"), and in the federal government's promotion and development of nuclear fission before crucial elements of the nuclear program were assured (notably, a safe means of permanently storing radioactive wastes). Part of the recent controversy turns on the question of what prerequisites should be mandatory before more radioactive wastes or reactor systems are produced, as well as on the question of what to do now that the investment of considerable time and money has apparently been offered as a substitute for fulfilling the prerequisites.

Besides these issues, which pertain more to management philosophies, economics, and public policy, there are several others which deal exclusively with the problems of and relationship between mechanical and human fallibility. These questions are treated extensively in Chapters 11 and 14.

During the preceding description of reactor systems and containments, it was pointed out that these and other components essential to safe operation are designed to function reliably and safely under *all* circumstances. The operative word here is "designed." Nuclear engineers stake their careers on promises that their systems will, in fact, work *as designed.* Their design philosophy is consistent, systematic, and stresses safety above all other considerations. To attain this objective, nuclear designers emphasize the concept of "defense in depth" to protect the environment and people from the radioactive material in the reactor core. This concept entails a series of barriers which "contain" the fuel and which we have already described in some detail. The first barrier—the zirconium sheath—is the *rod* enclosing the fuel pellets. The rods are secured inside the *reactor vessel* which features heavy steel walls. The reactor is protected simultaneously by the several *cooling systems* and by the *steel con-*

tainment shell which surrounds it. Finally, the entire plant is encased in a *reinforced concrete building.* These containment systems represent a series of concentric shells, much like Chinese boxes, to provide extra volume immediately outside the reactor, to reduce pressure inside of the drywell and reactor in case of accident, and to keep any radioactive material from reaching the outside atmosphere.

Fortunately, this philosophy has yielded a perfect safety record for the industry, and it has been necessary to use the outer defenses of the system (the containment and shield building) only a few times. The problems identified by nuclear critics pertain mostly to the nuclear plant cooling and electro-mechanical systems. These complex systems have an annoying tendency to go awry, occasionally because of human errors, which cannot be anticipated. Edward Teller's view of human fallibility suggests some of the consequences.

> So far we have been extremely lucky. . . . But with the spread of industrialization, with the greater number of simians monkeying around with things that they do not completely understand, sooner or later a fool will prove greater than the proof even in a foolproof system.[7]

Evidently, if major accidents are to be averted in the long run, we will need a disciplined cadre of nuclear priests who can be relied on to guard nuclear fission, reprocessing and waste disposal facilities for what might be millennia. These priests must be extraordinarily diligent and meticulous, and their care must extend beyond the operation of the nuclear systems to include the police and custodial functions. Until it can be demonstrated that this cadre exists and will be effective for many years into the future, readers may view with skepticism claims that nuclear systems will (nearly) always work *as designed.* To preserve its credibility, the nuclear industry hedges its bet by announcing that the chances of nuclear systems failure are one in 20,000 per reactor per year.[8] This is not a significant concession and the suspicion grows that someone is writing an escape clause into the Faustian bargain* while preserving the illusion of (near) infallibility.

* Alvin Weinberg, former director of the Oak Ridge National Laboratory, points out that

> We nuclear people have made a Faustian bargain with society. On the one hand, we offer—in the catalytic nuclear burner—an inexhaustible source of energy. . . . But the price that we demand of society for this magic energy source is both a vigilance and a longevity of our social institutions that we are quite unaccustomed to.[10]

Backup and redundant systems significantly reduce the chances of an accident due to design inadequacy or mechanical failure. To some extent, these systems may buffer safe, continuous reactor operation from human error as well. But complex systems tend to malfunction for complex reasons and in ways undreamed of by their designers.[9] In the end, it may be that man himself may govern accident probabilities that now appear favorably small because nuclear plant risks were calculated with primary attention devoted to the *engineering* aspects, about which a great deal is known, rather than to the *human* aspects, about which there remains a great ocean of truth to be discovered, and all we know may be compared to a pebble on the beach.

6
The Nuclear
Fuel Cycle
Simplified

Just as a private automobile is completely dependent on the availability of petroleum, a nuclear plant must have a dependable supply of fissionable fuel to produce electricity during its life span, now estimated between thirty and forty years. At present, the nuclear fuel most frequently used in American commercial reactors is uranium oxide (UO_2), which must contain enough uranium-235 (about 3 percent) to assure reactor heat of the proper intensity and duration. Figure 1 shows the sequence for manufacturing, reprocessing, and disposing of nuclear fuel.

The Front End

Step 1: Mining and Milling: In the United States, uranium ore is found principally in the Colorado Plateau and the Wyoming Basin in geologic formations known as Western Sandstone. Of every 2,000 pounds of mined high-grade ore, less than 0.2 percent, or four

The Nuclear Fuel Cycle

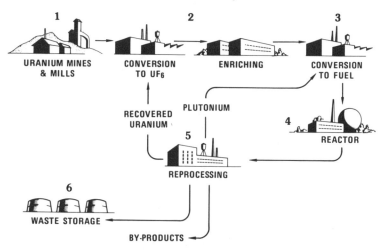

Figure 1

pounds, is pure uranium. Of these four pounds, 99.3 percent is the unfissionable isotope, uranium-238. The remainder, about 0.7 percent, or less than half an ounce, is fissionable uranium-235, which forms the basis for reactor fuel. Once removed from the mine and ground until it is as fine as sand, the raw uranium ore must be separated from the sandstone at a milling plant and refined chemically. The end-product of this refinement process is a uranium compound (U_3O_8) known as "yellowcake."

Step 2: Conversion and Enrichment: Plants in Illinois and Oklahoma convert this refined, natural uranium to uranium hexafluoride (UF_6), a dense gas, which is then shipped to an enrichment plant to boost its uranium-235 concentration fourfold, from 0.7 percent to about 3 percent. Enrichment plants now operating in the United States are located at Oak Ridge, Tennessee; Paducah, Kentucky; and Portsmouth, Ohio. At the enrichment plant, the UF_6 is passed through numerous porous membranes (diffusion cells) that allow the lighter uranium-235 molecules to pass through, blocking the slightly heavier gas molecules containing uranium-238. These plants, operated by the federal government, consume enormous amounts of electricity as the gaseous diffusion enrichment process propels considerable volumes of UF_6 through a series of high-pressure pumps and diffusion cells. Gradually, the concentration of uranium-235 increases and the fuel becomes sufficiently "enriched" to be useful in a nuclear reactor.

Step 3: Conversion and Fabrication. The enriched fuel, converted again to solid uranium oxide (UO_2), is shipped to another plant where it is "fabricated" (baked and molded) into pellets. These cylindrical pellets are inserted into thin zirconium alloy fuel rods between twelve and fourteen feet long. The rods are then gathered into fuel assemblies each containing up to sixty-three rods. Finally, the fuel assemblies are shipped to a nuclear plant where they are loaded into the reactor core. After the plant receives its operating license, control room operators carefully withdraw control rods from the reactor core. As the control rods are withdrawn, less boron carbide in the rods is available to absorb neutrons expelled by fissioning uranium-235 atoms. Instead, these unabsorbed neutrons collide with other uranium-235 atoms and sustain the "critical" chain reaction described in the previous article. The heat generated during the fission process converts water to steam, which turns the turbine-generator to create electricity.

Each reactor requires a surprisingly large quantity of uranium ore. Calculations from Morgan Huntington's data[1] imply that one 1,000-MWe reactor operating for forty years will require as much as 11,400 tons of uranium oxide (U_3O_8), assuming a 70 percent capacity factor. Geologists with ERDA have estimated that only 780,000 tons of proven yellowcake reserves are available in the United States, provided that buyers are willing to pay up to $30 per pound production cost.[2] This estimate does not include ores recoverable at a higher production cost or recoverable uranium ores from *potential* reserves at $30 per pound production cost or less.

Simple division (780,000 tons divided by 11,400 tons per reactor) demonstrates that we have enough domestic uranium for fewer than 70 nuclear plants. The Republican administration under President Ford urged the construction of 700 reactors in the 1,000-MWe* range by the turn of the century.[3] To fuel these reactors for their forty-year lifetimes would require as much as 7,980,000 tons of U_3O_8, or more than ten times as much ore as our government has determined exists in proven reserves. Not surprisingly, the inter-

* MWe: Megawatts-electric. 1 megawatt = 1 million watts. A watt is a measure of power; one watt is that amount of power required to lift 2.2 pounds 4 inches per second.

national uranium market has responded to the anticipated fuel shortage with escalating prices. Since 1972, uranium has risen from approximately $8 to $40 per pound, an increase of 400 percent.[4]

To obtain a single pound of uranium-235, over 70,400 pounds (35 tons) of high-grade ore must be mined and processed. But regions from which high-grade ore is mined may be soon exhausted, and we may have to depend on ore with a much lower uranium yield. This will require open-pit or strip mining of hundreds of square miles. The ratio of mined ore to fissionable fuel also indicates that the enrichment process is vulnerable to runaway costs, for the leverage built into the ratio could compound labor or transportation cost increases. Not only are thousands of man-hours involved in the front end of the fuel cycle, but the fuel must be shipped between mine and conversion, enrichment, and fabrication plants on its way to the reactor. Even if enrichment costs remain stable, enrichment-plant capacity has lagged nuclear fuel demand in recent years.[5] Since new plants for fuel conversion, fabrication, and enrichment will not be ready for some time, the industry will be short of enriched fuel if new power plants are granted operating licenses in the near future.

The costs of environmental damage and physical injury must also be considered in determining real fuel costs. The most serious of these arise from radioactive pollution by the material left over after the milling process. Approximately 99.8 percent of the total mass of mined material is rejected during this phase of the fuel cycle and, over the last thirty years, about *100 million tons* of this material has been accumulating in "tailings" piles near milling plants in the Southwest.[6] According to Dr. Robert Pohl, a Cornell University physicist, the considerable amounts of thorium-230 present in these tailings may produce as many as *5 million* cancer casualties over many centuries.[7] Because thorium-230 has a half-life* of 80,000 years, it must be isolated from the environment for at least a million

---

* The time in which half the atoms of a radioactive substance disintegrate—or decay—into other nuclear forms. If a radioactive element has a half-life of ten years, the radioactivity of any amount of that element will be reduced by half when ten years have elapsed. During that ten-year period, half the atoms in the sample will change spontaneously into a different element, or into a different energy state of the same element. As each atom decays, it may emit alpha particles, beta particles (electrons), or gamma rays, all of which may damage living things.

A conservative rule of thumb for determining when the radioactivity of a sample will decline to a safe level involves multiplying the half-life of the sample element by twenty. At the end of the period calculated, the radioactivity of the sample will be reduced to one-millionth of its original potency.

and a half years. And because so much thorium-230 will be produced if the international nuclear program continues to grow, enough will remain even after several hundreds of thousands of years to assure millions of early deaths. The survivors will be grateful that our uranium reserves were limited.

The only unassailable evidence of cancer related to uranium ore that we have is the alarmingly high rate of lung cancer among uranium miners. Because carcinogens and radioactive particles leave no calling cards, there is no way we have of absolutely determining which of the unstable isotopes—radium, thorium-230, radon-222, and others—caused these cancers. The most recent, thorough study of these cases is included in *The Nuclear Fuel Cycle*, published by the MIT Press.*

The Back End

Once loaded into the reactor core, the UO_2 pellets remain there for about three years. The energy released by fissioning uranium-235 atoms begins to decline as the percentage of this isotope decreases, and fission products such as xenon, strontium, cesium, and iodine build up to interfere with efficient heat generation. In addition, neutrons released during the fission process convert a fraction of the uranium-238 atoms into plutonium-239. A portion of this plutonium is consumed during fission, and the remaining unfissioned plutonium and uranium-235 are available for reprocessing when removed from the reactor.

Step 4: Temporary Storage. Each year approximately one-third of the depleted fuel is replaced. The removed fuel assemblies are stored in a nearby spent-fuel pool of water so that their radioactivity may decline in isolation.

Step 5: Reprocessing. After several months in the spent-fuel pool, the depleted fuel may be sent to a reprocessing plant where residual uranium-235 and plutonium-239 are extracted. Although their concentration in the spent fuel is diminished, both isotopes may be used again as fuel. The recovered uranium and plutonium may be shipped to conversion plants and thus return to the *front end* of the fuel cycle where they may eventually be refabricated into reactor fuel. The commercially usable by-products are distributed outside of the fuel cycle.

* See Bibliography.

Step 6: Long-Term Storage of Wastes. Waste materials from the reprocessing plant are classified according to their penetrating power. High-level wastes must be solidified and cased in ceramic before being shipped to waste storage locations. (Possible permanent storage locations include geologic bedded salt formations.) Low-level wastes, usually in liquid form, may be stored in tanks similar to those on the Hanford Reservation in Richland, Washington. For maximum protection, however, low-level wastes should also be stored in "permanent" geologic formations. Though the penetrating power of these wastes may be low, the hazardous products may persist for some time.

The back end of the commercial fuel cycle stops today at the spent-fuel pool. The Nuclear Regulatory Commission has not licensed any reprocessing plants and, as of October, 1976, there were no commercial reprocessing plants operating in the United States. The only plant that ever operated is now closed permanently.[8] The West Valley facility in northern New York State, owned by Nuclear Fuel Services, Inc., closed following radiation-safety standards violations between 1966 and 1972 and will never be reopened.[9] General Electric built a $64-million reprocessing plant in Illinois that never reached operating status; the company recently concluded that its reprocessing system just would not work.[10] A third plant, in North Carolina, is almost finished but it was designed to produce reprocessed plutonium in liquid form. For safety reasons, plutonium must now be converted into a solid state before shipment.[11] The owners, Allied Chemical and General Atomic, are worried that the cost of reprocessing may be higher than the value of the fuel extracted from spent rods and are hoping that the federal government will take over the operation.[12] If reprocessing costs are not subsidized and turn out to be higher than anticipated, utilities will have little incentive to reprocess their fuel and will tend to leave it in spent-fuel pools until uranium becomes so expensive that reprocessing represents a financial advantage.

No one knows when the balance will tip toward reprocessing. Few commercial investors, however, may be willing to sit by the wayside with unused reprocessing plants waiting for the volatile uranium market to further skyrocket, thereby creating a demand for reprocessed spent fuel. The absence of reprocessing facilities assures that spent-fuel pools across the country will reach full capacity soon.[13] These are designed to hold between 900 and 1,000 metric tons of spent fuel and, of this capacity, only a small fraction remains available as of March, 1977. Some devices to increase spent-fuel pool storage capacity are now being marketed;[14] these stopgaps may be helpful in the short run but will not solve the broader problem of assuring an adequate uranium supply and reasonably safe radioactive waste storage.

The fuel cycle remains open-ended for a number of reasons: poor planning, cost overruns, delays and shutdowns suffered by reprocessing plants, environmental restrictions, unresolved technical problems with reprocessing, and the long lead-time required for construction of reprocessing plants. Failure to close the cycle results in two serious predicaments: reactors are deprived of an important source of fuel as long as reprocessing facilities are not available, and spent-fuel pools soon may have to be used as permanent storage sites, which may set an upper limit to the amount of spent fuel that reactors will be allowed to generate. The utilities are trying to adapt to a "throwaway" fuel cycle instead. The rationalizations offered to justify these conditions reflect the confusion and environmental insensitivity of the nuclear industry.

7
The Rise and Fall of Nuclear Power

A temporary halt to liquid metal fast breeder research occurred in 1966 when the Enrico Fermi Plant suffered a partial meltdown (See "We Almost Lost Detroit" for details of this accident.) That same year, a coalition of reactor suppliers, electric utility companies, and government agencies successfully launched a program for building hundreds of light water reactors throughout the United States. Between 1965 and 1966, twenty-six nuclear units were purchased by utility companies as the first step of an ambitious plant construction program. The long-range plan provided for one thousand large operating reactors by the turn of the century.[1]

Industry and government leaders who developed this aggressive schedule were faced with several important prerequisites. If the long-range plan was to succeed, it was necessary to assure stable construction costs; minimum licensing delays; available capital subject to moderate interest charges; a long-term supply of uranium at predictable prices; a permanent, safe system for reprocessing and storing radioactive wastes; and ample experience with and tests of reactor behavior and emergency systems. Not all of these prerequisites were perceived as critically important at the time, and it is now clear that industry and government leaders made several overly optimistic assumptions, especially about licensing delays, fuel, and construction costs. Other prerequisites entailed unknown variables, such as emergency systems behavior and waste storage. A prudent course would have been to postpone nationwide construction until the more important unknowns were resolved and the critical prerequisites met. Instead, the construction program commenced at the same time that efforts were being made to meet these prerequisites.

Aggressive marketing characterized the first stage of this program. Promotion by the AEC and the reactor manufacturers stampeded normally conservative utility companies into plans for purchasing nearly 170 reactors by the mid-1970s, though the utilities had very little nuclear expertise[2] themselves and less than solid evidence that the fuel, waste disposal, and safety problems would be solved. To stimulate enthusiasm in private industry, the federal government initiated a system of direct and indirect subsidies that, together with the reactor manufacturers' "loss-leader" approach to sales, attracted dozens of electric utility company orders.[3]

From 1965 until 1974, the plant construction boom continued, and the utilities kept ordering new plants despite rising interest rates and unproductive research into the waste and safety problems.[4] The Atomic Industrial Complex dismissed as obstructionist the technical controversies raised by AEC scientists and the strenuous objections by citizen groups intervening during the licensing process.[5] Several successful round trips to the moon during the Apollo space program assured many American citizens that a technically advanced society should be able to solve its important technical problems. This presumption was reinforced by those who were in a position to make vast sums of money. Given America's predicted electrical demand, the limitations of other electric power sources, and the 1973 oil embargo, nuclear energy appeared to be the most profitable option. Those on the nuclear bandwagon were the ones most likely to reap substantial profits, and their vision of an imminent bonanza distracted many from the problems that stalled industry progress by 1976.

Signs of a bust were unmistakable as early as 1974. Over 60 percent of all nuclear plant construction projects were postponed or cancelled that year,[6] and only five domestic plants were purchased in 1975. By January of 1976 the industry realized that it was facing more than just a temporary lull in sales, and that several converging trends might defeat earlier plans for a nuclear network across the United States. Several prominent analysts have studied the economic causes of this bust[7] and in the next article, "The Uneconomics of Nuclear Power," David Comey summarizes their conclusions.

Evidently, unresolved technical and safety problems compounded these economic difficulties. The earlier decision to construct plants while simultaneously conducting prerequisite research was, by 1976, reduced to an absurdity. By then, many of the safety studies begun years before had failed to produce conclusive results.[8]

Daniel Ford of the Union of Concerned Scientists suggests why the nuclear program lost momentum in spite of enormous subsidies and promotion efforts:

> Nuclear plant performance, for the plants that had begun to operate, also proved to be disappointing and, to some safety reviewers, disconcerting. Breakdowns and malfunctions at nuclear plants seemed to be endemic, with many involving failures of safety-related equipment. Safety deficiencies were discovered that led, from time to time, to across-the-board restrictions on nuclear plant operations, to requirements for the backfitting of new equipment, and to shutdowns for emergency inspections of components. All of these problems, combined with the usual run of operating difficulties for large central-station power generating equipment, greatly reduced the dependability of nuclear plants as a source of electricity. On the average, the nuclear capacity installed in the United States has spent about as much time idle as it has producing electric power.[9]

Nuclear plant performance reached a low point in 1976 when the NRC reported that the forced outage rate* for all operating U.S. plants was 18.8 percent for the first five months of that year.[10] Although some plants have demonstrated excellent operating records,† industrywide performance has been dismal from the beginning, and attempts to focus attention on these data as a basis for

---

* The average rate at which plants were shut down for unscheduled repairs and off-normal conditions, expressed as a percentage of the total hours that the plants should have operated.

† Especially San Onofre 1, Connecticut Yankee, Monticello, Nine Mile Point 1, Point Beach 1, Point Beach 2, Vermont Yankee, Yankee Rowe, Robinson 2, and Turkey Point 4.

reform have been met with reprisals, ridicule, and other extraordinary industry reactions.[11] Industry sensitivity to criticism and its bureaucratic approach to the myriad economic, environmental, and technical problems tended to erode public support for nuclear power at the same time that its critics were gaining credibility.

As opposition to the nuclear industry consolidated in 1975 and 1976, and as the moral arguments against mortgaging the resources and safety of future generations began to capture public attention,[12] the industry found itself hemmed in from many sides and reacted by launching still another public relations campaign,[13] funded largely by private corporations, to defeat antinuclear initiatives in several states. Toward this end the industry may win several battles but lose the war. In a contest that is likely to be repeated several times, the California antinuclear initiative was defeated in June of 1976 in the wake of a pronuclear campaign with expenditures that exceeded $3 million. A month after California voters rejected the initiative, the U.S. Court of Appeals issued three plant-licensing rulings that led the NRC to announce a temporary moratorium on full-power operating licenses and construction permits.[14]

That the court's conservative view is at odds with voter enthusiasm for nuclear power is evidenced by the state initiative results. But there was a substantial difference in the arguments and data presented. In California, the industry appealed to the voters' job security and standard of living.[15] The Court of Appeals, on the other hand, had the opportunity to deliberate over arguments presented by several expert attorneys, and both judges and clerks were familiar with the considerable body of government data and prior court decisions. The landmark decision that followed makes it clear that the judiciary will not be swayed by vague, undocumented, and optimistic expert testimony favoring nuclear power[16] and will, instead, reject industry efforts to steamroller projects that could damage the environment. A close reading of the decision confirms that earlier licensing procedures tended to insulate inadequate technical testimony (favoring the industry) from challenge and cross-examination,[17] and acknowledges that members of the licensing boards are not always hospitable to intervenors.[18] To that extent, the government discouraged open review of public safety issues.

The decision and the lengthy opinion sustaining it no doubt pleased the small band of dedicated public interest advocates (especially John Gofman, Henry Kendall, Daniel Ford, Anthony Roisman, Ralph Nader, David Pesonen, Richard Ayres, Gustave Speth, and Myron Cherry) who had led the fight since the early 1970s to protect the public from radioactive pollution. These men devoted half a decade and considerable time and resources to testifying against the nuclear industry before agencies that frequently obstructed their efforts in the public interest.[19]

The industry will undoubtedly continue to win citizen votes by investing enormous sums to defeat state initiatives (as in California), but may find that courts will repeatedly reject industry and government attempts to bend the law to suit their own purposes. Regardless of what happens in the judicial and political arenas, the future of nuclear power may depend more on how many more electric utilities decide to buy reactors. They will do so only if electric generation by nuclear power appears profitable in the long run and if demand for electricity increases.

Both criteria are doubtful. Following the oil embargo of 1973, electricity prices increased sharply and demand dropped as citizens and private industry found ways to reduce electric consumption. This behavior stunned utility companies,[20] many of which reexamined their plans for nuclear plants that had assumed *increasing* use of electricity. If electricity prices continue to climb as fuel, construction, and other utility costs increase, it is likely that per capita electricity consumption will level off, or even decline, especially if alternative energy sources and conservation practices gain national acceptance.

Even if electric demand is sustained, capital cost overruns and plant unreliability may convince utilities that nuclear fission is not a profitable way to manufacture electricity. Coal and conservation may become the primary means of carrying us through until the more benign technologies of solar and wind power emerge as mainstays of a steady-state economy. Our coal reserves are substantial— 433 billion tons of proven reserves and 1.5. trillion tons of total known reserves[21]—and are enough to supply our energy needs for centuries. What has been lacking are the political decisions to invest enough funds toward research that would limit coal pollution problems and diligently explore conservation and alternative energy sources.

At the same time that conservation groups were urging a shift to coal and conservation, a series of articles in national newspapers and industry newsletters suggested that the nuclear industry was in serious trouble.[22] According to the executive director of the Union of Concerned Scientists:

> The economic collapse of the nuclear industry, from the vantage point of the economy as a whole, can . . . be accommodated without harm. In fact, while an unhappy event for the nuclear industry, the slowdown . . . is a desirable adjustment for the U.S. economy. The simple economics of the situation is that business collapse of a selective nature is a healthy and necessary thing for an economic system. A rationally operating economy cannot afford to allow its resources to be incompetently managed or to be devoted to projects that do not serve its real purposes. It is desirable . . . that the "Invisible Hand" that Adam Smith told us takes charge of a free market economy would, as the need presents itself, put its fist down to discourage uneconomic activity. The nuclear program, it would appear, has been given some not-too-gentle pushing about by the Invisible Hand—for our economy's own good.[23]

In retrospect, the billions of dollars wasted on unproductive nuclear research and plant construction would have been better invested in developing alternative energy sources (such as solar and thermonuclear fusion) and solving the pollution problems associated with coal.

8
The Uneconomics of Nuclear Power
David Dinsmore Comey

The early promoters of nuclear power promised that electricity from nuclear plants would be so cheap in the near future that no one would bother to meter it. Instead, nuclear power now appears so expensive that it may be, to use that most pejorative of terms, uneconomic. In short, the pot of gold has disappeared. Ten years ago, however, the rainbow was substantial enough to attract billions of dollars of investment capital and today the fiscal momentum is enormous: the 1977 fiscal year budget allocated 65 percent of ERDA's entire research and development funding to the nuclear effort.

Electric utilities are now beginning to realize that the promised profits were illusory.[1] These doubts, together with the escalating costs of construction, operation and fuel, have combined to halt the ambitious program of nuclear plant development that began about 1965. We can only guess what the future holds. An educated guess, nevertheless, depends to a great extent on these cost factors. In fact, future nuclear industry profits depend almost entirely on construction, operation and fuel costs and we might examine them more closely.

Construction costs have gone up for a variety of reasons, one of which was that the early plants were not realistically priced. They were built by the reactor manufacturers as so-called "turn-key" projects:[2] the utility company bought the entire plant as a package for a specified price and, when the plant was finished, the utility merely had to "turn the key" and it was in business. Cost overruns were absorbed by the reactor manufacturer. Several reactor manufacturers deliberately set their prices low, knowing they would lose money on the turn-key plants, in order to attract initial orders. After enough utilities were hooked to provide a production stream, manufacturers ceased accepting any more turn-key project orders, and thereafter plants were subject to cost escalation during construction.

The firms that construct plants invariably work on a cost-plus basis, which provides little incentive to keep costs down. Furthermore, these firms discouraged their utility-customers from adopting a less-costly, standardized plant design and the utilities played into their hands by insisting on special features. Today, standardized plants are being explored by only a handful of utilities despite years of government urging.[3]

Over the years, the utilities realized that profits depended on building as large a plant as possible. The reactor manufacturers obliged, only to have the government step in and insist on sophisticated safety systems to protect the public against the consequences of a major accident. By the time they were installed and tested, these systems turned out to be far more expensive than expected.[4] At the same time, environmentalists began intervening during plant licensing hearings, protesting that safety systems were defective and discovering hidden defects in plant design that cost millions to remedy. As public pressure for better designs increased, the AEC regulatory staff began cracking down on the industry. Intervenor efforts and government regulation combined to delay plant licensing and increase plant costs due to inflation and interest payments on nearly completed, unlicensed plants.[5]

Shortages in skilled construction labor developed; the supply of welders in particular became a problem.[6] Qualified welders are in such demand that they reap extraordinarily high wages, work long overtime periods at double or triple pay, and earn up to $40,000 per year. Expensive overtime must also be paid to skilled electricians, boilermakers, pipefitters and metalworkers who are also in short supply. Moreover, the complex design of a nuclear plant makes it difficult to keep all the "critical paths" of the construction process in synchronization. Regiments of skilled welders are frequently kept waiting while other construction crews finish a segment that must be completed before welding can begin.

As a result of all these factors, the cost of a nuclear plant has gone from less than $100 per kilowatt of generating capacity in 1964 to more than $1,000 per kilowatt for a plant ordered in 1976, a rise of more than 900 percent.[7] During these years, the Consumer Price Index rose "only" 77 percent. This phenomenal rise in cost represents a market adjustment which reflects some of the real costs of nuclear power, as opposed to the artificially low costs that attracted utilities when the plants were first marketed.

With a cost of $1,000 per kilowatt of generating capacity, a typical new nuclear plant with two units of 1,200 megawatts each represents a $2.4-billion investment. Much of that money must be borrowed because utility common stock prices are depressed; most are selling below book value and raising capital by floating new issues of common stock merely dilutes the present stockholders' equity. But borrowing is expensive: although the cost of raising money used to be about 5 percent for a utility, it is now about 14 percent. Many utilities are finding the investment banking community unwilling to underwrite enormous loans for facilities which have not been a reliable source of power in the past and which will tie up substantial capital in the future.[8] To compete successfully for nuclear plant loans against other projects, the utilities will have to convince the lenders that nuclear fission is a profitable way to generate electricity.

The profitability of a nuclear plant depends on the amount of electricity it generates. Because plants have turned out to be less reliable than originally forecast, they have not generated as much electricity as expected.[9] Consequently, profits have lagged. In the first place, the reactor manufacturers and the AEC predicted that nuclear plants would generate electricity at 80 percent capacity factor, but that has not turned out to be the case. The capacity factor of a power plant is simply a measure of the total electricity produced by a plant during some period of time, compared to the electricity it would have produced during this same time period had it operated at its full design power level.

In order to conform to the requirements of the National Environmental Policy Act, the federal government prepares environmental impact statements for each plant. In each statement, the results of a cost-benefit analysis of the plant are published to justify its construction. The cost-benefit analysis of nearly every nuclear plant operating today assumed that each plant would produce electricity over its lifetime at an average capacity factor of 80 percent or better.[10]

The actual performance data do not support this claim, however. The AEC began compiling nuclear capacity factor data in 1973. During that year, the nation's large nuclear plants showed an average capacity factor of 56.1 percent. This dropped to 51.7 percent in 1974, rose to 54.9 percent in 1975, and leveled off at 55.1 percent in 1976 (see Table 1). The average capacity factor for the years 1973–1976 was 54.6 percent, an average that is weighted by size of the plants operating during that period. The cumulative-to-date capacity factor for all plants from first year of operation through 1976 was 53.7 percent. Moreover, the plants show a decline in performance as they get older. During the first two years of commercial operation, they average 54 percent, rise to an average of 63 percent during the next six years, and thereafter decline to an average of 39 percent, less than half of the design level of 80 percent (see Table 2).

There are a number of reasons for this disappointing performance.[11] During the first two years of operation, there is a carryover of construction-related quality assurance problems. In addition, utility personnel are fine-tuning plant operating procedures that would, under ideal circumstances, be perfected the day the plant received its license. Once these problems are resolved and the bugs eliminated from the plant's systems, the capacity factor rises to a peak. Thereafter, wear-related problems set in, the most serious of which are corrosion and (for PWR's) the accumulation of highly radioactive "crud" in the primary cooling system. For these plants, any repair work on cooling systems will consume enormous amounts of time and people in order to avoid excessive radioactive exposure to workers.[12]

A worker can receive his maximum permissible exposure after working on the primary system for as short a time as a few minutes, thus "burning him out" for the next three months. This has meant in some cases that thousands of workers have had to participate in

TABLE 1

Nuclear Power Plant Capacity Factors

Unit Name	Reactor Mfr.	Design Power Net MWE	Age Yrs.	Cumulative to Date	1976	1975	1974	1973
Arkansas 1	BW	850	2	58.80%	52.1%	65.5%	—	—
Browns Ferry 1	GE	1065	2	14.15%	13.9%	14.4%	—	—
Browns Ferry 2	GE	1065	2	15.75%	16.8%	14.7%	—	—
Brunswick 2	GE	821	1	34.48%	34.5%	—	—	—
Calvert Cliffs 1	CE	845	1	84.93%	84.9%	—	—	—
Connecticut Yankee	W	575	9	74.04%	79.7%	81.8%	86.4%	48.1%
Cook 1	W	1090	1	71.07%	71.1%	—	—	—
Cooper	GE	778	2	54.90%	53.3%	56.5%	—	—
Dresden 1	GE	200	16	50.56%	54.2%	39.8%	20.1%	32.0%
Dresden 2	GE	809	6	50.78%	61.5%	41.9%	47.7%	70.8%
Dresden 3	GE	809	5	50.92%	56.8%	30.9%	45.2%	52.1%
Duane Arnold	GE	538	2	50.75%	52.7%	48.8%	—	—
Fitzpatrick	GE	821	1	57.63%	57.6%	—	—	—
Fort Calhoun	CE	457	2	53.35%	54.7%	52.0%	—	—
Ginna	W	490	6	60.78%	47.9%	70.8%	48.9%	79.2%
Hatch 1	GE	786	2	52.50%	59.9%	45.1%	—	—
Indian Point 1	BW	265	14	37.79%	0.0%	0.0%	52.9%	0.0%
Indian Point 2	W	873	3	45.67%	29.6%	63.9%	43.5%	—
Kewaunee	W	560	2	68.45%	68.8%	68.1%	—	—
Maine Yankee	CE	790	4	62.63%	85.4%	65.1%	51.6%	48.4%
Millstone Point 1	GE	690	6	54.80%	61.9%	64.5%	64.5%	31.1%
Millstone Point 2	CE	828	1	62.46%	62.5%	—	—	—
Monticello	GE	545	5	69.56%	83.3%	60.3%	61.2%	68.5%
Nine Mile Point 1	GE	610	7	58.31%	76.8%	57.0%	61.7%	65.4%
Oconee 1	BW	887	3	56.97%	51.3%	68.1%	51.5%	—
Oconee 2	BW	887	2	59.15%	54.3%	64.0%	—	—
Oconee 3	BW	887	2	62.95%	61.0%	64.9%	—	—
Oyster Creek	GE	650	7	64.97%	67.6%	55.2%	64.5%	63.0%
Palisades	CE	821	5	26.48%	39.5%	33.8%	1.1%	33.5%
Peach Bottom 2	GE	1065	2	57.00%	59.5%	54.5%	—	—
Peach Bottom 3	GE	1065	2	60.65%	64.7%	56.6%	—	—
Pilgrim 1	GE	687	4	46.08%	40.0%	43.0%	33.6%	67.7%
Point Beach 1	W	497	6	70.52%	78.0%	67.1%	72.2%	63.0%
Point Beach 2	W	497	4	78.45%	86.2%	85.9%	73.0%	68.7%
Prairie Island 1	W	530	3	60.23%	70.2%	79.6%	30.9%	—
Prairie Island 2	W	530	2	62.80%	57.2%	68.4%	—	—
Quad Cities 1	GE	809	4	56.90%	47.7%	60.3%	50.3%	69.3%
Quad Cities 2	GE	809	4	58.03%	60.6%	34.9%	63.1%	73.5%
Rancho Seco	BW	913	2	22.05%	27.5%	16.6%	—	—
Robinson 2	W	707	5	72.46%	78.5%	67.3%	77.7%	60.8%
San Onofre 1	W	450	9	68.13%	62.6%	82.3%	79.8%	57.7%
Surry 1	W	823	4	52.38%	60.9%	54.3%	46.0%	48.3%
Surry 2	W	823	3	50.97%	46.3%	70.1%	36.5%	—
Three Mile Island	BW	819	2	68.75%	60.3%	77.2%	—	—
Trojan	W	1130	1	21.19%	21.2%	—	—	—
Turkey Point 3	W	745	4	59.80%	66.0%	67.0%	55.5%	50.7%
Turkey Point 4	W	745	4	61.50%	57.6%	61.1%	65.8%	—
Vermont Yankee	GE	514	4	61.68%	72.2%	79.1%	55.1%	40.3%
Yankee Rowe	W	175	15	71.31%	81.4%	77.8%	59.5%	68.2%
Zion 1	W	1050	3	47.60%	51.6%	53.4%	37.8%	—
Zion 2	W	1050	2	51.40%	50.3%	52.5%	—	—
		37,725	209	53.69%	55.1%	54.9%	51.7%	56.1%

BW—Babcock/Wilcox CE—Combustion Engineering GE—General Electric W—Westinghouse
(Averages weighted by megawatt capacity)

TABLE 2
Performance VS. Age of Plant
(3-year Moving Average)

Year of Operation	Capacity Factor
2	54.9%
3	57.5%
4	61.3%
5	63.3%
6	67.4%
7	65.7%
8	65.9%
9	66.2%
10	54.4%
11	53.4%
12	39.1%
13	41.5%

the repair of a single nuclear plant; the results have been long outages and lowered plant capacity factors.[13]

The AEC prediction that nuclear plants would operate at 80 percent capacity factor for 30 years is unlikely to be fulfilled, according to two Swedish engineers who have developed a realistic model forecasting U.S. nuclear performance.[14] Calculations based on their model over a thirty-year plant lifetime yield an average capacity factor of 42.7 percent (see Figure 1). If this model is correct, then, in order to meet the Energy Research and Development Administration's projections of electric generation by nuclear plants in 2000 A.D., we will have to build nearly twice as many nuclear plants as would be needed if the government's 80 percent capacity factor were realized.[15]

Low capacity factors raise the price of electricity another way. The lower a plant's capacity factor, the greater its generating costs per kilowatt-hour will be. This is because "fixed costs" (e.g. taxes, interest payments, depreciation) do not change, regardless of how much or how little electricity the plant generates. For example, the cost of electricity produced by a plant operating at a 35 percent capacity factor is just *double* that produced by a plant operating at an 80 percent capacity factor because the same fixed costs are distributed over much less electricity.

FIGURE 1
Nuclear Plant Capacity Factors Vs. Age of Plant

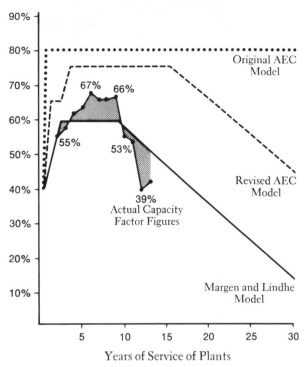

Years of Service of Plants

In addition to construction and operating costs, the price of nuclear fuel significantly affects the profitability of nuclear power. One way of viewing fuel costs is to determine how much electricity was generated, say, last year by using nuclear fuel. Then, calculate the cost of nuclear fuel used to generate that electricity and compare this cost with what we would have had to spend *if* we had burned fossil fuel instead. The industry is fond of this rather narrow comparison because it ignores *all other* costs. For if you do not include these "other costs," if you assume that the only cost of generating electricity is the cost of fuel, then the costs of generating electricity by nuclear fission appear to be less than the costs of generating electricity by burning fossil fuel.

A recent industry survey showed that fuel costs for nuclear plants were indeed far lower than fuel costs for coal-fired plants.[16] The nuclear industry ran advertisements based on these statistics that proclaim how many millions of dollars have been saved.[17] If you read the advertisements very carefully, however, you will see that they may be referring *to fuel costs only.* Not included in these cost calculations are the fixed charges for taxes, interest and depreciation, which are greater by far for nuclear plants.[18] If you calculate all the costs—fixed, fuel and operating—first for fossil and then for nuclear plants, and then compare the total combined costs, coal turns out to be cheaper than nuclear in almost every case.

By emphasizing fuel cost savings, the industry views nuclear plant economics from too narrow a perspective. Before long, however, advertisements of fuel cost savings may disappear along with the statistics. The nuclear industry assumed until recently that uranium would be inexpensive and plentiful.[19] Indeed, for more than 20 years, uranium prices stabilized between $6 and $8 per pound. In 1973, however, the price began skyrocketing and, by early 1977, had reached $41 per pound.[20] At this rate, the price will exceed $100 per pound in a few years, eliminating any advantage that uranium had over coal.[21]

A recent casualty of this astronomical price rise was Westinghouse Corporation, which manufactures pressurized water reactors and, earlier, had captured a fair percentage of the nuclear reactor market. During the 1960s Westinghouse made the mistake of assuming that uranium prices would remain stable and signed contracts with utilities that bought its reactors, guaranteeing them a steady supply of uranium at pre-inflation prices. Once uranium prices started climbing, Westinghouse regretted this bit of merchandising. Faced with a $2.5 billion loss due to the spread between promised and actual uranium prices, Westinghouse recently announced that it would not honor these fuel contracts, a move that shocked the industry and left at least twenty-seven utilities out in the cold. Most of the utilities promptly sued and the matter is now before the courts.[22]

The rise in uranium prices reflects shortages that are bound to get worse. All of the known, low-cost uranium reserves have been contracted for and the medium-cost reserves that are known to exist in the United States will not be sufficient to supply the demands of plants currently planned for the 1980s.[23]

Several studies by respected geologists and economists conclude that the amount of recoverable uranium in North America may be limited.[24] But to provide for just our own nuclear plant requirements to the year 2000 A.D., we will need to discover new uranium lodes equal to at least 9 times as much as exists in the Colorado Plateau,[25] a four-state area encompassing much of New Mexico, Arizona, Utah and Colorado. Whether these discoveries will occur remains to be seen. If the domestic uranium supply turns out to be inadequate, we may have to compete on the world market with foreign nations, the nuclear programs of which are also developing a voracious appetite for "yellowcake." If this does not drive uranium prices sky-high, we may still have to contend with the risks of importing uranium from politically unstable countries where prices and export policies fluctuate with the whims of the party in power.

The need for restoring some stability and control over uranium supplies and prices may have been foreseen as early as 1971 by members of the Uranium Institute in London. Confidential papers taken in 1976 from the files of Mary Kathleen Uranium Ltd. of Australia reveal that Uranium Institute members met to set prices, rig bids, and establish a market quota system.[26] The existence of this cartel was confirmed in October of 1976 when Westinghouse charged in a countersuit that twenty-nine uranium suppliers illegally rigged U. S. uranium prices.

Both the cartel and the international free-for-all it attempts to remedy pose severe problems for the United States nuclear industry. In the first place, neither arrangement gives utilities much basis for making economic forecasts. As one major Wall Street investment banking firm put it recently, "Uncertainty represents the one thing that no market can stand—the stock market or the utility purchasing agent, whose utility plant is about to run out of fuel."[27]

Grim as this situation sounds, there are four additional jokers lurking in the economic deck of cards which may further aggravate present difficulties. First of all, the federal government has recently increased by 20 percent the amount of uranium each utility must supply to an enrichment plant in exchange for a shipment of fuel for its reactor.[28] The utilities are already protesting that they cannot obtain this extra uranium due to the present tight supply. What is worse, the government plans in a few years to boost the amount of uranium required for enrichment *by another 20 percent.*[29]

Second, all commercial fuel reprocessing plants in the United States are either shut down, or under construction with unresolved environmental questions haunting them.[30] Consequently, there are no commercial reprocessing facilities today and none of the uranium-235 in hundreds of tons of accumulating spent fuel is available as recycled fuel. Unless the government allows the plutonium reprocessed from spent fuel to be recycled and used in reactors at a later date, the economics of reprocessing make it extremely unlikely that any of this leftover uranium-235 will be recovered from the spent fuel in the future. But industry projections assumed that this recycled fuel *would* be available. If not, international uranium requirements will be boosted by another 25 percent.

Joker number three is that currently operating nuclear plants have not been able to convert as much of the uranium into electricity as expected. These plants were designed to produce 9.2 million kilowatt-days of electricity per metric ton of uranium but, in fact, have produced less than 3.7 million.[31] Because of this difference between the expected and actual "burn-up" rates of uranium, we will need at least *40 percent more* uranium to produce the required amount of electricity.

Government and industry knew that uranium might be in short supply—not as short as it turned out to be, but nevertheless limited. They were counting on the breeder reactor—the LMFBR—which can convert the more plentiful uranium-238 isotope into plutonium, a potent reactor fuel. However, both environmental and technical roadblocks may prevent the breeder from becoming a commercial reality. The environmental questions have resulted in uncertain government support[32] and there still remain a great many unresolved technical questions[33] to be answered before the breeder becomes a reliable means of electric power generation. Its promise as a solution to the uranium shortage is clouded because its doubling time (the amount of time it takes to double the original inventory of plutonium) may be *up to five times longer* than originally planned.[34]

The effect of these four jokers (increased fuel required for enrichment, no uranium-235 recycle, bad burn-up and no breeder) will be to increase substantially the uranium requirements of the nuclear industry at a time when even the presently forecast demands cannot be provided for.

This ensures that uranium costs will climb, reaching at least $100 per pound and possibly going two or three times beyond that level. If the cost of nuclear-generated electricity increases as a result, other means of electric generation may appear more profitable. For example, to remain competitive with coal at $30 per ton, uranium prices must not exceed $62.50 per pound.[35] As uranium prices continue to rise, so will the profits of companies supplying the fuel, although at some point they run the risk of pricing nuclear power out of business. When that point of no return is reached, however, the private nuclear industry will undoubtedly ask for massive federal government assistance.

Several recent developments foreshadow a government bail-out.[36] First, after huge cost overruns, ERDA may take over the Barnwell, South Carolina, fuel reprocessing plant[37] now under construction by Allied Chemical and Gulf Oil. Second, President Ford proposed a Nuclear Fuel Assurance Act which provides federal guarantees of up to $8 billion to assure that private industry will not take a loss if new uranium enrichment plants are built. This proposal pales by comparison with his $100 billion Energy Independence Authority, which would create an independent agency beyond the control of Congress and result in a major expansion of the country's commitment to nuclear power.[38] These proposals are indicative of the rather desperate measures that the nuclear industry is concocting in order to stay in business.

Nuclear power costs more than alternative sources of electric generation, the plants are less reliable and now their fuel is in danger of running out. No relief for any of these problems is visible on the horizon and, unless the federal government begins pouring immense subsidies into the nuclear industry, it may very well be done in by its economics.

9
The Silent Bomb: Radioactive Waste

If you leave a glass of water undisturbed long enough indoors, the water will evaporate and the airborne water molecules will distribute evenly throughout the room, slightly raising the humidity. Similarly, when a ship dumps nondegradable, toxic or radioactive wastes in deep water, the materials eventually disperse throughout the ocean. The air surrounding the earth and the water covering part of it tend to scatter suspended substances regardless of toxicity, diminishing their concentrations but also maximizing their dispersion. Eventually, perhaps millions of years after release, an equilibrium point is reached when tested water samples everywhere will show identical, infinitesimal traces of the dumped wastes.

These dispersion processes are one way that the ecosphere has protected living things in the past by diluting all suspended substances over time. But we have developed extremely potent poisons in the last century which, if released, are especially dangerous in small quantities because they can be more easily inhaled or ingested in that form. The dispersion process thus aggravates pollution problems by scattering microscopic particles of plutonium and other toxic substances released by human errors.[1] Because many of these substances are long-lived and may well evade our best efforts to contain them, these dispersion processes will eventually scatter lethal quantities of them throughout our fragile ecosphere.

Plutonium, thorium, and other elements produced by the nuclear industry remain potent for 250,000 to 1,500,000 years depending on their half-lives, the quantity of the material, and the degree of safety required. Detoxification and transmutation of plutonium and other radionuclides are feasible, but may involve higher costs than anticipated. Meanwhile, the decision to continue plant operation and construction assures that radioactive wastes will gather in prodigious quantities. This decision will commit thousands of future generations to exercise the greatest care and diligence to ensure that under no circumstances will plutonium and other long-lived, toxic materials be released to the biosphere.[2] According to Hannes Alfvén, the waste repositories must be located in areas immune from

> riots or guerrilla activity, and no revolution or war—even a conventional one—[must] take place in these regions. The enormous quantities of extremely dangerous material must not get into the hands of ignorant people or desperados. No acts of God can be permitted.[3]

Of course, it may be possible to bury canisters containing these wastes in wells thousands of feet underground. Deep geologic disposal, whether under stable land masses or far under the ocean floor, meets Alfvén's requirements. But while it offers one hope of safe, permanent disposal, it cannot be "tested" in the conventional sense, and it may involve unforeseen geological and technical problems, such as safe canister transport thousands of feet from sea level to a point on the ocean floor. Whatever disposal system is finally agreed upon, we will be asked to put our faith in nuclear science and technology whose spokesmen imply, paradoxically, that their disposal systems are infallible even though the people who design, develop and describe them may not be.*

---

* For this reason, safety assurances by members of the technological priesthood must be viewed skeptically. Government and industry lack enough data from which to calculate reliable predictions of human error, catastrophic nuclear systems failure, and their synergistic effects. Even if they possessed enough data, risk analysts who

Geologic burial does not solve all environmental problems posed by radwastes, however. Before they are buried, plutonium and other toxins may escape because of catastrophic accident, terrorism, theft or mere chance. The precise amount of plutonium lost during normal industry activities is unknown. The margins of uncertainty involved suggest that as much as 2 percent of plutonium-239 (and equivalent amounts of reactor plutonium) may be lost, and that enormous efforts will be required to reduce this.[5] If the standard of care improves *by a factor of 100,* and if AEC's and ERDA's estimates of 1.6 million pounds of plutonium produced by the year 2000 is reduced by *a factor of 1000,* we calculate that enough plutonium will be released to provide *minimum potential cancer doses for 108 million people.* * The possibility of continuing plutonium losses was emphasized by Dr. Ralph Lumb, chairman of a 1967 AEC advisory committee on safeguarding nuclear material, who, referring specifically to the future plutonium problem, said it is "like any other business . . . the more you have of something, the more you're going to lose."[6]

The amount of plutonium that will be produced and the fraction that will inevitably be released are unknowns for the time being. What is absolutely certain, however, is that plutonium must be guarded essentially forever, that it is "fiendishly toxic" for this span of time, and that some of the plutonium that we produce will inadvertently or deliberately be released. Minute particles of

derive predictions therefrom would inevitably introduce certain blind errors into these calculations, for the analysts are as fallible as the people and the systems they are studying. Indisputably, they are unable to identify or guard against their own infallibility, which tends to be intrinsic and unique for each person. Consequently, one may assume that these calculations include certain errors, *the identity and degree of which are unknown.* If so, then any statistical prediction of human or systems failure, together with derivative accident probabilities, are at best defective. At worst, they are meaningless, especially if these predictions are developed from an inadequate data base in the first place. If would be helpful, but unfortunately impossible, for risk analysts to make quantitative, detailed, "and infallibly correct and complete calculations, not only of engineering variables (such as effectiveness, reliability and relevance of safety devices), but also of human malice *and of human fallibility, including their own.* Even the most generous enthusiasts of risk analysis are not in the habit of [claiming that they can achieve these results. Instead, they settle for identifying] the designer's omissions; but who is to discover their own?"[4]

* These calculations are based on the assumption of micron-sized particles dispersed in the air and lodged in the lung, and also on Dr. John Gofman's figure of 338 million cancer doses per pound of plutonium.

plutonium will disperse and penetrate living things and some of these living things will die or mutate after the lodged plutonium undergoes radioactive decay.

The exact chain of events leading to mutation, cancer, or death as the result of radioactive exposure is not clear, nor do we know exactly what forms of mutation or cancer are the result of exposure *only* to plutonium. This is because a great many chemical and radioactive substances are mutagenic and/or carcinogenic,* and corporations worldwide will continue to release these materials until exact traceability to cancers and mutations can be proved for discrete amounts of these toxins. It is utterly irresponsible to continue dumping these materials into the biosphere until someone can prove beyond a reasonable doubt that specific people have suffered death, illness, or mutation as a result. This proof may be scientifically impossible, partly because of the time scale involved in mutation, and partly because of the astonishing amounts and variety of other toxins already dispersed throughout the environment.

Plutonium is only one of the thousands of poisons we have created since 1900. Dozens of radionuclides and poison gases, hundreds of defoliants and insect sprays, and tens of thousands of other chemical compounds ranging from the "violently lethal" to the "merely toxic" have been manufactured for agricultural, industrial, research, or military purposes in the past seventy years.[7] Tons of these materials are dumped into the environment annually simply because we have nowhere else to deposit poisons that have lost their usefulness. More importantly, *thousands of tons* of radioactive material released by nuclear explosions and spills are now dispersing through the ecosphere.[8] The accumulating toxic material released from all these sources will increase daily for the indefinite future. It remains to be seen just how much longer we can keep using the earth's environment as an unlimited sink without catastrophic consequences.

There may be a point where the effects of these releases on the biosphere and on human beings will become irreversible. Someday, the critical saturation point of one of these toxins may be reached.[9] It may be plutonium that tips the balance, or arsenic, or perhaps it may be only sulphur dioxide which will reach, say, one part in ten thousand everywhere on earth. At that point, a great many living

* Mutagenic substances cause genetic mutations. Carcinogenic substances cause cancer.

things will begin to die. It will be too late to scrub the seas or decontaminate the air, for the cleansing solutions we will use may be contaminated as well. By clever technical fixes in limited regions, the human species may sustain life for a generation or two. And that will be all.

The dispersion process described earlier is only one side of the coin. Another process works in an opposite manner: it tends to concentrate toxins, especially in organisms along the human food chains. Certain plants, insects, and animals tend to accumulate poisons before they are consumed by other living things. This concentration process frequently leads to our dinner tables. For example, strontium-90 scatters to the four winds after open-air weapons tests and eventually falls on grazing lands and crops. The isotope is consumed by humans and animals, and its concentration may be increased if humans regularly consume tissues or other foods contaminated with strontium-90.[10] Further concentrations of this bone-seeking isotope in your body will increase significantly your chances of early death due to leukemia or bone cancer.[11] Nor is this the only danger we face: iodine-131 tends to concentrate in the human thyroid gland; cesium seeks muscle fibers; and other isotopes regularly released by industry accumulate with effects we will not perceive for years.

What is especially tragic is that economic imperatives deflect our efforts to protect ourselves, our families and future generations from the consequences.[12] Assume that you ingest unsafe quantities of industry-produced iodine-131 over the years. Assume further that your thyroid gland develops an inoperable, malignant tumor, an adenocarcinoma, ten years from now. If you die and if an autopsy confirms the diagnosis, your survivors will neither be able to collect damages nor convince government or industry to shut off the radioactive discharges that were directly responsible for your death unless they can prove beyond a reasonable doubt that *your* cancer was caused *exclusively* by the defendant's discharges. As long as your survivors cannot prove this, the discharges will continue. The government will not interfere as long as the radiation levels of these discharges are within limits set by government committees, presumably representing the public interest.

The burden of proof is on you, the injured citizen, or on your survivors, to show who damaged you and how, not on the industry which poisoned you.[13] Its innocence is presumed until proved otherwise. Of course, proof of guilt is usually impossible to establish, unless a great many people are poisoned at once. This occurred recently in Minamata, Japan, where elevated mercury concentrations in fish and other ocean edibles followed many years of dumping toxic compounds into the bay. Over three hundred people died and several thousands were crippled or made ill. The Japanese court ordered the responsible company to pay a huge sum in damages to the survivors,[14] but only after a bitter, prolonged fight.

In our country, industries that limit their poison releases according to government standards enjoy all kinds of defenses—legal buffers—built into our socioeconomic system. Corporations that release toxins are never compelled to prove that "small" amounts of these releases are safe because it would be uneconomic and therefore inconvenient to do so, and because our government has co-opted that responsibility. Safety is presumed, and this presumption insulates industry from a multitude of legal actions.

The Inhuman Experiment

Of the many agents capable of causing cancer, radiation has been given special attention lately, and enough evidence has accumulated to convince the most skeptical scientist that radiation does, in fact, cause cancer.

During the last forty years, a great many people have received substantial radioactive doses. These occurred voluntarily during medical x-ray diagnosis and treatment, unknowingly during uranium mining and commercial activities, and involuntarily as the result of accidents and of military explosions over Japan.[15] A number of studies show a relationship between these radiation exposures and many forms of cancer.[16] Several times in recent years, these medical data were used as the basis of insisting on more stringent government regulation to protect the public.[17] Only limited reforms followed. The plain fact is that not enough people have died from radiation to force federal agencies to pay more attention to people and to disregard industry vested interests. As matters now stand, the deaths of "only" a few thousand overexposed human beings are the sole *conclusive* evidence we have to support a case for reduced radioactive pollution.

Our government maintains a permissive radiation policy that balances economic interest against human damage. Consequently, all living things have become subjects of a universal experiment to determine how much radioactive material can be released before unacceptable results terminate it.[18] The industrial sector will undoubtedly persist in releasing as much material as is allowed (or more) until public pressure becomes great enough to force significant reforms. As long as people continue to die two or three stages removed from the radiation or chemical release that caused the disease, and mutants continue to be born for reasons their attorneys can't precisely identify, the experiment will go on.

Meanwhile, the release of toxins may have synergistic effects which may call attention to the experiment earlier than expected. For example, recent research by Dr. Edward Martell of the U.S. Atmospheric Research Center indicates that low-level alpha radiation from insoluble radioactive lead-210 particles in cigarette smoke may be the primary agent responsible for bronchial cancer in smokers and that low-level radiation may play a major role in arteriosclerosis and heart disease.[19] This is another instance of hidden dangers that frequently surface after the fact from a permissive approach to toxin release or exposure. Smokers who grasp the implications of these data may become enthusiastic supporters of stricter radiation standards.

The Bad News for Today

We can guard ourselves only in limited ways from radioactive wastes that have been released to the environment. The radioactive waste effort is not concerned with these; instead it focuses effort on guarding materials that can still be isolated or have yet to be created so that releases are held to a minimum.

Several assertions have been made by nuclear critics regarding the federal radioactive waste program:

- *High-level radioactive wastes may not be disposable.* At present, they can only be guarded.* Once created, some of them must be guarded for hundreds of thousands of years.[20]

* Government scientists dispute this. However, as of this writing, no proved, workable safe disposal system has been developed and tested. The issue turns on whether our disposal systems should be only "reasonably" safe. Our view is that safety standards imposed on other hazardous industries are inadequate when applied to the nuclear industry, given the extremely long time that nuclear toxins remain potent.

- Most reactor wastes are so highly radioactive that they pose serious risks to all forms of life.[21]
- In addition to the known risks, there may be unknown genetic consequences of releasing radioactive materials.[22]
- The federal government has not yet developed and adequately tested a safe, permanent means of guarding radioactive wastes for hundreds of thousands of years.[23] After thirty years and millions of dollars spent on research, it claims to have developed only a general plan.*

Each of these problems will be examined more closely. But first, it may be helpful to define a few terms.

What *is* radioactive waste? What forms does it take and where does it come from? "Radwaste" is the general term applied to all the unusable, radioactively contaminated by-products of the nuclear fuel cycle and U.S. weapons programs. It also includes "hot scrap" and rubbish, decommissioned plants and their components, radioactive materials left over from military programs, and reactor and reprocessing "offgasses" containing radioactive iodine, carbon, tritium and krypton. Wastes are classified "high level" or "low level," depending on their penetrating power. Our main focus will be on high-level wastes, for these pose the most serious health and environmental dangers, although a few cases of low-level waste management will be mentioned.

If nuclear industry plans are fulfilled, reactors alone will create about 42,000 tons of nuclear waste annually by the year 2000.[24] Because these wastes will generate considerable radioactivity, they must be shielded from the environment for millennia. The wastes also generate a great deal of heat due to radioactive decay. Storage in a small area is out of the question. Instead, wastes will have to be placed in small containers and placed several feet apart for cooling by convection.

Recent industry announcements have misled the public by spotlighting only the radioactive waste generated by *nuclear reactors.*[25] An honest assessment would consider all waste sources, including (but not limited to) mill tailings and fuel reprocessing leftovers, detritus remaining from military nuclear programs, and the millions

---

* The installation of solidified plutonium in deep, stable geologic formations such as bedded salt is a reasonable approach if carefully implemented. But the first disposal sites selected recently turned out to be defective. See p. 124.

of contaminated components from decommissioned plants. As of 1976, there are more than 85 million gallons of liquid high- and low-level wastes in temporary storage tanks. By the year 2000, the government anticipates that between 60 and 200 million gallons of high-level wastes *alone* will be generated.[26]

Radwaste contains a galaxy of potent radioactive elements. One of the most toxic by-products of reactor fission is plutonium, of which a minimum dose of 1.4 micrograms could cause lung cancer in a nonsmoking male. Dr. John Gofman calculates that one pound of plutonium dispersed throughout the environment is equivalent to 338 million lung cancer doses.[27]

The same amount could trigger lung cancer in one hundred times as many smoking males, or ten times the earth's present population. Besides strontium-90 and iodine-131, radwaste also contains cesium-137, plutonium-239 and tritium; the half-lives of these isotopes range from twelve to 24,360 years. Although all these are carcinogens, plutonium is by far the most lethal, despite its low penetrability, and is classified as a high-level waste. It may be pulverized to a very fine dust and carried moderately far by the wind, and must be isolated from the environment for at least 250,000 years. Not only are commercial reactors now manufacturing plutonium in substantial quantities, but ERDA has also proposed that, in the future, plutonium be mixed with uranium and used as reprocessed "mixed oxide" reactor fuel, thereby creating additional opportunities for release during the reprocessing cycle and subsequent burnup.[28]

Finally, if Seaborg's estimate of 1.6 million pounds of plutonium production (including plutonium manufactured by breeder reactors) by the year 2000 is realized,[29] the chances of the human race surviving will reach a new low. Dr. Gofman has suggested that even with 99.99 percent storage reliability, the nuclear industry will release 160 pounds of plutonium each year. This will provide minimum potential lung cancer doses for about fifteen times the earth's present population.

Can these lethal substances be guarded for millennia? What percentage will actually be released despite industry promises? As long as no one can answer these questions authoritatively, prudence demands that we approach the unprecedented magnitudes of toxicity and half-life with utmost caution. This is essential, especially since the federal government may not implement the best available means of guarding these wastes, as pointed out recently by Lash, Bryson and Cotton of NRDC.[30]

One basis we have for predicting how carefully radwaste will be guarded in the future is the government's waste management record to date. A quick rundown of recent incidents suggests room for improvement:

- Tanks at the Hanford, Washington, radwaste storage reservation are corroding and leaking. Approximately 70 million gallons of high-level wastes are stored there, some so hot that they boil spontaneously and continuously. At one Hanford site, twenty leaks totaling more than 500,000 gallons have occurred since 1961.[31]

- The AEC dumped so much plutonium into a concrete-capped earthen trench at Hanford not long ago that scientists were concerned about the possibility of a chain reaction. Migration of high-level waste in the soil has been studied, and some scientists worry that significant amounts of Hanford waste may eventually reach the water table.[32]

- A tractor-trailer truck carrying thirty-two drums containing low-level radioactive waste slammed into the rear of a state road vehicle near Ashland, Kentucky, in January of 1976. Of the fourteen drums that rolled onto the road, six ruptured, scattering radioactive material over the highway.[33] Although the material was cleaned up, the accident suggests that releases during transportation may pose significant hazards.

- More than 47,000 concrete-lined steel cans of radwaste were dumped into the Pacific Ocean in 6,000 feet of water off San Francisco between 1946 and 1970. Many of the cans cracked or imploded due to water pressure.[34] Recent underwater photographs show what appear to be giant sponges attached to the barrels. Some measure three feet in height.[35]

- A 1970 program to study radwaste burial in salt mines at Lyons, Kansas, was dropped in 1973 after the possibility of water intrusion into the salt formation was raised by researchers.[36]

- A second site at Carlsbad, New Mexico, was rejected when test drilling indicated underlying brine contaminated with toxic gases.[37]

Besides permanent burial in salt mines, in deep geologic formations, and in deep holes in the ocean floor, several ingenious proposals to dispose of radwaste have been studied. They include:

- *Launch rockets loaded with radwaste into the sun.* This is unfeasible at the present as well as expensive ($5 to $10 billion). One satellite containing 900 grams of plutonium-238 disintegrated during reentry and the toxin was dispersed. This incident could be repeated with disastrous results unless every radwaste rocket works perfectly.
- *Sink radwaste barrels in ocean trenches.* Our knowledge of the dynamics, timing and variables associated with underwater geologic drifts may never be enough to provide reasonable assurance that sunken barrels will be safely buried. Furthermore, to be sure that the barrels are dropped accurately, we must have reliable information on all deep ocean currents. If dumping takes place on the basis of inaccurate or incomplete information, the barrels may remain unburied long enough for them to implode and leak.
- *Bury radwaste barrels under antarctic ice.* Transporting the barrels there may pose serious hazards because of storms and high winds. Our knowledge of antarctic substrata is not much better than our understanding of what takes place at the ocean's bottom. For example, there may be water under the ice cap. Hot barrels would eventually melt through, reach sea water, implode or corrode, and contaminate the ocean.
- *Bury wastes in underground caves formed by exploding a nuclear bomb.* Underground bomb explosions have fractured surrounding rock. Rock fractures might allow radwaste to migrate to the surface or reach subterranean water.

Thomas Hollocher analyzes these and other proposals in *The Nuclear Fuel Cycle.** An earlier (now rejected) ERDA strategy was to place solidified wastes in a retrievable surface storage facility under constant surveillance until an acceptable disposal method was found. Millions of dollars of research funds and thirty years of study have failed to develop a permanent, safe method for absolutely guarding radioactive wastes for periods up to a quarter of a million years. At this point, it is fair to ask if we ever will. Lash and Cotton of NRDC point out:

> Either the federal bureaucracy is to a large degree incompetent or the radioactive waste disposal problem is considerably more difficult than has been publicly admitted by the nuclear power

* See Bibliography.

industry. To a large degree both explanations are supported by a careful examination of the record.[38]

Lash and Cotton also recommend that:[39]

- Radioactive waste production should be slowed until management problems are studied further.
- No fuel reprocessing to extract plutonium from fuel rods should be permitted until questions regarding health and terrorist diversion are carefully studied.
- Immediate efforts should focus on deciding explicitly the goals of waste management and developing precise criteria for determining the likelihood that options under consideration will achieve these goals.
- The decision-making process regarding waste management nuclear power, and energy policy in general should be changed to include scientific peer review, citizen and legislator participation.
- We must examine the moral dilemma posed by creating long-lived and potentially lethal hazards to future generations.[39]

In July of 1976, the U.S. Court of Appeals decided two cases[40] in which the Natural Resources Defense Council was plaintiff. The Council, accompanied in one case by the Consolidated National Intervenors, insisted that NRC had neglected the problems of radioactive waste and fuel reprocessing when issuing nuclear power plant operating licenses. The NRC contended, in part, that reprocessing and radwaste issues should be resolved when reprocessing and radwaste facilities are licensed, not when nuclear plants are licensed. The court rejected this argument and the NRC subsequently announced a temporary moratorium on full power operating licenses, construction permits, and limited work authorizations.[41]

The court contradicted the Nuclear Regulatory Commission's earlier assertion that environmental effects of the fuel cycle, including waste disposal, were "relatively insignificant." After rejecting arguments that radwaste was a minimal problem, it reversed earlier findings that this problem should be considered separately from plant licensing. Attorneys for the environmental groups perceptively identified the government's effort to postpone radwaste consideration as a transparent tactic to avoid jeopardizing plant construction and operation. Their counterargument, implicitly accepted by the court, maintained that all adverse environmental effects should be considered before industrial commitments acquire a momentum of their own.

Of special significance was the court's detailed and critical analysis of the government's overall approach to radwastes. Several oxen were gored in the process of identifying a number of instances where misleading and inadequate standards were applied,[42] where procrastination was rationalized on the grounds that radwaste issues were "too speculative,"[43] and where an expert's effort to point out the risks of terrorist attacks was met with arrogant condescension.[44]

In the end, the government failed to convince the court that licensing hearings allowed adequate ventilation of the radwaste issues and rested its case on "confident assertions . . . that long-term waste management is feasible" without submitting adequate documentary evidence to sustain these assertions.[45] As of this writing, the NRC has completed a staff analysis of radwaste and conservation issues prior to new rule-making proceedings. If future hearings permit the cross-examination that the court identified as lacking, the "vague assurances" upon which NRC has relied, and which form the basis of the government's conviction that radwaste problems as yet unsolved will be solved, may be subject to rigorous probing. This may bring to light what intervenors have been pointing out for years: that the government's radwaste plan is insubstantial, that prospects for coming up with a permanent, safe, and tested system for storage are dim, and that one cannot "claim to have solved these problems by pointing to all the efforts made to solve them."[46] Congressional action may be soon spurred by suits brought by intervenors aimed at stopping industry production of wastes for which no eternally safe, tested disposal system exists at present.

10
Theft and
Terrorism

Denis Hayes, of the Worldwatch Institute, identifies several spectacular examples of nuclear threats in his forthcoming book *Rays of
Hope: A Global Energy Strategy:**

In November of 1972, three men with guns and grenades
hijacked a Southern Airlines DC-9 and threatened to crash it
into a reactor at the Oak Ridge National Lab if their ransom
demands were not met. In March of 1973, Argentine guerrillas
seized control of a reactor under construction, painted its walls
with political slogans, and departed carrying the guards' weapons.

A former official in the U. S. Navy underwater demolition
program testified to Congress that he ". . . could pick three to
five ex-underwater demolition, Marine reconnaissance or Green
Beret men at random and sabotage virtually any nuclear reactor

* See Bibliography.

in the country. . . . The amount of radioactivity released would be of catastrophic proportions."[1]

One 1974 visitor to the San Onofre, California, nuclear plant displayed a smuggled steel table knife and vitamin bottle—both with neat, typewritten labels—when his tour was next to the control room. Guns and nitroglycerine brought in the same way could wreak havoc; the "incident" demonstrates how easily the reactor could have been disabled . . .[2]

Werner Twardzik, a parliamentary representative in West Germany, joined a tour of the 1200-megawatt Bilbis-A reactor carrying a two foot "panzer-faust" bazooka under his jacket. He toured the world's largest operating reactor with the weapon undetected and presented the bazooka to the plant's director when the tour ended.

Threats to destroy a reactor in such a way as to release much of the radiation in its core are truly terrifying. Yet two French reactors were bombed by terrorists in 1975, and several other facilities were bombed in 1976. Between 1969 and 1976, 99 separate incidents of threatened or attempted violence against licensed nuclear facilities were reported in the United States alone. A nearly completed nuclear plant in New York was damaged by arson. A pipe bomb was found in the reactor building of the Illinois Institute of Technology. The fuel storage building of the Duke Power facility was broken into. Seventy-six additional incidents took place at government atomic facilities.

If the radioactive iodine in a single light-water reactor were uniformly distributed, it could contaminate the atmosphere over the lower 48 United States at over 8 times the maximum permissible concentration to an altitude of about six miles. The same reactor contains enough strontium-90 to contaminate all the streams and rivers in the United States to the maximum permissible concentration 12 times over. Such an even distribution of these materials would be impossible, but the figure serves to indicate that every reactor is a veritable Pandora's Box. . . .[3]

Guarding against terrorism requires foresight. But it also requires "2020" (A.D.) vision. Who in 1975 would have expected a group of southern Moluccan extremists to hijack a train in the Netherlands in order to bargain for Moluccan independence from Indonesia? Protecting ourselves against future terrorism means nothing less than building a nuclear system able to withstand the tactics of future terrorists fighting for causes that have not yet been born.

Denis Hayes' implicit claim that terrorism is impossible to guard against is underlined by Dr. Theodore B. Taylor's testimony before a U.S. Senate subcommittee on March 12, 1974:

Nuclear weapons are relatively easy to make, assuming that the required nuclear materials are available. All the information, non-nuclear materials, and equipment required for designing and building a variety of types of fission explosives are readily available throughout the world.[4] The technical skills and resources required would depend on the desired efficiency, predictability, total weight and yield of the explosives. Under conceivable circumstances, a few persons, perhaps even one person working alone, who possessed about 10 kilograms of plutonium, uranium-233 oxide or two dozen kilograms of highly enriched uranium oxide and a substantial amount of high explosive could, within several weeks, safely design and build a crude, transportable fission bomb—one that would be very likely to explode with a yield equivalent to at least 100 tons of high explosive, and that could be carried in an automobile. This could be done using materials and equipment that could be purchased in a hardware store from commercial suppliers of scientific equipment and materials. All types of plutonium, highly enriched uranium, or uranium-233 now used or contemplated for use by the nuclear industry could be used for this purpose. The explosion of such a device could kill at least tens of thousands of people and destroy hundreds of millions of dollars worth of property.

Smaller quantities of plutonium or uranium-233 than are required for fission explosives could also be used for relatively simple dispersal devices for contaminating very large volumes of air with lethal concentrations of suspended, small particles of these substances that are exceedingly toxic if they are breathed. Air dispersal of a few grams of the type of plutonium now being produced in power reactors could kill most of the occupants of a large office building or enclosed industrial facility.

The use of nuclear energy to generate electric power at rates now projected by the AEC would result in very large domestic and foreign flows of materials that could be used to make nuclear weapons. The annual rate of extraction of plutonium from reprocessed nuclear fuels from light water reactors in the U. S. would be about 30,000 kilograms by 1980. By the year 2000, light water reactors and fast breeder reactors in the U. S. would produce more than 300,000 kilograms annually. Worldwide, annual plutonium production would be more than 60,000 kilograms in 1980, rising to roughly a million kilograms in the year 2000. In addition, worldwide flows of highly enriched uranium and uranium-233 associated with high temperature gas-cooled reactors or other types of reactors that use the uranium-233-thorium cycle may reach several hundred thousand kilograms or more by the end of the century.

Taylor then turned to the problem of theft of nuclear material. He stated:

Important changes affecting opportunities for nuclear theft can occur as the nuclear industry develops. It happens, for example, that flows of fission-product-free plutonium and highly enriched uranium for civilian power plants are now practically at a standstill and are not expected to reach large interfacility flow rates for at least several years. Since mid-1972 . . . no plutonium has been extracted at commercial reprocessing plants in the United States. . . . [This] hiatus . . . will probably last for several more years, to be followed by rather sudden increase in the flow [of concentrated commercial plutonium and highly enriched uranium]. . . .

Without effective safeguards to prevent nuclear theft, the development of nuclear power on the scale now in prospect will create substantial risks to the American people and people [throughout the world]. Individuals or groups may attempt to steal nuclear weapons materials for money or to engage in political blackmail. It would be relatively easy for a small group possessing nuclear weapons materials to use them in any number of ways to threaten other groups within society, governments, or entire communities. The frequency and character of nuclear theft attempts in the future is likely to be influenced greatly, not only by the nature of physical safeguards against theft, but also by the general political climate and by prevailing attitudes toward violent behavior within societies where opportunities exist for such theft.

The U.S. system of nuclear material safeguards is incomplete at this time. Although recent regulatory actions have strengthened requirements substantially, some basic issues pertaining to physical protection and materials accountability measures have not been resolved. . . .

[Another important unresolved issue pertains to the question of] who should pay the added costs of new safeguards measures, and by what mechanism? . . . How can the added costs of highly effective safeguards be equitably distributed across the entire fuel cycle so that they are finally borne by the consumers of nuclear electric power?

A system of safeguards can be developed that will keep the risks of theft of nuclear weapons materials from the nuclear power industry at very low levels without interfering significantly with the continued development of nuclear power. . . . The overall objective of a safeguard system would be to detect attempts at theft of nuclear materials from authorized channels as early as possible, and to place sufficient impediments in the way of people attempting a theft to assure that [the theft operation will be controlled. The following measures could be used:]

1. Containment of nuclear materials at fixed sites within physical barriers designed to prevent unauthorized penetration long enough to allow on-site or reserve guard forces capable of dealing with *any credible type* of attempt at theft to arrive at the scene before the theft is completed.
2. Shipment of nuclear materials in massive containers in vehicles designed to resist penetration, transfer of the shipment to another vehicle, or commandeering of the vehicle itself long enough for accompanying or reserve guard forces to be able to deal effectively with any credible theft attempt.
3. Provision of automatic alarms that will immediately detect attempts to remove materials from authorized channels and sound alarms at several specified control points.
4. Provision of on-site and in-transit guard forces for the purpose of denying access of unauthorized people at places where nuclear materials exist.
5. Provision of on-call law enforcement forces that can be brought to the scene of an attempted theft before the theft can be completed, along with secure communications be-

tween the control points where the alarms are sounded and the off-site forces.

The present . . . approach to safeguards is . . . along these general lines but simply does not go anywhere near far enough to prevent thefts by groups of people with at least the skills and resources that have been used for major thefts of other valuables in the past.[5]

Taylor is recognized as one of the most technically competent people to comment on the problem, having designed A-bombs for the AEC at Los Alamos. He is also very much of a realist.

Whether or not the institutional and political obstacles confronting efforts to implement such effective safeguards against theft, and also against national diversion of special nuclear materials, can be overcome within the next few years, however, remains to be seen.[6]

To the "institutional and political obstacles" we might add the following exerpt from John G. Fuller's *We Almost Lost Detroit:*

[These considerations] did nothing to squash the enthusiasm of the supporters of fission power who countered rational questions with flimsy answers. Was there no possible way to guarantee an emergency cooling system for the light water reactors? Build them anyway, and hope for the best. Was the breeder reactor erratic, dangerous, unproven, and untamed? Forget Fermi, and plan another one, a bigger one, in Tennessee. Was there absolutely no solution whatever for the safe and eternal burial of radioactive wastes? Worry about that later. Was there no safe way to transport spent fuel by truck or train or air? Ship it anyway, and take a chance.[7]

It is precisely because of this arrogant carelessness that many thoughtful people distrust government and industry promises. If their careless approach is the rule rather than the exception, then plant executives, with the passive cooperation of NRC, may in the future pay only lip service to stringent procedures and security systems. This is especially likely if the massive security systems proposed by Taylor turn out to be more expensive than originally predicted. Finally, because the industry has had so little experience in dealing with the unprecedented security and quality assurance problems of nuclear power, in the future it may continue to *react* to problems as they occur rather than devote much time or resources to anticipating and preventing them.

This is not to say that Taylor's analysis or proposals are unrealistic or presumptuous. We are only pointing out that it's a very long distance and quite some difference between the attitude in a Senate hearing room, on the one hand, and the "business-as-usual" mentality that may be commonplace at reactor and fuel-processing sites, on the other. If Dr. Taylor could somehow be cloned and his equivalents placed in charge of plant security systems throughout the United States, his proposed security network might very well be standardized and uniformly successful. Since cloning is impossible, we must accept that plant executives may considerably dilute the effectiveness of Taylor-designed security programs by the time they are put in effect in locations where thefts can occur. One of the chief problems facing the industry is that there are not enough funds nor people to cover adequately every vulnerable plant and fuel cycle location. Thus, we can rely on people interested in stealing nuclear material to strike at the weakest points and thereby provide us with the first, only, and very possibly last proof that nuclear security systems are no more effective than safeguards protecting any other commercial activity.

Terrorism and the Nuclear Fuel Cycle

Dr. L. Douglas DeNike is a clinical psychologist and a former faculty member at the University of Southern California who has published and lectured extensively on the inadequate level of security in the American nuclear power industry. On November 19, 1975, Dr. DeNike, a technical consultant to Californians for Nuclear Safeguards, testified before a California Assembly committee on the problems of terrorist attacks. What follows is a brief summary of his testimony:[8]

- The General Accounting Office, Washington, D.C., stated on October 16, 1974, that a security system at a licensed nuclear power plant could not prevent a takeover or sabotage by a small number—as few as perhaps two or three—of armed individuals.[9]

• In France, six terrorist explosions have occurred at four nuclear-related facilities since May, 1975.[10]
• The results of successful nuclear power plant sabotage could be even more severe than the effects of the maximum credible accident analyzed in the federal government's Rasmussen report (WASH-1400). Those effects are given in Appendix VI of the final draft of the report: 3,300 deaths, 45,000 cancers, 45,000 persons made acutely ill, 240,000 cases of thyroid nodules, and total property damage costing $14 billion.[11]
• The seven nuclear power reactors operating or being built in California will annually discharge over 400 intensely radioactive spent-fuel assemblies, which must be shipped to the East Coast for reprocessing. The massive shipping casks used in shipping these spent-fuel elements could be blown up by sophisticated saboteurs utilizing shaped explosive charges. The resultant dispersal of particulate radionuclides could jeopardize many square miles downwind.

Neither the AEC nor the NRC has ever proposed power-plant security regulations sufficient to withstand an attack force as large as that involved in the 1973 Argentine takeover, described earlier. That group consisted of about a dozen heavily-armed guerrillas, and it escaped unscathed.

DeNike went on to state that:

I consider it most improbable that any government agency can assure a satisfactory degree of protection for nuclear facilities and materials. We are concerned here with substances such as plutonium, of which one and one-half ounces spread over a square mile would require evacuation and thorough clean-up, and of which less than 20 pounds would be sufficient to make an atomic bomb. Absolute prevention of theft is therefore necessary.[12]

In testimony that followed, DeNike listed a number of examples where federal and nuclear authorities were unable to prevent assassinations and fires in spite of all precautions. His conclusion: it is probably impossible to create invincible domestic safeguards for the civilian nuclear industry.

In a supplemental statement to the Assembly committee, Dr. DeNike identified several vulnerable nuclear plant locations where a terrorist group might strike:[13]

- *Spent-fuel pools.* The General Accounting Office has identified the relative insecurity of spent-fuel pools, which are designed neither to resist forced entry nor to prevent radionuclides suspended in the air from escaping. Timed, waterproofed charges, using fifty to one hundred pounds of high explosives, if armed and dropped into such an uncovered pool, would probably explode with sufficient dispersive effect to make an entire $1 billion plant unusable thereafter.

- *Control room.* In certain plants and under certain circumstances, a control room explosion can result in loss of control of the nuclear reactor. In the absence of controlled coolant flow, the radioactive fuel core may heat up and lead to an irreversible meltdown condition, even if the control rods are inserted and automatically cut off the chain reaction, and even if no pipe ruptures occur accidentally or deliberately. Fission-product after-heat produced by residual radioactivity in the fuel is sufficient to overheat the core unless necessary electric pumps and valves operate in coordinated fashion.

- *Cable-spreader room.* At all American nuclear power plants, vital control wiring with typically flammable insulation is ducted through the cable-spreader room located directly beneath the control room. An ordinary chemical bomb tied to a can of gasoline could start an uncontrollable situation in the cable-spreader room, leading to a fire similar to or more severe than the one which occurred at Browns Ferry.

- *Reactor containment structure.* Assume that saboteurs force entry into the containment area, either by coercion or assisted by explosives. A possible object would be to set time bombs simultaneously to destroy all four "cold leg" pipes which return vital primary-coolant water to the core. Since the plant's emergency core cooling system is designed to cope with complete severance of only one such pipe, multiple rupture would make core cooling impossible and meltdown would follow.

- *Shipping casks.* The extremely rugged casks used to transport spent-fuel elements provide a protective layer consisting of about two inches of steel and eight inches of lead. However, they are penetrable by one or more shaped explosive charges developed to pierce military armor. A hemispheric charge made of a few pounds of high explosive lined with any dense metal will penetrate two feet of solid steel.

Dr. DeNike's testimony presented several facts for the first time, none of which pleased industry promoters. A careful reading of his and Dr. Taylor's suggestions yields several tentative conclusions. Precautions to prevent release of radioactivity as the result of theft or sabotage sound ominous and expensive. No one has yet determined exactly how much it will cost to achieve adequate plant security or to maintain a federal force to guard vulnerable points in the fuel cycle. It's reasonable to assume that this kind of security will be very expensive, perhaps more than the industry anticipates. Equally important will be the effect that tight security precautions and ubiquitous police have on our quality of life[14] and the impossibility of measuring, in dollars-and-cents terms, the restraints associated with stringent nuclear security in every affected community. Is it possible that we are reaching a point where the nuclear option yields too much trouble and expense?

11
Reactor
Safety
California Assembly Committee on Resources, Land Use and Energy

A. The Controversy

The fissioning of the atom produces materials which are potentially very hazardous and must be carefully confined in the reactor.* To ensure safety, an agency of the Federal Government (the Atomic Energy Commission, now the Nuclear Regulatory Commission) was created to regulate the use of nuclear power to ensure adequate protection. The basic design philosophies for safety include multiple barriers to contain radioactive material and emergency systems to maintain the integrity of the barriers. A recent Federal Government study concluded that, if these safety devices were breached, a reactor accident could cause 3,300 fatalities, make 45,000 people ill, damage $14 billion worth of property, and contaminate a 3,200 square mile area. But the probability of such a catastrophic accident was estimated to be about one in 10,000,000 per year (with 100 reactors

* This report has been edited slightly by Peter Faulkner.

in operation), less than the chances of a major dam failure, airline accident, fire or earthquake.

Critics challenge the validity of this conclusion on several grounds: (a) the methodology was inappropriate; (b) human fallibility was not properly accounted for; (c) recent unexpected malfunctions indicate component failures or design inadequacies more likely than assumed; and (d) safety systems have not been tested and may not work when needed. The critics believe that the study created overconfidence in present systems and approaches and was based on optimistic assumptions. Nuclear power proponents note that, even though there have been some malfunctions at reactors and some safety devices have failed, there have been enough back-up systems so that no radioactive material has ever been released which imperiled the public. Proponents attribute this good safety record to vigorous regulation and good design An independent study of reactor safety by the American Physical Society concluded that not enough was known about certain reactor conditions and emergency system performance to verify that the design assumptions were in all cases conservative.

The hazards of nuclear plants are potentially large. While most concern focuses on the nuclear power plants themselves, it is argued that the public is at risk in one way or another in all parts of the nuclear fuel cycle.

An operating 1,000-MWe light water reactor—the size now commonly being built—contains a large amount of radioactive material.[1] All authorities agree that the release of large quantities of radioactivity could have serious effects. High levels of radiation exposure can cause almost immediate death. Low exposures can lead to various cancers (thyroid, lung, bone) or genetic damage which show up only after long latent periods. One critic notes that release of a quarter of a reactor's inventory of radioactive iodine would contaminate the atmosphere across the United States to an altitude of 10 km. to twice the maximum permissible concentration (established by the National Committee on Radiation Protection), and half of the strontium-90 inventory could contaminate the total annual fresh-water runoff of the same area to six times the maximum permissible concentration.[2] *This is not to suggest that such dispersion could occur, but only to provide some perspective on the volume of hazardous material contained in one reactor and the extreme caution it merits.*

Most of the radioactivity in the reactor is due to the fragments produced in the fissioning of uranium. The remaining radioactivity is the result of the capture of neutrons by the reactor structural material or the portion of the fuel which is not fissionable. Once the reactor shuts down, the generation of radioactive material ceases and the quantity of radioactivity decreases, initially at a very rapid rate, due to the decay of very short-lived fission fragments. This decay process itself generates heat, however. Immediately after shutdown, a 1,000-MWe (3,300 MW thermal) reactor generates 225 MW (thermal) of after-heat, or about 7 percent of the heat at full power. The need to remove this residual heat after shutdown leads to concern over many reactor safety features.

At other points in the nuclear fuel cycle, radioactive materials are also present. The greatest amounts of this material outside the reactor are handled during the transportation of spent fuel, reprocessing, and nuclear waste disposal, all requiring extreme caution.

A program of safety regulation by the Federal Government was instituted to assure public protection. Because of the potential hazards of the careless use of nuclear power, in 1946, the government created the Atomic Energy Commission, to control the use of the atom, to promote its peaceful use, and to safeguard the public. In 1954, Congress and the President took steps to create and encourage a private nuclear industry under the scrutiny of the AEC. Over the years, the AEC established an extensive set of criteria and standards controlling the commercial use of nuclear power and required that the detailed design of every reactor be closely examined before authorizing its construction and operation. In 1975, owing to complaints that the functions of promotion and regulation were incompatible in one agency, the functions were split. The Nuclear Regulatory Commission (NRC) was given the job of being the impartial judge of safety, and the Energy Research and Development Administration (ERDA) was assigned the task of promoting nuclear power along with a number of other energy options.

At every stage in the fuel cycle, a small portion of the radioactive materials leak out. The NRC is now responsible for putting maximum limits on the levels of these normal operating releases and the levels are now set so low that most believe that these releases are of minor consequence.* NRC regulations also limit the amount of

* However, the present low emission levels were established after a controversy, which at its height lasted two-and-a-half years, was set off by criticisms that the

radioactivity to which plant workers may be exposed. It is also the NRC's duty to see that there will not be serious accidents at reactors or elsewhere in the fuel cycle. The fear of high exposure-levels from accidents is the basis for most of the critics' concerns.

The industry and Federal Government regulators have designed a concept which they feel will protect the public from the release of large amounts of radioactive material from the reactor during both normal and abnormal operation. This concept involves three levels of safety:

- Almost all of the radioactive material is contained within the ceramic-like uranium fuel pellets which are themselves enclosed in metal (zircaloy) rods.
- The fuel rods are sealed inside an extremely strong steel pressure vessel and coolant piping.
- The entire reactor coolant system and pressure vessel is enclosed in an air-tight containment structure designed to confine and trap radioactive material.

A large fraction of the radioactive material can only be released if the fuel itself melts. It becomes crucial, then, to ensure adequate cooling to prevent the fuel core from melting.

The critics are skeptical about the actual adequacy of the three levels of safety. Major concern has been centered on a postulated loss-of-coolant accident (LOCA) and its consequences.[3] The cause of a LOCA might be the rupture of the main cooling system involving either a coolant line or the pressure vessel itself. The water in the system is under high pressure and would be rapidly expelled if a large break occurred, leaving the core dry. Residual decay heat would then begin to melt the core unless additional cooling water was supplied. If not enough coolant is available in time, the core might melt into a mass hot enough to burn its way through the pressure vessel and, possibly, through the containment building. By breaching all three safety levels, radioactive materials would escape and potentially could cause considerable damage.

To guard against the occurrence of a LOCA, nuclear power plants are designed to provide what the federal regulators term

---

original level posed too great a hazard. In 1972, the AEC dropped the permissible level by a factor of 30, while cautioning that, overall, medical x-rays were probably of more concern.

"defense in depth." The first level of safety involves quality assurance for materials and components, as well as conservative design which allows for errors and malfunctions. The second level consists of engineered safety features and redundant backup systems. The third level provides additional safety features designed to handle failures in the first two lines of defense. One of these safety features is the emergency core cooling system (ECCS). The ECCS is a reserve water supply and delivery system designed to activate seconds after a LOCA and to prevent the core from melting.

The reliability and adequacy of the ECCS has been one of the major debates involving critics, the industry and the federal regulators. However, no major loss-of-coolant accident has ever occurred. In the few cases in which portions of the ECCS were called upon to operate, they generally worked, although there have been occasional malfunctions.

The nuclear industry and the AEC (now NRC) have expressed confidence that the ECCS will indeed work when needed. Their confidence is based on a substantial amount of research which used scale models in tests and which developed mathematical models of reactor operations. Critics claim that some of the tests demonstrate the inadequacy of the ECCS, while the industry uses the same test results to confirm their confidence in the computer codes (models).

Critics also maintain that the mathematical simulations of reactors have consistently failed to predict the course of events in laboratory tests[4] and therefore are inadequate. Proponents respond that the errors are usually on the side of conservatism and in any case, as new data becomes available, the codes are refined. They argue that more expertise is being applied to the task by involving several national labs.[5] Proponents also note that tests and data collected from actual reactors confirm parameters estimated in the computer codes.[6] However, the study of reactor safety conducted by the American Physical Society concludes that:

> Despite qualitative indications of general conservatism within the ECCS Acceptance Criteria, we feel that the experimental data are not adequate to demonstrate convincingly that the integrated ECCS systems effects are conservatively prescribed, even if all of the individual pieces were demonstrated to be independently conservative (which they have not been). Therefore, any meaningful quantitative evaluation of system effective-

ness, or of the ECCS safety margin, must depend upon the adequacy of the system analysis codes. At this time, none of us has been convinced that the current generation of codes is adequate to this purpose.[7]

The computer codes grow in significance because they formed the basis of a set of acceptance criteria adopted by the AEC for judging the adequacy of ECCS systems in proposed power plants. The controversy lead to an extensive set of rulemaking hearings before the AEC, requiring 125 days of hearings and generating 22,000 pages of transcript. At the conclusion of these hearings, the AEC did modify its interim criteria but the critics were not totally happy with the outcome. They charge that the AEC arranged procedures to deny critics the rights of subpoena, cross-examination, and discovery in addition to narrowing the scope of the hearings exces sively.[8] The critics still contend that the "NRC is using shaky and unproven computer predictions as a basis for answering such vital questions as the effectiveness of reactor safety systems."[9]

Defenders of present safety regulations argue that, at any one time, the NRC has done the best job it could and that better regulations have evolved over time and continue to be.[10] In 1963 the AEC set out to build a loss-of-fluid test (LOFT) facility to erase doubts about the adequacy of the ECCS and the computer codes. Over the years the mission of LOFT has continually changed and the program has been delayed in order to include tests on new safety features rapidly evolving in LWR's.[11] Full tests are now slated to begin in early 1977, using PWR fuel bundles in a reactor which produces 55 MW of heat (1/60th the scale of a commercial reactor). Critics maintain that LOFT is coming too late to be useful, since by the time results are available, over 100 reactors would already be in operation in this country. The chairman of the APS study stated he "deplored the rate at which the LOFT tests have been brought to fruition," and notes that it may be 10 years before convincing results are available.[12] The critics feel a lot is at stake with LOFT and that the computer codes may turn out to be surprisingly inaccurate. ERDA, the agency which will conduct LOFT, feels the tests are not nearly so vital, and will only enhance the confidence in computer codes by providing data for incremental improvements. More importantly, the APS report finds that the LOFT series has inherent limitations and will not mimic the behavior of an actual reactor in several major respects.[13] Accordingly the LOFT results will not be conclusive of ECCS performance in nuclear power plants.

Until recently, most of the attention on major accidents was directed to the LOCA. New government studies raise the possibility that other unusual reactor conditions could lead to a meltdown. One category of these unusual events, called "transients"—abnormal, non-equilibrium core conditions, hot spots and the like—have been singled out as likely causes of a partial meltdown.[14] Transients may occur as a consequence of operator error or equipment malfunction or failure. Ordinarily a serious transient would trigger a reactor shutdown. Following shutdown, cooling systems would normally operate to keep the core from overheating. Failures in either the shutdown system or the decay heat cooling system potentially could cause a core meltdown. Critics and proponents disagree on the implications of this finding. Proponents feel that the sequence nec-essary to produce serious consequences from a transient is very unlikely. The critics point to the problem as one more area, only recently discovered, about which we know very little. ERDA began a series of tests in 1974 in their "power burst facility" (PBF), in part to gain more information about fuel core behavior during a transient.

Reactor accidents may also be initiated by disruption of piping and control systems as a result of a severe earthquake. One witness indicated that California was one of the worst places in the nation for this type of problem.[15] The industry is confident that nuclear plants can withstand the forces of an earthquake without endangering the public. Nuclear Regulatory Commission criteria require a minimum distance between a reactor and an active fault as well as a design which will withstand ground accelerations between 0.5 and 0.66 g. Critics are skeptical for two reasons. First, they do not believe that all active faults are known. Second, they point out that, during the moderately severe San Fernando earthquake (6.6 on the Richter scale), the ground acceleration was measured at 1.25 g. Given the limited amount of scientific data, critics believe that siting reactors in California is too risky. Proponents argue that sufficiently stable areas exist and that seismic hazards are not insurmountable.

Proponents admit the *possibility* of several courses of events which could lead to a major release of radioactive material from a reactor but contend that the *probability* of any of these events is exceptionally low, especially with the safety features which are a part of every U.S. plant. Critics believe that the probabilities are much higher than the industry is willing to accept, and that the consequences could be of unparalleled proportions.

B. The Reactor Safety Study

To settle the dispute over reactor safety, the AEC in 1972 commissioned a comprehensive Reactor Safety Study (RSS) to analyze a large number of potential accident sequences and the probabilities of each accident type.* This study, released in draft form in August, 1974, and published in final form in October, 1975, concluded that the consequences of a reactor accident were no larger, and, in some cases, much smaller than the consequences of many other accidents to which the public is exposed. The major surprises were that meltdown accidents were more likely than previously imagined, but that their consequences were much less than expected. The study's main conclusions were:

- The risks posed by nuclear reactors to both individuals and society are smaller than the risks of fire, air crashes, explosions, dam failures, and the release of toxic chemicals.[16]
- With 100 nuclear plants in operation, among the 15 million people living within 25 miles of a nuclear plant, two fatalities and 20 injuries are expected each year due to reactor accidents. In the same population, 560 would be expected to die in fires and 4,200 in automobile accidents.[17]
- The worst possible reactor accident would produce 3,300 fatalities, 45,000 early illnesses, and 14 billion dollars in property damage, and would require decontamination of a 3,200 square mile area, and relocation of residents from a 290-square mile region.[18] In the 10- to 40-year period following such an accident, cancers induced by the radioactive materials released would claim the lives of 1,500 people per year, thyroid nodules requiring treatment would show up in 8,000 people per year, and 170 children born each year of parents exposed in the accident would

---

* The RSS was not the first time accident sequences were studied. In some areas work has been going on since 1947. But the RSS tied much of this research together and it now serves as something of an upper boundary on present understanding of reactor safety. Accordingly, the discussion is organized around the points brought up in the RSS and this organization may appear to overemphasize the importance of the RSS.

be affected with genetic disorders.[19] With 100 reactors operating, the report maintains that the chances of all this happening are one in 10 million per year.

• The most common accident involving the release of radioactive material has a one-half percent chance per year of occurring (with 100 reactors in operation). In this accident, the chances that even 1 person will be killed, injured, or suffer cancer are less than 100 percent—conceivably no health problems could arise. Less than $100 million of property would be damaged and on the order of 50–60 acres of land would have to be decontaminated.[20]

In two previous studies in 1957 and 1965, the AEC tried to estimate the maximum conceivable consequences of a reactor accident. The RSS results are in the same range as those in previous studies—if anything, RSS is a little higher. The major contribution of RSS was assigning probabilities to and looking at less serious but more frequent accidents.

The critics of the Reactor Safety Study go right to the methodology employed and claim that the numbers generated are meaningless. One witness stated that (1) RSS had used an incorrect and thoroughly discredited methodology, (2) the methodology was incorrectly applied in a number of instances, and (3) given the foregoing, the effort in the RSS was misdirected into a concerted attempt to generate low estimates of accident risk.[21]

The Reactor Safety Study used techniques previously employed by NASA and the Defense Department, designated "event trees" and "fault trees." The analyst using these techniques attempts to trace a sequence of events leading to an accident, specifying the components involved and the rate of failure of each. RSS uses this approach to combine the component failure rates and event chains with an estimate of the frequency of accidents. Critics claim that, in analogous situations in the space program, the predicted accident frequencies turned out to be wide of the mark. For example, the Apollo 4th stage rocket engine had a predicted reliability of 0.9999 (1 failure per 10,000 missions). The highest reliability actually achieved by this engine after thousands of tests was only 0.96 (4 failures per 100 missions).[22] The General Accounting Office also notes that in the space program, where the cost of improved reliability was no hurdle, only 85 percent of the launches were successful;

for the moon rockets (the Saturn V) the launch reliability was 92 percent. Therefore, the critics view the nuclear power plant system reliability of 0.999999 predicted by the RSS as being completely at odds with comparable experience. The critics note that the inaccuracy of fault-tree analysis in predicting absolute reliability was realized 10 years ago in the space program and the tool was relegated to tasks of identifying potential design weak points and making relative comparisons of reliability between two systems.[23] Critics contend there have been a number of subsequent techniques developed in the aerospace field for estimating absolute reliability which RSS does not even discuss. This contrasts with statements in the RSS that the "latest" aerospace methodologies are being used.

The major weaknesses of the event-tree/fault-tree approach pointed out by critics are:

- *Not all accident modes can be identified.* The analyst, unless omniscient, misses sequences of events which significantly reduce reliability. In the aerospace field, for new systems, the actual causes of failures often turned out to have been things completely overlooked.
- *All relevant failure rates are not available.* The reliability of each component in a system must be known with some precision in order for the fault trees to be meaningful in an absolute sense. RSS notes, however, that the data from nuclear reactors were not sufficient nor detailed enough. Consequently, data from analogous industrial situations was used. The critics note that the data from non-nuclear situations may not be applicable in the more demanding conditions in a nuclear reactor, leading to gross overestimates of system reliability since the uncertainty in the failure rate may be as much as a factor of 1,000 or 10,000.[24]
- *Component failure interdependencies are not included.* RSS assumes the failure of one piece of equipment is independent of failures elsewhere. However, fragments from an explosion in one area may damage other pieces of equipment.
- *Design deficiencies are not analyzed.* In the aerospace missile program, a large percentage of the test failures were due to design error, rather than to equipment malfunction.[25] RSS assumes no design problems and that accidents occur only through equipment malfunction.
- *Human error not adequately considered.* According to the critics, operator errors are probably the most likely cause of reactor

accidents, but the errors are hard to anticipate and quantify.[26] RSS does consider human error and tries to estimate "human failure" rates based on experience elsewhere, but the critics maintain that human errors are not so easily dealt with. They ask how one "quantifies the risk of stupidity." RSS estimates of the accident probabilities appear misleading when compared with the probabilities of other (non-nuclear) accident situations in which human error plays a big role and with which we have direct experience.

In looking over the RSS, the APS study concludes:[27]

> . . . we recognize that the event-tree and fault-tree approach can have merit in highlighting relative strengths and weaknesses of reactor systems, particularly through comparison of different sequences of reactor behavior. However, based on our experience with problems of this nature involving very low probabilities, we do not now have confidence in the presently calculated absolute values of the probabilities of the various branches . . .[28]

In response to these criticisms, the RSS authors maintain that the fault-tree/event-tree methodology *is* appropriate.* They also cite several letters in defense from NASA, the GAO and a British firm specializing in reliability techniques. The British firm notes that they have been able to predict failure rates within a factor of 4 of observed rates using fault trees in analysis of some 50 systems.[29] All the letters caution, however, that fault trees are good only where there is enough accurate data on individual component failure rates.

The RSS also points out that the low possibility of serious consequences is only due in part to the system failure rates estimated through fault trees. The low probability of adverse weather conditions (inversion layers, no strong winds) *and* a breeze carrying any radioactive plume over the most densely populated area around a reactor are equally significant factors. Therefore even if the fault-tree analysis were off by a factor of 100, the probability of a large accident would increase only from one in a billion to one in ten million. But

---

* The RSS authors note that fault trees as conventionally applied do have deficiencies, but that they have attempted to eliminate these deficiencies by using event trees which work backwards from the accident to the sequences of events that could have caused it, rather than starting with a failure in one component and working through to see if an accident results.

some have stated that under a similar setting, experts have disagreed by a factor of 10,000 on the probability of any one event.[30] Were this range of uncertainty to hold for the RSS, its results would become meaningless. They point to a statement made by Clifford Beck, Deputy Director for Regulation of the AEC, in 1965 which contradicts the current official optimism in the RSS techniques:

> There is no objective quantitative means of assuring that all possible paths leading to catastrophe have been recognized and safeguarded, or that safeguards will in every case function as intended when needed. . . . There is not even in principle an objective and quantitative method of calculating the probability or improbability of accidents or the likelihood that hazards will or will not be realized.[31]

After the draft RSS was released in August, 1974, several organizations undertook extensive evaluations of it. Several of the criticisms they made led to changes in the final version. Several criticisms[32] still stand, however:

- *Deliberate sabotage is not considered.* The RSS authors reply that the probability of successful acts of sabotage cannot be estimated and that the consequences would not exceed the largest calculated for other accident events.[33] Therefore, consideration of sabotage would add nothing to the study. The critics feel unsatisfied by this response especially in light of the recent announcement by NRC that at 99 times since 1969 incidents of threats or violence were directed at licensed nuclear facilities[34] and two explosives charges were detonated in a reactor in Europe by saboteurs.[35]
- *Treatment of earthquake risk flawed.* The RSS extrapolates past earthquake experience to generate probabilities of serious seismic damage. According to the critics, too little is known about previous earthquakes, because of lack of instrumentation, to make this extrapolation valid. They point to the comments of geologists and seismologists that every new large earthquake has been surprising in terms of what was expected.[36]
- *Evacuation estimates unrealistic.* Critics allege that the fatalities and injuries stemming from a reactor accident are low in the RSS because an unrealistically rapid and efficient evacuation was assumed. The RSS authors counter that their estimates were

based on an examination of actual evacuation experience under similar conditions. Furthermore, they assure evacuation of only a small population 25 miles or less from the reactor in the direction of the passing cloud. The evacuation is done to minimize early exposure and prevent substantial doses to individuals near the reactor. The remainder of the affected population is not assumed to be evacuated.[37]

- *Only LWR's of the present design, conventionally sited, were considered.* Critics note that reactor concepts other than current vintage BWR's and PWR's sited on land were not included in the RSS. They contend that proposed offshore, floating nuclear plants, for example, pose a much greater risk. The possibility that other reactor designs, such as the HTGR or the Canadian CANDU reactor, present a much lower risk than LWR's is also overlooked, according to the critics.[38] The authors of the RSS reply that data on these other systems were even less adequate than for LWR's, and that the BWR and PWR would be the backbones of the reactor business for some time to come.

- *Steam explosion analysis slighted.* While the uranium in a reactor does not contain enough fissionable material to create an atomic explosion, recent research indicates that molten fuel dropping into water could release the equivalent of 19 tons of TNT of energy in a steam explosion within the pressure vessel. If only 20 percent of the core melted, the resulting force equivalent to 3.8 tons of detonating TNT might still be enough to wrench the pressure vessel apart.[39] The RSS admits there are no definitive experiments on steam explosions to evaluate what the hazard really is, or how likely an explosive interaction between molten fuel and the pool of water at the bottom of a reactor would be. Nevertheless, the RSS predicts that in only one in ten fuel-melt accidents will an explosive interaction occur, and in only one in ten of such cases will the explosion yield enough energy to breach the containment and release a significant amount of radioactive material. In the absence of any experimental verification of the probabilities used in the RSS, critics wonder if the chances aren't quite a bit higher. The RSS authors believe they have been suitably conservative and note that initial experiments are now underway.

C. Reactor Safety Problems

There has never been an accident involving the excessive release of radioactive materials at any of the 55 nuclear power plants now in operation in the U.S. However, the critics point to a number of unusual occurrences and unexpected problems as evidence that the operation of these plants has not been smooth and could have led to serious accidents:[40]

- On June 5, 1970, the Dresden-2 BWR near Chicago suffered a partial loss-of-coolant when the pressure control system malfunctioned. In trying to correct the problem, operators allowed the water level in the reactor to increase and exceed safety margins. The emergency core cooling system was called upon, but the high-pressure portion was shut down for repairs following earlier damage and the low-pressure system was inhibited because of the 350 psi pressure in the reactor.[41]
- During a routine check at a power reactor, abnormal radioactivity was observed in the plant drinking fountains. The contamination was found to have arisen from an inappropriate cross-connection between a 3,000-gallon radioactive waste tank and the water system.[42]
- Repeatedly, important valves have failed to operate properly. In one case, an entire core spray system was found inoperative because check valves had been jammed during assembly.[43] Elsewhere, 63 out of 191 containment isolation valves failed to operate when tested.[44] Their failure apparently resulted from rust, oil, and water present due to lack of preventive maintenance. Plant valve failures not involving nuclear systems have resulted in the deaths of two technicians; other valve failures have led to the uncovering of a reactor core.[45]
- Small hairline cracks were found in the four-inch bypass pipes in the primary coolant recirculation loop at the Dresden-2 plant. The AEC then told utilities with similar BWR's to inspect for bypass line cracks in September, 1974, and similar cracks were found on four additional reactors. Later, on January 28, 1975, small cracks in two 10-inch diameter core spray pipes were found on the Dresden-2 reactor. The new NRC then gave the utilities 20 days to inspect for cracks in this new location. No cracks were revealed at other installations.[46]

- The Millstone Point reactor scrammed twice in three days, once when an automatic valve failed in the feedwater line and the water level in the reactor dropped, and again when a defective pressure regulator allowed the pressure to build up in the reactor and increased reactivity in the core. The valve malfunction was attributed to excessive wear due to vibration. The pressure regulator failed because of a poor solder connection.[47]
- The Vermont Yankee reactor accidentally reached a critical level of reactivity during a refueling, with the reactor vessel head off and the containment building open. A scram occurred automatically.[48]
- Shrinkage or "densification" of fuel occurred in a number of pressurized water reactors, leading to collapse of fuel rods, partial blockage of coolant flow channels, and localized hot-spots in the core. The AEC temporarily reduced the permissible level of power generation of a number of plants by 5 to 20 percent until fuel rods could be replaced. The AEC and Westinghouse have said that the difficulties of the densification have been eliminated through design changes.[49]
- Damage to fuel channel boxes used for directing coolant flow around the fuel rods was caused by excessive vibration of instrumentation tubes against the channel boxes. The situation could have led to overheating and damage of the fuel rods, or blockage of coolant flow.[50]
- A four-by-six-inch connector in the feedwater line at the Quad Cities nuclear power plant broke and 8,500 gallons of reactor cooling water escaped from the severed line. Flow-induced vibrations were thought to have caused the problem.[51]
- All three shaft seals on one of three main coolant pumps in a PWR failed, resulting in a leak from the reactor coolant system to the containment building that could not be isolated. Approximately 132,500 gallons of radioactive primary coolant water leaked out and the normal reactor coolant water makeup system was unable to maintain the coolant level. The leak greatly exceeded the postulated leakage for pump seals of that design. The reactor power was quickly reduced from 100 to 36 percent and the leaking pump was shut down. The reactor shortly thereafter completely shut down automatically, a safety injection system provided reactor coolant, and the fuel was adequately cooled.[52]

• In probably the most talked-about accident, a worker using a candle flame to check for air leaks in cable penetrations through a containment wall set fire to foam packing at the Browns Ferry-1 and -2 nuclear plants in Alabama, March 22, 1975. The fire spread to cables in nearby cable trays, burning for seven hours and damaging 2,000 electrical cables. Both units lost power but were eventually shut down. The emergency core cooling system was disabled, along with several other safety features; the Unit-1 water level dropped at one time to 48 inches above the core (200 inches is normal) and the pressure climbed rapidly at another time. The control room was full of smoke and the operators were wearing air packs. The control room also lost its phones and public address system to the rest of the plant, making it difficult to dispatch men to open and close valves manually that no longer could be operated from the control room. Finally the fire was brought under control after several unsuccessful attempts to use chemical extinguishers. Only minor injuries to plant personnel resulted.[53]

Since no member of the public was exposed to excessive radiation as a result of these incidents, nuclear power proponents look on these occurrences as vindication of the adequacy of the safeguards and conservative design of modern plants. The critics are not equally impressed. They indicate that these errors and malfunctions demonstrate an inability to construct and operate plants with the degree of perfection required. The proponents admit that occurrences such as these confirm human fallibility, but contend that through such experiences, corrections can be made and designs improved. The critics wonder when one of these "incidents" will get out of hand.

D. Tentative Conclusions

The critics testified that these unexpected problems cast a pall over the validity of the Reactor Safety Study. They point to the number of times that breakdowns were caused by improper installation of equipment, inadequate quality assurance,[54] defective components, design errors,[55] and outright errors, compared to the RSS assumptions of essentially perfect operation until a component failed, and of perfect design and construction. The critics do not necessarily contend that the use of nuclear power is inherently unsafe. Some do point out that they believe nuclear power could be made acceptably safe. However, they do not feel present U.S. programs will result in adequate protection.[56]

In this regard they see the RSS as instilling a false sense of security in present systems and present approaches.[57] They point to the annual letters from the AEC's independent Advisory Committee on Reactor Safeguards [ACRS] listing generic unresolved safety issues. In 1975, the ACRS listed 27 safety items as still unresolved, including pressure buildup in the containment structure following a LOCA, rupture of high-pressure lines outside the containment building preventing operation of critical safety components, stress-corrosion cracking in BWR piping, common-mode failures, and pressure-vessel failure from post-LOCA thermal shock. The critics believe that the issues raised by the ACRS should be dealt with *before* more reactors are built. The NRC, however, takes the position that the ACRS items are important areas for more studies but they are not crucial for the safe operation of nuclear power plants.[58] NRC and ERDA contend that, through a vigorous safety research program, improvements can continually be made and any problems which require modification of existing facilities can be handled without too much inconvenience. Critics argue that the present reactor safety research program will not provide the studies which are needed.

The American Physical Society study, for example, reaches the conclusion that the research program presently planned by ERDA —the loss-of-fluid test and the power burst tests, specifically—gives no assurance of actually being able to resolve the serious questions which have been raised about reactor safety.[59] The uncertainty is the result of the difficulties encountered in extrapolating the results from scale models which may not be entirely accurate mimics of large commercial power reactors. APS recommends that alternative emergency core cooling systems be designed and tested, so that we do not rely exclusively upon present ECCS concepts which are in dispute.

Furthermore, the APS study recommends that much greater effort go into (1) improving reactor containment, (2) mitigating the consequences of a large release of radioactive material, and (3) learning how to clean up afterward. The APS report mentions underground siting and the design of a "core catcher" to prevent melt-through by the core as areas for serious re-examination. Undergrounding, as also advocated by Dr. Edward Teller, is also suggested as a means for providing an extra level of safety. Berm containment, a concept originally advocated by the Oak Ridge National Laboratory, involving the placement of power plant components in a scooped-out hole and backfilling over all the structures with the removed dirt, is another option which provides much of the release-control advantages of undergrounding, but more cheaply. Examples of possibilities for mitigation of reactor accident effects include preparation of serious evacuation plans, and distribution of iodine pills to block uptake of radioactive iodine by the thyroid. At present only about $1 million is allocated to these tasks out of a total reactor safety research budget of roughly $70 million; almost all of the money is presently going to studies of ECCS performance.[60]

Beyond possible modifications to existing reactor designs, some critics note that other completely different reactor designs may be inherently safer than those we use now. The High-Temperature Gas-Cooled Reactor (HTGR), installed at the Fort St. Vrain nuclear power plant, offers some advantages because it does not require elaborate coolant piping systems. Unfortunately, General Atomic Company, which manufactured the HTGR, is no longer in the reactor business. The other reactor most often mentioned is the Canadian CANDU reactor. The CANDU uses natural, unenriched uranium and heavy water.* Its major advantage in terms of safety is that it does not use a massive pressure vessel with a small number of coolant lines, but instead incorporates hundreds of pressure tubes within which are placed the fuel bundles. Complete loss-of-coolant from a single break becomes practically impossible. The CANDU is also attractive because it can be adapted to a thorium breeding cycle that will not have some of the safety problems identified by critics of the U.S. liquid-metal cooled fast breeder reactor.

Proponents do not feel the safety situation is so bleak as to warrant these major changes. Nor do they believe other reactor designs[61] or siting concepts offer significant safety advantages.

Issues and Comments

Having studied both sides of the safety debate, the reader is ready to explore some implications and related topics. A recent controversy pertaining to nuclear plant safety focuses on the reliability of

---

* Heavy water occurs naturally in all water but is rare. Separating it out involves a technical process which is energy consumptive.

reactor systems, as indicated by how much electricity they generate and how much time they are out of commission for unexpected reasons.

A few industry leaders have been candid about low-capacity factors and extended forced outages.[62] Implicit in these data is evidence that plant operators are being extraordinarily careful—witness the fact that there have been no major accidents in spite of the substantial number of forced outages and abnormal occurrences over the past several years.* It is reasonable to assume that when an unsafe condition emerges, most plant operators immediately shut down the reactors rather than take a chance of compromising plant safety.

If this practice is universal, it also imposes certain limitations on the industry. First, it may be impossible for a nuclear plant to maintain a perfect safety record and at the same time deliver a steady supply of electricity. Second, if the public can be protected adequately only by shutting the plants down a significant amount of time, that may tell us something about their intrinsic safety, since plant safety depends very much on how well they were built in the first place. If they were constructed without proper quality controls, this weakness in the form of sloppy workmanship is bound to show up later during plant operation—pumps breaking down, valves leaking, instruments that don't work when they should. If these malfunctions are chronic, if plant personnel spend a significant amount of time repairing things that should have been designed, fabricated or installed properly when the plants were being built, then we may have grounds for suspecting that the plants may not be adequate in a broad sense.

If their overall adequacy is open to question, if the plants *were* constructed without proper quality controls, and if sloppy workmanship was literally "built into" each plant, then we may be faced with a generation of plants that will suffer forced outages at an unsatisfactorily high rate. If so, we may have evidence of *inadequate safety* that courts will regard as persuasive.

The editor's nuclear-industry experience convinced him that plant safety could be substantially improved and that safety deficiencies are reflected in less than predicted capacity factors. These deficiencies may be traced to the industry's failure to provide a rigorous system of quality assurance (QA) involving third-party inspection of nuclear systems and components while they are being manufactured and before they are shipped to the plant site.[63] This failure is especially evident in the NRC's practice of auditing quality assurance on a sampling basis and delegating total responsibility for

* Except in the case of the Browns Ferry fire. See pp. 3–22.

QA implementation to the licensee—the utility—which seldom has the personnel or the resources for carrying out this responsibility with meticulous attention to detail.

The data presented by David Comey in "The Uneconomics of Nuclear Power" confirms that original predictions of 80 percent capacity factors have failed by at least 25 percent. The number of forced outages and abnormal occurrences is a matter of record,[64] although their significance is disputed. The *Wall Street Journal,* nevertheless, went so far recently as to identify nuclear plants as "atomic lemons," and there are plenty of data to sustain this point of view. To determine if it is justified, we might view reactor operations from the perspective of the people who operate these systems.

Because of frequent component and systems problems, the role of plant reactor operators has undergone a subtle shift. *On fossil-fueled plants* their job has traditionally been to keep the plants running properly, to make adjustments in the flow of coolant, steam and electricity. By comparison, a significant amount of *nuclear*-control-room-operator training time is spent in simulators where trainees are drilled in procedures to prevent what is often the result of poor quality control from getting out of hand. Of course, it's comforting to know that these conscientious and competent men are in charge, but this is beside the point. What is relevant is that *nuclear plants are far more complex than fossil-fueled plants and that many nuclear plants have suffered disappointing operating records.* The control-room operators at these plants must suspect why their reactor systems aren't running up to par. At least some of them must realize that *there is a relationship between the number of forced outages and abnormal occurrences after a plant is built on the one hand and, on the other, the degree of quality assurance that was enforced while it was being constructed.*

However, it is very difficult to prove that this relationship exists, because many things other than poor quality assurance can cause plant failures (e.g., human error, defective operating procedures, acts of God). As long as intervenors and the public cannot prove that poor quality assurance is the major cause of below-par plant performance, the industry enjoys the benefit of the doubt—that the present system of fabricating and constructing plants is satisfactory, that quality assurance controls are adequate.

A careful analysis of the causes of plant breakdowns, and of the correlations between these breakdowns and evidence of inadequate quality assurance is the first step in developing a case for substantial reforms. Efforts in this direction have been suspended in favor of litigation of more concrete issues. As can be seen from recent court decisions, the environmental threats posed by fuel reprocessing, transportation of spent fuel, and long-term storage of wastes are more likely to receive attention in the future as long as the industry's record of electric productivity deteriorates no further.

The debate over plant reliability and safety will remain lively, nevertheless; and properly documented, a very good case can be made for questioning intrinsic safety on the basis of poor plant performance. However, until intervenor attorneys petition for injunctions against plant operation on the basis that unreliable plant performance is the only conclusive evidence we have that plants are not safe, the issue of how much evidence is needed to be persuasive will not be reached.

That we may have enough was suggested by none other than the author of the Reactor Safety Study, MIT Professor of Nuclear Engineering Norman Rasmussen:

> Probably one of the most serious issues that the [nuclear power plant] intervenors can raise today, with good statistics to back their case, is that nuclear power plants have not performed with the degree of reliability we would expect from machines built with the care and attention to safety and reliability that we have so often claimed.[65]

12
Insurance and Nuclear Power Risks
Senator Mike Gravel

Congress and the nuclear power industry face a basic and irresoluble paradox in trying to insure nuclear power, because the kinds of risks we take in generating nuclear electricity are simply not insurable.

There are several reasons why nuclear power is uninsurable.

The most obvious is the scale of the possible accident. Here are the government's own various estimates of possible damages in a nuclear catastrophe:[1] $6 billion, $7 billion, $17 billion, conceivably even $280 billion; outright deaths of 3,400 to 43,000; contamination of an area the size of California or Pennsylvania.

Another factor making nuclear power uninsurable is the far-reaching and long-term nature of the damage that could be done. If great quantities of radioactive materials escaped into our environment, there would be no way to judge the extent of the cancer cases or birth deformities which might be caused. The health consequences would be incalculable and recompense would thus be impossible.

Cancer, as in the case of Hiroshima, might appear as late as 20 years or more after the event, in spite of the 10-year statute of limitations provided by the Price-Anderson Act [renewed twice since 1957 to provide insurance in case of a nuclear plant accident]. And the range over which the damage would be inflicted could not be judged. Radioactive carcinogens would be distributed throughout the biosphere and concentrated by some plants and animals in ways completely beyond the control of man.

"How shall we insure nuclear power?" Those familiar with the issue already know that nuclear power is uninsurable. If it were insurable, nuclear power would be subject to normal liability laws and there would be no need for Price-Anderson. Perhaps the real decision is, put very bluntly: "Who's going to be left holding the bag?"

If one of the Indian Point reactors, twenty-four miles north of New York City, suffers a meltdown and a breach of containment, who's going to be left holding the bag? If the San Joaquin Project contaminates this nation's largest food-producing area, who's going to be left holding the bag?

Back in the mid-1950s, when the insurance problem was first tackled by Congress, the nuclear industry made it very clear who was *not* going to be left holding the bag. They let us know that if normal liability were allowed to apply, they would not take the nuclear gamble. The public, on the other hand, was not wary. And today, it is the public which really carries the burden of nuclear power risks, even though few citizens are aware of it.

The Congress decided that, in the worst nuclear catastrophes, citizens' constitutional right to just compensation would be suspended. An arbitrary sum—$560 million, a fraction of the possible damages—would be shared on a pro-rated basis. And even if the victim were to get only pennies to the dollar under this scheme, Price-Anderson would prohibit him from recovering further from the nuclear industry.

Furthermore, the $560 million is to be paid mostly by the taxpayer, not by the industry or its insurers. At the present time, $125 million is guaranteed by insurers and $435 million by the federal government.[2] This federal indemnity is provided to industry at a cost far below what would be charged for private insurance. In the meantime, insurance companies have added to their homeowner and other insurance policies a clause excluding coverage of losses following a nuclear catastrophe.

So, in a word, the public is holding the bag.

Or, as a Columbia University study of Price-Anderson declared in 1974: "The decision to limit liability represents a determination that a major share of the costs of an accident should be borne by its victims."[3]

I believe that we have failed in our obligation to the public in permitting the limitation of corporate liability. The law indicates that, should damages exceed $560 million, the Congress may provide disaster relief. But another Columbia study has shown that congressional action in such instances can be expected to be too little, too late. And, if congressional action is needed to cover a nuclear catastrophe, then what purpose does Price-Anderson serve? Nothing in the Act guarantees extra relief.

The only conclusion is that Price-Anderson exists to protect the utilities and the nuclear manufacturers. Period.

Even with those provisions which supposedly assure $560 million for victims, there are loopholes which protect the industry at the expense of the public. For one thing, industry's costs for investigating and settling claims are to be subtracted from the $560 million. In other words, the funds to pay utility lawyers to challenge victims' claims will come from money supposedly designated for paying those claims. In addition, a utility's property outside of its reactor is covered by the $560 million. This means that the utility will recover from its own "public liability" insurance.

It is also interesting to see how much private insurance the utilities have managed to get for damage to their reactors after an accident. While they claim that only $125 million in coverage can be found for public liability, their reactors are covered up to $175 million.

Admirers of the Price-Anderson Act point to its assurances that money will be available to victims on a no-fault basis. In fact, however, the guarantees made by Price-Anderson are rather illusory. To begin with, in many states, utilities would be held strictly liable even without Price-Anderson.

In addition, victims must accept the settlements offered to them or else go to court much as though Price-Anderson's no-fault provisions did not exist. How many people are likely to accept their pro-rated share if the damages exceed $560 million?

It should also be noted that the $560 million provided by Price-Anderson is now worth much less than when the Congress approved it 20 years ago. And in that same period the possible consequences of an accident have grown because reactors are larger and the population densities near them is greater. The Congressional Research Service calculated that $3 billion would be more appropriate coverage, and even that calculation is more than three years old.[4]

Finally, there is the problem of proving damages. Even under the no-fault provisions, victims must prove that their injuries were caused by the accident. But a cancer does not grow with a flag in it identifying the cause. It is impossible to prove that any cancer has been caused by radiation. And the industry is not going to give victims the benefit of the doubt.

The case of Edward Gleason, a truck dock worker, is instructive. In 1963, Gleason handled an unmarked, leaking case of plutonium that was being shipped to a nuclear facility. Four years later he developed a rare cancer: his hand, and then his arm and shoulder, were amputated. He sued, but even though plutonium is among the deadliest of radioactive poisons, company insurers argued that his plutonium accident could not be proven as the cause of the cancer.

Gleason's suit was dismissed in 1970 on statute-of-limitations grounds, although a settlement was later made. He died of cancer in 1973 at the age of 39.

Finally, even if the victim can prove his case, how much of the $560 million is going to be left for those who suffer cancer years after the accident? Only by restoring the citizen's right to recover relief beyond the no-fault amount can Price-Anderson claim to treat the public fairly. And the nuclear power industry should be exposed to such suits, not isolated from the consequences of nuclear accidents.

We should also be aware that Price-Anderson has acted as an industry subsidy. As such, it has made nuclear power look more attractive economically than it really is. And this has meant pouring tens of billions of dollars into nuclear fission and keeping those dollars from the benign and renewable energy sources that we should have taken seriously long ago.

The most important point remains to be made. The reality of a nuclear catastrophe exceeds the capacity of the imagination: a hundred thousand persons each suffering damages of $60,000 in the case of the $6 billion estimate; or a million persons each suffering damages amounting to $17,000 in the $17 billion case. There has never been a peacetime disaster on such a scale.

As long as we rely on nuclear power, we court this kind of a disaster. There is only one way to protect the public from the consequences of a nuclear power catastrophe and that is to stop building nuclear fission reactors.

Harold Green, an attorney and law professor, has stated the Price-Anderson paradox most precisely. He wrote in the *Michigan Law Review:*

> The fact that the technology exists and grows only because of Price-Anderson has been artfully concealed from public view so that consideration of the indemnity legislation would not trigger public debate as to whether nuclear power was needed and whether its risks were acceptable. . . .
>
> It is remarkable that the atomic energy establishment has been so successful in procuring public acceptance of nuclear power in view of the extraordinary risks of the technology that are so thoroughly and incontrovertably documented by the mere existence of [the Price-Anderson Act].[5]

If utilities choose to gamble with nuclear power, let them share the risks along with the rest of us. If the odds are really as favorable as the industry advertises, the risk should not bother the utilities.

If utilities will not build nuclear plants under these conditions, then it is time for us to confront the *meaning* of their refusal.

In the interests of launching the nuclear power industry, the constitutional right of the citizen to just compensation was suspended. This abridgment of rights was not justified in the industry's early days—it is even more offensive today.

Issues and Comments

The historical background of Price-Anderson is provided by a 1973 AEC study:

> Shortly after the 1954 [Atomic Energy] Act opened the nuclear industry to private enterprise, it became apparent that the problem of potential liability to the public was a substantial roadblock to investment of private funds in nuclear power projects. The insurance industry, with virtually no experience in assessing nu-

clear risks, was unprepared and unwilling to write third party liability insurance contracts against the astronomically high potential liability that could result in the event of an accident. In addition, the very size of the potential liability was beyond the capacity of the insurance industry. Faced with the inability to acquire adequate insurance protection, the utilities and the nuclear equipment manufacturers served notice that they were unwilling to proceed with nuclear power development, which could result in bankrupting public liability claims . . . unless some mechanism were found to provide adequate financial protection.[6]

A program wherein the federal government underwrote these risks was developed by the AEC and the Joint Committee on Atomic Energy.

Their consideration resulted in enactment in 1957 of an amendment to the 1954 Act commonly referred to as the Price-Anderson Act. This legislation rested on two justifications: first, that industry, on which reliance was placed for development of nuclear power, should be protected against potential bankrupting liability thereby enabling industry to proceed with the development of this technology; and second, that assurance should be provided that, in the event of a serious accident, funds would be available to compensate the injured public for its losses.[7]

The Price-Anderson Act has been renewed twice since 1957 and a national controversy has developed concerning its provisions. A recent report by the California Assembly Committee on Resources, Land Use and Energy summarizes the main issues:

Critics [of the Price-Anderson Act] claim that industry support for limited liability is evidence of the industry's own uncertainty about reactor safety. Proponents argue that, without limits on liability, the potential losses would deter investors. Critics reply that such investor decisions are all a part of the market operating to halt hazardous activities and that the market should be allowed to operate. Proponents claim that the limits make other benefits of Price-Anderson possible—rapid recovery, no-fault payment of a claim, and an immediate fund of liquid assets—but critics doubt that the benefits, in reality, are much different than those provided by state law otherwise applicable. The major benefit of Price-Anderson appears to be in providing a level of indemnification not available in the private insurance market. Critics propose ways in which this indemnification can be provided without the liability limit. The industry is divided on whether the limit is still essential.[8]

As of 1977, the controversy remains unresolved and focuses especially on two related questions: Why can't the liability limitations be eliminated? and To what extent should the nuclear industry be required to *internalize* the costs of nuclear power associated with catastrophic accidents?

Insurance coverage of $560 million may comfort many people but it is important to realize that this would absorb *less than 3 percent* of possible losses that may exceed $17 billion. Of this $560 million coverage, more than 89 percent, or $500 million, was originally provided by the federal government.* According to a Columbia Law School study:

> One rationale advanced for limiting the government's obligation to $500 million was that a claim for that amount would not significantly disturb the government budget. In the end, the figure of $500 million was chosen because it was small enough to "not frighten the country to death" but probably large enough to protect the industry.[9]

Whether $560 million affords adequate protection is a question we will leave to the reader. If the government eliminates liability limits, the industry will have to internalize the fiscal risks created by nuclear technology. These risks are large enough to bankrupt several major corporations if a major accident occurs. Unless these liability limits remain in effect, the civilian nuclear industry risks insolvency and will probably stop building reactors if the limits are eliminated. One of industry's arguments for retaining them is that private insurance companies are either unwilling or unable to provide adequate coverage even if the industry was able to pay the premiums. If so, then perhaps it is better in the long run to let the normal market forces rather than the government determine if nuclear power risks are too hazardous. As pointed out by the 1974 Ford Foundation's Energy Policy Project:

It is our view that the nuclear power industry is now sufficiently mature for a revision of the Price-Anderson Act, so that the marketplace will reflect the potential social costs of nuclear

---

* Recent legislation decreases the government's indemnity share to $435 million and adjusts private industry coverage to $125 million. Additional provisions provide for the gradual phase-out of government indemnity, a system of retrospective premiums of $2 million to $5 million per reactor payable by the utility in the event of an accident, and gradual increase of the $560-million liability limit.

power. If nuclear power is indeed as safe as the AEC and the industry claim, the risks are small enough to be borne by the enterprises involved. If the utilities are unwilling to buy new plants on certain sites, or to buy reactors of certain designs, without the shield of Price-Anderson, then those locations and those plants are too risky to be built.[10]*

* An equally conservative view was presented by Herbert W. Yount, vicepresident of Liberty Mutual Company, before the JCAE in 1956. An excerpt from his testimony and a brief summary of congressional reaction to his recommendations may be found in "The Dawning of the Nuclear Age," p. 221.

THROUGH
THE MAGNIFYING
GLASS

13
Quality Assurance: Industry Watchword

This is a good time to define just what we mean by "Quality Assurance," henceforth referred to as "QA." The airline industry provides a fine analogy. Let's say that you've bought a ticket for a flight to New York on a new jumbo jet. You're vitally interested in the safety of the plane you're going to fly in—your life depends on it. While the plane was being manufactured, several QA programs were implemented to assure that all systems were safe. The aircraft has flown about 4,000 hours since then and, as would be expected, the engines and other systems have required repairs from time to time. Some of the repairs were made after trouble showed up in flight, many others were made after careful preventive maintenance inspections (prevention is an important QA function) revealed that a problem was "about to happen."

The day before your flight, the aircraft lands at San Francisco International Airport. On the maintenance write-up forms the captain reports only minor problems. The plane is taxied into the hangar where the maintenance night shift converges on it with prepunched work order cards which assign to each man a portion of the inspection task. This QA system is computerized as much as possible, but the records kept on file pertain to specific employees and the specific task each has accomplished. The system is refined and exact to the point that careful track is kept of the types of maintenance each man is qualified to execute. This assures that the tasks are properly matched to each man's qualifications just before the jet is inspected.

One man is assigned to inspect the fuel system on the number three engine on the right wing. He follows a printed procedure and traces fuel lines and couplings on the main fuel line inlet to the engine accessory section, all the way to the engine ignition chamber. During his inspection, he notices a slight leak in the primary fuel pump. He notes this on his maintenance form and turns it in after the inspection is complete. Another mechanic, qualified especially in fuel pump replacement, goes to the parts shop and checks out a replacement. The parts shop forwards a packet of documents accompanying the fuel pump to the aircraft's master file in maintenance operations after the pump leaves his shop. The packet provides a complete history of the pump, including assembly, inspection, test, acceptance, and shipping; it consists of more than twenty pages.

Another mechanic installs the pump and conducts a preliminary test to make sure it is working properly. Another inspector checks *his* work, and finally, before the aircraft leaves the bay with a new pump on board, the entire fuel system is rechecked by a maintenance supervisor. Each man signs his name to the maintenance order, and the plane is not cleared for the flight line until all steps are "signed off."

This procedure assures that, if anything goes wrong with that pump on any subsequent flight, the malfunction can be traced to the responsible people—all the way back to the time that it was manufactured. Of course, this creates a lot of paper work, but by obliging each person to authenticate his responsibility by signature the system assures that everything will be done properly. Each man knows he may be held responsible later for any error.

QA for nuclear plants is carried out in much the same way as airline maintenance QA. However, our airline example is limited to an application of QA during the aircraft's operational phase (maintenance, installation, and check-out). It's important to remember that QA controls must be uniformly applied during design, development, procurement, fabrication, processing, assembly, inspection, test, operation, maintenance, packaging, shipping, storage, site installation, and check-out. Our fuel pump went through all but two of these phases before being installed in the engine accessory section. Remember, though, that *pump quality is only as adequate as the phase where QA controls are weakest.*

Compared to QA applications in the nuclear industry, airline maintenance QA applications are more similar to those encountered in NASA, mainly because of the time element: the operating life of an aircraft or spacecraft may be shorter than that of a nuclear plant, and inadequate quality assurance may show up earlier there than in the nuclear case. As a result, traceability may be easier in the airline and aerospace industries, for intervening causes can be limited if the time interval is shorter.

Nuclear QA programs have been imposed on the industry only recently by the AEC whereas nonnuclear QA was developed over many years *within* other industries. This partly explains why nuclear industry QA controls may have only recently come to management's attention and received their support. Another reason: nuclear utility managements only recently began enforcing QA systems rigorously partly because they may not have been under the same pressure as the airlines and NASA to keep accidents to a minimum. Witness the sweeping reforms that were imposed on the aerospace industry by NASA management following the Cape Kennedy fire in which astronauts Grissom, White, and Chaffee lost their lives.

Though this introduction to QA is adequate for a general understanding of this management control system, you can learn more from Title 10 of the Code of Federal Regulations, Chapter 50 (abbreviated 10CFR50) by consulting Appendix B of that chapter. Universally accepted as the QA "bible," Appendix B is short and concise, if difficult. More important, any businessperson interested in a systems approach to quality may gain several insights by reviewing this important document.

Bear in mind, though, that Appendix B refers to only the first of several levels of control. To implement a good QA system, a company needs at least two more documents: a corporate QA program detailing the specific procedures required for carrying out Appendix B, and a body of procedures derived from the corporate QA program which spell out in detail the method by which each quality objective will be obtained. Our airline analogy, where procedures for checking fuel lines as well as installing and inspecting fuel pumps were followed, applied only to this third level.

Public safety depends very much on how well the nuclear industry develops, understands, and applies its QA programs. To be effective, QA should be uniformly stringent, beginning with earliest design phases right on through the operation, maintenance, and decommissioning of nuclear power plants. The next article is a case study of General Electric's program during the design and marketing of its most recent boiling water reactor system, the BWR/6, and the containment system, designated Mark III, with which it mates.

14
Shakeup at General Electric

In April, 1972, the Nuclear Energy Division of General Electric Company, San Jose, California, unveiled a new product line for nuclear power stations. Designated BWR/6 (boiling water reactor/ sixth model), it allegedly incorporated major design improvements.[1] In addition, General Electric announced that it was developing a new pressure-suppression containment configuration, designated "Mark III," and a new turbine building layout. Under the terms of subsequent contracts, General Electric controlled all aspects of BWR/6 design, manufacture, and development. As for the Mark III, General Electric assumed responsibility for the containment design *concept* and let the electric utility company which purchased this system work out the details, prior to construction, with the architect-engineer hired to build the plant.

Before describing the technical difficulties that General Electric encountered with the Mark III containment, it might be helpful to review a diagram of the design (see Figure 1). Note the location of the suppression pool of water, which runs all the way around the base of the containment in the form of a square with the drywell and the reactor in the center. The drywell, which surrounds the reactor and primary coolant system, is a pressure boundary that will channel steam during a loss-of-coolant blowdown* past the weir wall and through the vents into the suppression pool water. As you can see from Figure 1, the rest of the containment and the reactor building are open to the suppression pool. When pressure builds up in the drywell following a major reactor leak or pipe rupture, the water in the vent annulus is forced down and outward by the steam pressure, thereby uncovering the horizontal vents. The steam then flows outward through vents into the suppression water. This flow allows the steam to condense, and reduces pressure in both the drywell and reactor by providing extra volume for condensation and expansion. Thus most of the steam from the drywell dissipates as it escapes into the extra volume provided by the containment, along with various noncondensable gases.

The Mark III shown in Figure 1 identifies few of the actual system details. An important feature, omitted in order to avoid cluttering the illustration, is the series of platforms located one above the other and running around the outside of the drywell. These are positioned *above* the suppression pool and are sandwiched, like the pool itself, between the containment and drywell walls.

In an early Mark III design, various mechanical and electrical equipment essential to reactor operation was located on these platforms.[2] After releasing the drawings of the Mark III design concept to its customers, General Electric encountered during design development certain problems with potentially destructive waves that might be generated in the suppression pool during a LOCA. During subsequent design development, General Electric decided to relocate several components originally destined for installation above the suppression pool.[3] In addition, GE either included "blowholes" in these platforms or substituted metal gratings where solid, reinforced

---

* If a pipe feeding coolant to the reactor, or if a main steam line, breaks close to the reactor vessel, a loss-of-coolant accident (LOCA) will occur. During a LOCA, the reactor experiences a "blowdown" during which the coolant and steam in the reactor blow out through the ruptured pipe into the drywell. The steam that will be generated by continuous deluge of the reactor core by the emergency core cooling system will also be vented into the drywell.

MARK III Containment

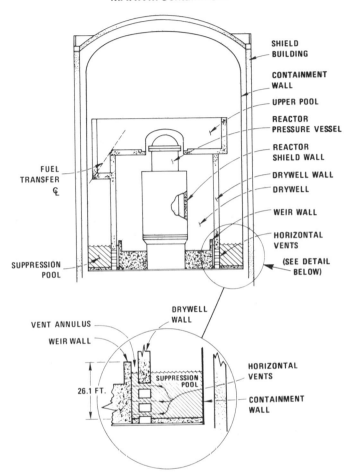

Figure 1

concrete had been intended. The significance of these changes will soon become apparent.

The problems became evident after the Mark III system was released for marketing, and after a number of customers had purchased the BWR/6 product line, intended for mating with the Mark III containment system. A good QA program at the *design* stage might have caught these problems much earlier. Readers will want to judge for themselves whether these are "routine" changes that "normally" occur while customer and designer work together to develop a new concept, or whether these changes could and should have been identified earlier. If you wondered why they were not, you may be especially interested in what three former General Electric engineers have to say about their employer's approach to design verification (see Appendix A).

One of these engineers, Dale Bridenbaugh, commented on General Electric's decision to postpone detailed design until after a system is marketed. According to Bridenbaugh, it has been common nuclear industry practice for the "buy" decision to be awarded to the company capable of demonstrating the lowest *apparent* initial capital cost. In most cases, the company winning the contest tends to be the one marketing the system that appears to cost the least. Fierce competition drove General Electric to bring out a new product line to solve its sales problems at the time. Marketing the Mark III system before it was designed was "a stroke of creative salesmanship," in Bridenbaugh's words.

Once the contract to furnish the reactor system was signed, General Electric was free to make design changes in the normal process of translating a general concept into a workable, safe reactor system. That these design changes will increase final reactor system costs well beyond the *apparent* initial cost may be expected by both parties. Both need a fiscal escape hatch and the freedom to make various required changes after the contract is signed. Nuclear plant construction is often on a cost-plus-fixed-profit basis, and consequently the architect-engineer has little incentive to limit design changes. The utility often does not have adequate staff to monitor the architect-engineer, and the escape hatch often turns into a huge loophole; after the contract is signed, thousands of design changes occur which contribute to enormous cost overruns ultimately paid for by the electric customer.[4]

Several electric utilities became interested in the BWR/6 product line and Mark III containment design, and by January of 1974 a number had signed contracts or were negotiating with General Electric for purchase of these systems. One of these utilities, Potomac Electric Power Company (PEPCO), serves the Washington, D.C., metropolitan area and surrounding counties. Its management had appointed as vice-president of its nuclear group an outstanding nuclear engineer, Paul Dragoumis, now recognized as one of the most aggressive utility executives in the nation. Dragoumis lost no time in assembling a strong engineering staff prior to breaking ground for the Douglas Point Nuclear Generating Station, due for commercial operation in the 1980s.

During 1973, General Electric forwarded drawings of the Mark III containment to Dragoumis and his staff for review. PEPCO's embryonic nuclear engineering and construction group included fewer than thirty engineers, but they took on the review of General Electric's drawings with gusto. Dragoumis's group identified several serious Mark III problems, some of which may have been overlooked by General Electric.[5] In 1973 and early 1974, Dragoumis met with General Electric to discuss these problems and hammered away at three of them: blowdown contamination, suppression pool swell, and containment stress.

Blowdown Contamination

During normal reactor operation, enormous amounts of water and steam move through the reactor vessel. A sudden change in load or operating conditions may cause a rapid change in reactor power level or pressure. This is defined as a "reactor transient," an occurrence which can lead to the opening of one or more reactor relief valves and the venting of radioactive steam to the suppression pool. This type of "blowdown" is neither dangerous nor unusual; a typical operating reactor will experience a transient, leading to the opening of the relief valves and subsequent blowdown, about once each year. Routine blowdowns were not significant problems in earlier containment designs, but in the case of the Mark III, essential equipment requiring maintenance during operation was located inside the containment. Three important questions confronted Dragoumis: When would the blowdown occur? For what period of time would the relief valves lift? Would the blowdown occur while workers were inside the containment?

Dragoumis was concerned that workmen would not be able to escape from the containment in time and might be overexposed to the radioactive steam or gas. General Electric was optimistic and assured PEPCO that if everyone escaped within ten minutes after the blowdown commenced, they would receive a radiation dose not to exceed 40 percent of that allowed by the federal government for each calendar quarter. General Electric offered to let PEPCO review and verify its calculations; Dragoumis was skeptical nevertheless, and directed his staff to investigate the problem themselves. This proved to be only the tip of the iceberg; the PEPCO engineering group may have anticipated other Mark III problems as well.

On January 8, 1974, General Electric contracted BWR/6 and Mark III customers and notified them of delays that might arise while GE studied the blowdown contamination problem. In a separate communication, General Electric notified Dragoumis that PEPCO's Douglas Point plant might be delayed by a more serious problem, suppression pool swell,[6] and other General Electric customers may have been similarly notified.

Suppression Pool Swell

Dragoumis had been negotiating with General Electric about the problem of suppression pool swell as early as June of 1973. Both parties were concerned about the consequences to the Mark III containment of a potential problem that had been anticipated earlier. What if a pipe were to rupture, spewing enormous quantities of pressurized steam into the drywell? (See Figure 1). It was determined that this might cause a blowdown so rapid that the water level in the suppression pool might move upward with tremendous force. In 1973, while the Mark III was being marketed and after a number of utilities had signed or were negotiating contracts for the BWR/6 product line and Mark III containment design, General Electric began building a one-third scale Mark III mock-up to test the adequacy of the design concept. According to the three former General Electric engineers, in analyzing the resulting Mark III test data, General Electric determined that the rapid discharge of air through the horizontal vents into the water resulted in a high vertical swell of the pool surface, which created large hydrodynamic forces on structures and components above the suppression pool. This led to structural modifications in the Mark III design as well as to embarrassing questions about the possibility that similar problems applied to Mark II plants and to the thirty-eight Mark I plants in operation or under construction all over the world.

While the one-third scale Mark III containment was being readied for tests in 1973, General Electric assured PEPCO that suppression pool swell could be a problem but that it was probably not a significant one. Unfortunately for everyone, tests on the Mark III model showed otherwise in December of 1973. The test results persuaded Dragoumis that the suppression pool water level might indeed jump upward with enough force to damage equipment that General Electric had located above the suppression pool, and that the reactor's safe operation might be affected.

The series of platforms located directly above the suppression pool (omitted from Figure 1 for the sake of simplicity) now become important. Dragoumis identified reactor cleanup pumps and hydraulic control units (for control rods) located there as being especially vulnerable to damage in the event of suppression pool swell. If these systems were damaged, the possibility of control rod malfunction would be significantly increased. The consequences of common-mode failure of control rod function—a possible core meltdown and containment rupture—were unacceptable.

Dragoumis and his staff were also concerned that General Electric may have failed to consider adequately that the tremendous force of the rising suppression pool water would strike a limited surface area at the top of the pool. Although important cleanup pumps and hydraulic units had been located even higher, other equipment was to be installed near the top of the pool, presumably on or below the first platforms. Dragoumis determined that unless the limited surface area was reinforced, the impulse force of thousands of gallons crashing into this area might disintegrate this equipment, creating missiles that could damage nearby components. On the other hand, if the surface area above the suppression pool *was* properly reinforced, it would withstand the impulse force of the rising suppression pool water but would present what amounted to a restricted orifice for this water column. The principles of fluid dynamics assure that water velocity through this narrow "orifice" would increase, and that fast-moving water would reach higher altitudes than it would if the equipment and surface above the suppression pool disintegrated on impact.

During 1974 discussions with Dragoumis, General Electric informed him that a General Electric general manager J.J.W. Brown, had been replaced. Whether General Electric's or Dragoumis's staff were the first to identify these problems is known only by the principals. There is no question, however, that these events must have been embarrassing to General Electric and that they occurred *after* General Electric had marketed the Mark III system. In February of 1976, when the specifics were released to the press, General Electric played down these problems as either originally exaggerated, routine, or solved.[7] General Electric's implicit suggestion that the problems reviewed here were incidental and not especially significant in 1976 is misleading, for according to 1974 internal PEPCO memoranda, both General Electric and PEPCO regarded them earlier in a much more serious light.[8]

Nevertheless, a later newspaper report on the controversy includes allegations by General Electric executives that the problems of blowdown contamination and suppression pool swell have been solved. Joseph Showalter, a General Electric manager, asserted that personnel caught inside the containment during a blowdown would receive only a "very moderate" dose if they escaped in "seven or eight minutes," and that blowdown contamination had "turned out to be 'merely a calculation problem.' "[9] According to Showalter, the other problem, suppression pool expansion, was dealt with by moving a steam tunnel, hydraulic control units, and the reactor water cleanup pump room to new locations.

These allegations, together with Showalter's assurance that General Electric spent millions of dollars testing the Mark III concept, may reassure the public, but they missed the main point. The issue is not how much money General Electric spent on tests or what technical fixes were applied after the Mark III problems were identified. Instead, we might ask if General Electric's reactor containment systems are as safe as claimed if problems similar to the ones described here keep cropping up after the systems are released for marketing. The timing of the technical fixes applied to these problems is also important: if General Electric had adequately tested the Mark III concept *before* marketing the BWR/6 and Mark III package, it might have identified much earlier the problems described here and would not have had to solve them retroactively. By building a one-third scale model of the Mark III *after* the system was announced as marketable, General Electric may have followed conventional nuclear industry practices; the company also took the chance that embarrassing systems problems might be identified *after* contracts to design and develop the system had already been signed.

Could General Electric have spent more time refining the Mark III concept so that no major flaws remained when utilities signed contracts for the BWR/6 and Mark III package? Perhaps, but then the company would have had to gamble that its competitors would not lure away potential customers during the testing phase. Apparently, General Electric decided that it was less risky to market the Mark III system before all the bugs were solved. Subsequently, General Electric announced through various industry trade journals and other media that it was currently offering the reactor-containment package. These announcements committed General Electric to marketing at least the general concept. A prototype of the Mark III concept had not been completely tested by the time that the system was announced publicly, and General Electric had to assume that the system would work and that system design problems could be resolved later. Unfortunately, General Electric soon found itself out on a limb, a victim of its own aggressive marketing.

Containment Stress

While General Electric was trumpeting its Mark III containment, the Atomic Energy Commission announced a new reactor and containment design requirement that put nuclear systems designers in a difficult position. During 1973, the AEC indicated that reactor and containment systems should be able to handle a dual emergency: an anticipated reactor transient *plus* failure of the reactor control rod drive systems (hereafter designated ATWS, for "anticipated transient without scram"). In other words, a sudden change in reactor operation, such as a load rejection, might occur requiring immediate insertion of the control rods to reduce reactor power and to stop the transient. But, asked the AEC, if you can't insert the control rods, what steps do you take to avoid a nuclear accident? Although the probability of ATWS occurring was very small, the question was legitimate.

General Electric examined several possibilities and replied that the most effective way to deal with ATWS was to redesign its systems so that the reactor recirculation pumps would be automatically shut off. This would reduce reactor power to about 30 percent, after which liquid boron would be pumped into the reactor. The boron would absorb neutrons created by the fission process and thus would safely shut down the reactor.

This wasn't a bad idea, except that while all this was taking place the safety and relief valves would be relieving excess pressure in the reactor vessel by venting enormous quantities of steam into the suppression pool. Dragoumis determined that it might take so long to shut down the reactor that more thermal energy would be vented during ATWS than in the case of a reactor loss-of-coolant accident where the reactor control rods *were* inserted immediately. The limiting internal design pressure in either case was 15 psig (pounds per square in gauge, i.e. in excess of 14 atmospheric pressure.) According to Dragoumis, an ATWS sequence might vent enough steam from the reactor vessel so that this design pressure would be exceeded and the containment overstressed; this could lead to containment cracks or fractures.

Regardless of how small the probabilities were of all this happening, it appears that consideration of the ATWS sequence raised fundamental doubts about the Mark III design. In a February 1, 1974, confidential internal memorandum to top PEPCO executives on the problem, Dragoumis pointed out that these doubts should be extended to include, as well, doubts about the Mark I and Mark II General Electric containments, of which there were dozens installed, operating, or under construction all over the world.[10] At least one other engineer (assigned to PEPCO but not an employee) felt that this statement, coming as it did from one of the most knowledgeable and responsible men in the nuclear industry, raised serious questions about the integrity of *all* General Electric reactors, and had grave implications for public safety, which took precedence over all other considerations. But not even Dragoumis was prepared for what happened next.*

The confidential February 1 memo, together with a number of other industry proprietary documents pointing out potentially serious industry problems, were leaked to *The New York Times,* the Nader organization in Washington, D.C., and several U.S. senators and congressmen. It is not clear what occurred during the next twelve months. However, the June, 1975, issue of *Nuclear News* included the following report, which highlighted the problem of suppression pool swell but altogether ignored ATWS:

NRC CONCERNED ABOUT GE CONTAINMENT ADEQUACY

The Nuclear Regulatory Commission has announced that it is requesting information from the operators of 20 General Elec-

---

* The nuclear industry may have held Dragoumis responsible for subsequent events. Shortly after his confidential memo was leaked, PEPCO announced that its nuclear station was "postponed." A few months later, Dragoumis switched jobs and assumed leadership of the FEA's Office of Nuclear Affairs under Frank Zarb. Immediately after, Congress abolished the office and with it, Dragoumis's job. Ambushed by this congressional move, Dragoumis remained under Zarb for several months and then returned to PEPCO as a vice-president for corporate planning. It is noteworthy that PEPCO did not rehire him in a position that would allow him to use his considerable nuclear expertise directly, and that every job change Dragoumis made was followed very closely by the industry in prominent periodical reports.

tric Company BWR's on the adequacy of their containment system structures.

The information is being sought as the result of new data developed by General Electric in tests on its proposed Mark III containment system. The data relate to previously unidentified hydrodynamic forces that could develop, in varying degrees, in the containment suppression pools of GE design. This would include the Mark I and the Mark II designs as well as in the design of the Mark III. (The containment suppression pool is used to reduce pressure in the containment in the event of a loss-of-coolant accident as well as to depressurize the reactor when the relief valves are opened.)

According to an NRC spokesman, the Commission is concerned that these additional pressures, which would be generated in upward waves during a large LOCA, could cause damage to system piping. The NRC has concluded, however, that these additional stresses would not cause serious impairment of the containment system.

Specifically, the Commission is asking that each containment structure be analyzed, that a description of the structural analytical methods used be provided, and that the evaluation either demonstrate that the containment design could withstand suppression pool dynamic loads with adequate safety margins or show the design modifications that would be required to meet design limits.

Accordingly, the NRC staff has sent a letter requesting such information from each of the utilities currently operating the 20 GE nuclear plants. Each plant has a Mark I or similar containment design. In addition, data are being sought from those utilities presently constructing GE plants. There are 17 of these, 11 of which are being constructed with the Mark II system and six with the Mark I system.

General Electric reportedly has made design modifications in the proposed Mark III containment system, and these are currently being reviewed the Commission.[11]

If you read this article carefully, you will notice that NRC did not initiate a *formal* investigation, including public hearings, of the problem. Instead, it appears to have let the utilities submit their own calculations, the entire matter being treated as routine. In this case, the effort seems to have been directed toward proving that the plants are safe enough for continued operation rather than to assess their safety openly.

During 1976, NRC was openly accused on three occasions of failing to regulate adequately so that public safety is assured. Testimony by the three former General Electric engineers (see Appendix A) and by former NRC official Robert Pollard (see Appendix B), as well as a resignation letter by an NRC reactor engineer Ronald M. Fluegge, all allege extraordinary dereliction by NRC. Fluegge's letter reads:

October, 20, 1976

Marcus Rowden, Esq.
Chairman, USNRC

Dear Mr. Rowden:

As a safety analyst for the Nuclear Regulatory Commission, I have been repeatedly frustrated in my effort to make the agency deal honestly with pressing nuclear safety problems. I have decided to resign from NRC, effective this Friday.

Let me put my underlying concerns "on the record."

I believe—in common with a substantial number of my colleagues on the NRC technical staff—that NRC is violating its public trust. This agency is supposed to protect the public interest in nuclear affairs. We are supposed to carry out objective, independent safety reviews that identify potential nuclear safety problems. We are supposed to see to it that such matters are convincingly resolved before nuclear plants are allowed to operate so that our fellow citizens will not be exposed to the awesome consequences of nuclear radiation accidents.

Yet time and time again, NRC has covered up and brushed aside nuclear safety problems of far-reaching significance. We are allowing dozens of large nuclear plants to operate in populated areas despite known safety deficiencies that could result in very damaging accidents. We are issuing Safety Evaluation Reports that are carefully censored to conceal major safety problems. We are withholding from the public NRC staff technical analyses of a wide range of unpleasant nuclear safety difficulties. We are giving the public glib reassurances about nuclear plant safety that we know lack an adequate technical basis. NRC management has let its commitment to the industry and Administration goals for nuclear power expansion compromise the agency's regulatory integrity.

Bob Pollard, our former colleague, decided earlier this year to resign and to speak openly about the current state of nuclear safety and nuclear safety regulation.* NRC's official response to Bob's assess-

* Editor's note: See Appendix B for Pollard's testimony.

ments is widely regarded in-house as whitewash because it concealed from the Congress the fact that he spoke the views of a fair segment of the agency's technical safety experts. I certainly share his concerns about Indian Point-2, which surely would be shut down by any prudent nuclear safety regulator, and have indeed advised the NRC management of safety problems that require the immediate shutdown of Indian Point-2 *and of all presently operating commercial PWR nuclear plants in the U. S.* Needless to say, despite acknowledgment that these safety problems are real and could result in unprecedented accidents, NRC allows the plants to remain in operation.

NRC employees, following Bob's resignation, refrained from speaking out for fear of harassment, reprisals or loss of their jobs. I have now found a new position in the Midwest, outside of the commercial nuclear power program, from which I can speak freely. I intend to speak out, not as an opponent of nuclear power, but as a proponent of this useful energy source who wishes to see its serious safety defects promptly corrected.

Sincerely yours,

/s/ Ronald M. Fluegge
Reactor Engineer
Reactor Systems Branch

The three former General Electric Engineers pinpoint NRC's failure to evaluate properly reactor safety as "the ultimate deficiency of our nuclear program." They identify the General Electric containment assessment program (for the Mark I design) as an example of this deficiency, involving nineteen operating plants and six under construction as well as thirteen overseas.[12] Their experience with the federal government's unwillingness to face honestly major safety problems[13] with the Mark I may be compared with the way that the NRC proposed to deal with Mark III deficiencies, as reported in the *Nuclear News* article above.

One is tempted to consider the important implications of the evidence presented so far. In its effort to expedite the marketing of the BWR/6 and Mark III package, General Electric appears to have slighted what engineers outside the nuclear and defense industries regard as elementary: a step-by-step *development* of design, including drawings, full-scale prototype, testing, and design revisions before a product is sold. It appears that General Electric abbreviated this sequence in order to get the jump on its competitors, sign as many contracts as quickly as possible, and capture a significant share of the market. Instead of prototyping the Mark III design before marketing it, General Electric, with the apparent approval of the AEC, overemphasized the theoretical rather than the practical approach in design verification. In other words, General Electric *did not* conduct extensive full-scale prototype tests or subject the design of the BWR/6 and Mark III package to extensive laboratory and field test verification before it was marketed. Instead, General Electric attempted to prove that the design was safe and workable *on paper* and postponed extensive work on these other prerequisites. As a result, the problems revealed by the Dragoumis document became more evident *after* the package was sold.

General Electric is probably no more guilty of shortcuts than are Westinghouse, Babcock & Wilcox, and Combustion Engineering— the other major nuclear steam-system suppliers. It may be that the entire nuclear industry has abbreviated the standard, careful approach to design verification evident in other industries not concerned with getting government contracts. The nuclear industry may justify this practice on the grounds that "everyone" does it and the government allows it. Although these points may be true, both avoid the main issue: that safety margins may be eroded by postponing or omitting adequate prototype laboratory and field test verification of nuclear systems. It is also fair to ask why the Nuclear Regulatory Commission allows abbreviation of design verification, or postponement of verification, until after the marketing phase. Has this been going on so long that it has become standard practice? Or is this a temporary phenomenon, albeit so widespread that regulatory agencies cannot clamp down without shaking up the entire industry?

A third possibility emerges from the 1976 Common Cause allegations of widespread industry conflict of interest in ERDA and NRC (see pages 206–207). By making sure that many executive positions in federal agencies are filled by people hired directly from companies having a keen interest in minimal government interference, the nuclear industry has reduced federal enthusiasm for rigorous regulation. Industry-trained NRC and ERDA officials must be aware that their chances of recovering civilian jobs when they leave the government might be hurt if they are zealous in regulating their future employers.

What implications for safety does the abbreviation of design verification suggest? Industry has taken a calculated risk by emphasizing theoretical verification of reactor and containment designs. This risk makes sense from a profit standpoint, for frequently there is much more time to develop prototypes and to obtain laboratory and field test data *after* a contract to build the system is signed. However, considerations of public safety may cast that risk in a far more ominous light. The problem turns on *how much* testing is needed to assure safety, and what recourse the industry has if later tests show serious safety hazards in plants that are already operating. Both questions are legitimate because testing theoretically can go on forever—you can never check out every possible accident mode or stress problem during the prototype stage. Eventually, business imperatives require that construction begin on the basis of "reasonably thorough" design verification. Fortunately, after construction starts, the testing *does* go on, but every now and then these later tests reveal problems that are already designed into operating plants. This puts the utility in an impossible situation—on the one hand, a billion-dollar plant, ready for service, but with serious defects; on the other hand, the general public, which needs the power. Too often, the solution is obvious—start it up and let the public bear the risk.[14]

In reply to industry and NRC assertions that this description of the situation is untrue or misleading, we have republished here testimony and a resignation letter by former NRC engineers Robert Pollard (pp. 315–319) and Ronald Fluegge (pp. 184–185), respectively. These men insist that NRC permitted plants to operate in spite of known safety hazards and who raise the strong possibility that this is the rule rather than the exception in the nuclear industry. Congressional testimony by the three former General Electric engineers (Appendix A) corroborates this conclusion.

On December 13, 1976, the U. S. Senate Government Operations Committee received testimony from four other NRC engineers who confirmed Fluegge's allegations. Demetrios Basdekas, Jose Calvo, Don Lasher, and Evangelos Marinos testified in concert that potentially unsafe plants are licensed to operate, that political considerations frequently win out when technical objections to licensing are raised, and that NRC's safety review process is demonstrably weak.[15]

The allegations of these four men converge with Fluegge's and Pollard's. If their views are correct and if, as they claim, a fair number of technical safety experts within NRC subscribe to these views, then something akin to mutiny is clearly brewing within NRC. Even if no additional NRC staff members come forward, the interlocking charges of the six outspoken government engineers tempt an observer to conclude that NRC's internal implementation of quality assurance during safety reviews is not much better than private industry's, and that both government and industry weigh economic considerations rather heavily when considering safety issues.

Issues and Comments

Rigorous enforcement of quality assurance (QA) programs from the beginning might have prevented or mitigated many of General Electric's Mark I, II, and III containment problems. If an industry accepts firsthand inspection by a third party of its products from the design through the installation and operation stages, the applied principles of QA should catch design errors and yield very high reliability later on. To be effective, though, the industry will have to take these principles seriously to the extent of subordinating short-term profit, and political and union imperatives to them.

If Admiral Hyman Rickover's naval reactor and shipbuilding program could achieve this on a smaller scale, the nuclear industry ought to have been able to do the same. The successful Apollo moon landings are other instances where strict QA enforcement yielded spectacular results. But both of these endeavors were *government* projects in which civilian industry played an implementing, but not an executive, role. In the case of the nuclear industry, the federal government limits its role to that of regulator and delegates responsibility for all executive activities to private industry. For this and a variety of other reasons, the civilian nuclear industry was unable to sustain rigorous QA enforcement by third-party personal inspection of components and systems at the vendor level and below.

H. Peter Metzger in *The Atomic Establishment** describes in detail the flaws in the federal government design for the nuclear industry that contributed to this failure. It is now apparent that some of these legislative design errors led to a situation where too much was expected of the electric utilities in too short a time. Specifically, the utility companies found themselves saddled with the job of implementing QA programs they neither knew how to design nor had enough qualified personnel to develop. With very few exceptions, these companies had no experience with QA whatsoever until 1972. By that time, those with nuclear plants nearing completion had their

* See Bibliography.

hands full with licensing, training, and staffing problems, and regarded AEC requirements for operating QA programs with either suspicion or resignation. In many cases, the task of developing QA programs for operating nuclear plants was farmed out to engineering and consulting firms.

Nevertheless, most nuclear plant staffs do not become involved in rigorous QA until the government requires them to. This involvement often occurs too late to assure adherence to QA principles by vendors while the plant is being built. Even if the government's requirements were changed to favor this innovation, the effort would fail. One reason is that the giant corporations that build reactor systems (e.g., General Electric and Westinghouse), and those that specialize in building the rest of the plant (e.g., Bechtel, Ebasco, Stone and Webster), would resist electric utility interference, as would many of the smaller subcontractors and vendors. All of these companies have been building reactors and plants for ten years or more and their modes of operation are not easily changed. Barring a major nuclear accident, significant QA reforms are unlikely this late in the program.

Of course, General Electric and Westinghouse have extensive quality assurance programs of their own, and these programs are implemented from design through installation and testing of reactor systems. In the Mark III case, we are still left with the question of why General Electric did not detect the blowdown contamination, suppression pool swell, and containment overstress problems earlier than it did. Were design quality assurance practices suspended until after the system was sold, or were they implemented early in the design phase, prior to sale, but not rigorously enough? Only General Electric knows the answer.

The management decision to postpone detailed design and testing of the Mark III until after the system was sold leads one to question General Electric's commitment to quality assurance. A company that assigned the *highest* priority to quality assurance would tend to insist on thorough design and testing before systems are released for marketing to avoid the embarrassment of having to make fundamental design changes after several customers have purchased the system. General Electric apparently did not insist on thorough pretesting before marketing, possibly because it was under pressure of time, or because of heavy competition, or because a significant percentage of its executives are brought up through the marketing ranks rather than through the engineering divisions (this was pointed out by Richard Hubbard, one of the three General Electric engineers who resigned). The executive decision to postpone thorough testing of the Mark III, therefore, may have reflected the marketing orientation of General Electric's top management.

The tendency to sell first and test later would not be so serious if it were not evident industrywide. The emergency core cooling system was designed and verified on paper before installation in nuclear plants throughout the world but was never given a test under actual operating conditions because of the accident risk.[16] Thorough laboratory testing of the ECCS under simulated operating conditions has not been successfully accomplished as of 1977, long after the ECCS was sold. Another case involves General Electric, which invested over $60 million in design and construction of the Morris, Illinois, nuclear fuel reprocessing plant only to discover, after it was built, that plant systems would not work as designed.[17] A third example is the waste storage system designed for the Hanford Reservation in Richland, Washington, which leaked hundreds of thousands of gallons of radioactive liquids.[18] Unlike the Morris plant, the Hanford Reservation cannot be shut down and written off for reasons of faulty engineering; the Hanford wastes will have to remain in temporary tanks until a better storage system is devised.

The ECCS, Morris, and Hanford examples span the full range of nuclear industry activities (operation, fuel reprocessing, and waste storage). In each case the same tendency appears: to postpone the testing phase until after substantial commitments are made, and to let the operation of the system offer proof of whether it will work *as designed.* This tendency may be widespread enough so that neither broad reforms, centralized quality assurance enforcement, nor restoration of the normal test-first, sell-later cycle can remedy the fundamental problem. This may be a good time to reassess other industry problems and, for the time being, to seek their causes before proposing solutions.

The suspicion grows that many of these problems may never be solved, not because we are lacking the skills or resources, but because they are unsolvable in the first place. If so, we should face the truth immediately. There may be limits to technological development, just as there may be limits to affluence, economic growth, and how far we can send astronauts into space. We would do well to recognize these limits before resource depletion or nuclear proliferation finally convince the scientific community that they exist.

15
A Moment
of Truth

After deliberating the human and environmental implications of nuclear power for several months, three management-level engineers from General Electric's Nuclear Energy Division resigned their posts on Monday, February 2, 1976.[1] Dale Bridenbaugh, Gregory Minor, and Richard Hubbard together had amassed fifty-four years of service for the company. Bridenbaugh, forty-four, had been chairman of an industry task force investigating the adequacy of General Electric Mark I containment design and was manager of performance evaluation and improvement for the company. Hubbard, thirty-eight, had been manager of the quality assurance section in the company's nuclear energy control and instrumentation department directly responsible for implementing quality programs and equipment to meet NRC specifications. Minor, also thirty-eight, managed the company's advanced control and instrumentation design.

Each man abandoned a position paying between $30,000 and $40,000 a year because he felt that "nuclear energy represents a profound and irreversible threat to life on the planet." Each cited different reasons for this resignation. Bridenbaugh recalled a recent sales presentation to Egyptian officials to whom he explained the operating benefits of the GE boiling water reactor design. "At about the same time," Bridenbaugh recalled, "Dick Hubbard was talking to the Israelis about buying a similar plant. I came back and asked my boss how we rationalize these sales to countries at war with each other. He said that wasn't our responsibility." Hubbard added, "My boss characterized my attitude as paternal. Who are we to question who we sell our products to? If they want to buy them, that's their business."

Minor told *The New York Times* that he began having second thoughts while in Hanford, Washington, when he first looked down through the dozen or more feet of water that separated him from the deep blue radiation given off by fuel elements below. "I looked through that ten or fifteen feet of water, the life-saving shield between me and that fuel, and I knew that if any one of those elements were to come up and hit me in the eye, that I was dead, just like that. Or if the water was gone, I was dead, just like that."

Minor's concern climaxed with the Browns Ferry fire in March, 1975. "I had responsibility for the people who designed the redundancy of those safety systems. We knew we were building the things well. I even thought it was overkill, But when (the fire) even precluded the operating of those safety systems . . . it was a very big shock."

Hubbard said, "It's a tunnel-vision kind of thing. I was responsible for making sure that neutron signals get converted into amperes. That's all you look at. People talk about radioactive waste storage and reprocessing and proliferation, and you say 'that's not my responsibility.' Even GE says 'some of those things aren't our responsibility.' . . . Looking at the broader picture, I conclude that nuclear power is neither safe, clean, nor economical."

He added, "We all picked the nuclear industry rather than defense because we thought we were making a contribution to mankind. I know that I felt personally attacked when someone told me something was wrong with nuclear power. . . . The funny thing is that I've had lots of calls and notes from my co-workers at GE [after resigning]. I've had much support for my decision and not a single criticism."

Hubbard was especially concerned about the responsibility that reactor manufacturers (vendors) have toward utilities who purchase a plant. "I know what goes on there," he noted. "Once the vendors sell a plant, they're through with it. There is no assurance that the utility will continue to buy good parts, or, if something goes wrong, that the vendor will help the utility fix it. It's just another commercial endeavor to them."

Bridenbaugh added his concern for what he called "the grandfather clause." "After you build twenty nuclear plants and you find some basic design flaw, you say you won't build any new plants with the same problem. But what do you do with the existing plants? Do you go back and redesign them? That's not NRC's policy."

The nuclear industry and NRC will have difficulty rationalizing Bridenbaugh's resignation; his role as manager of GE's review of the adequacy of Mark I containment design casts him unmistakably as an expert and credible spokesman. In April, 1975, sixteen electric utilities and GE formed, *sub rosa*, an organization to investigate the implications of some recent tests on Mark I containment structure adequacy in the event of a loss-of-coolant accident.[2] The tests had been performed in connection with GE's current containment design, the Mark III. "We knew," reported Bridenbaugh, "that the containments were not designed to absorb all currently known loads. Given the *most probable* course in a LOCA, would the thing function as intended or would it not? It was a high-priority, intense investigation with twenty nuclear plants, a vendor, several architect-engineers, and fifteen utilities. The logistics were almost impossible."

On January 27, 1976, Vermont Yankee Corporation shut down the 540-MWe Vermont Yankee plant pending resolution of the safety issues surrounding the Mark I containment problem. Operators were concerned about the integrity of the torus, a circular donut-shaped cylinder containing water around the base of the pressure vessel that condenses the high-pressure steam generated in a loss-of-coolant accident. Under accident conditions, steam pipes from the reactor pressure vessel rupture and steam is released into the dry-well. High pressures in the dry-well force the steam downward into the torus so that the steam is released underwater and condensed. This pool also serves as one source of water for the emergency core cooling system.

Said Bridenbaugh, "The question we were concerned with was whether the torus support structure was adequate. A phenomenon occurs in a LOCA which tends to push the torus upward. We wanted to know whether it could jump and still maintain integrity."

He added, "The decision to shut down Vermont Yankee was made by the operator. It was directly related to the Mark I containment problem. Given the situation at Vermont Yankee, I think his decision was correct. The loads which led to his decision do exist in some degree in all other, similarly designed plants."

Two weeks after resigning from their jobs, Bridenbaugh, Hubbard, and Minor flew to Washington, D.C., to testify before the Joint Committee on Atomic Energy. An edited version of their testimony appears in Appendix A (pp. 280–314): A slightly abridged version of the summary of their testimony follows:[3]

SUMMARY OF
TESTIMONY OF DALE G. BRIDENBAUGH, RICHARD B. HUBBARD AND GREGORY C. MINOR BEFORE THE JOINT COMMITTEE ON ATOMIC ENERGY

FEBRUARY 18, 1976

The three engineers have concluded that, despite continual assurances to the contrary by the industry, power plant owners, and the NRC, *nuclear power plants are not safe.* Their testimony noted many design deficiencies and unresolved regulatory issues that have a serious impact on the safety of existing nuclear plants. Flow-induced vibration in the cores of both Boiling Water Reactors (BWR's) and Pressurized Water Reactors (PWR's) caused serious damage to in-vessel components and resulted in the failure of feedwater spargers (a device designed to uniformly distribute water within the reactor) at a number of reactors, including Northeast Utilities' Millstone Point, Northern State Power's Monticello, Boston Edison's Pilgrim and Commonwealth Edison's Dresden and Quad Cities Plants. Flow-induced vibration also resulted in the failure of the Local Power Range Monitors (LPRM's) and fuel channels at Vermont Yankee and NPPD-Cooper Plants. Doubt exists whether the core spray, which is designed to distribute large quantities of cooling water over the hot fuel rods to cool the core, will work in a loss-of-coolant accident. Similar concern was expressed as to the effect of the "End-of-Cycle-SCRAM Reactivity Effect" which produces a power surge and higher pressure transient in the reactor primary system during a SCRAM than was originally anticipated.

Proper functioning of control rods is essential to control the reactor core, yet the life of a control rod is not known. A control rod can fail to function from either nuclear depletion or when it reaches the end of its mechanical life. Some rod materials, such as stainless steel, have cracked. Material failures in control rod mechanisms create greater uncertainties about the functioning of control rods in the reactor core and the likelihood of a rod drop accident.

Several problems were noted with respect to the integrity of the pressure vessel. Nozzle breaks between the vessel wall and the biological shield could result in buildup of pressure sufficient to tip the pressure vessel. Pressure waves in the suppression pool of the Mark I and the Mark II plants could create loads on the supporting structures greatly in excess of design values.

The engineers traced the evolution of the pressure suppression containment designs through Mark I, Mark II and Mark III stages. The just-completed preliminary study of the Mark I containment system indicated the possibility that greatly increased loadings in the torus could result in the failure of the primary containment system in the 19 operating Mark I plants. Other factors of the Mark I containment system which have resulted in constant erosion of design margin are corrosion allowance in material thickness, the failure of seals for electrical penetration into the containment vessel, and in the functioning of vacuum breakers between the wet-well/dry-well. The engineers conclude that because the potential consequences are great, the advisability of continued operation of the Mark I plants under these circumstances should be questioned.

Material failures in reactor systems have been a continuing problem; for example, it was recently determined that stainless steel, which is used so extensively in reactor construction, suffers from intergranular stress corrosion cracking.

NRC Quality Assurance Program requirements are inadequate. The NRC inspection of manufacturers of pressure bearing equipment is less stringent than is required in commercial industry by the ASME Boiler Code. NRC regulation of product qualification of electrical control systems is currently less stringent than would be required for commercial electrical appliances (such as toasters or hair dryers) by the Underwriters Laboratories. It is recommended that NRC regulations be amended to require a quality assurance program to minimize the likelihood of common-mode failures of safety-related items, and that an analysis of all field-failure data be required to be reported to the NRC.

Nuclear power plants today are not designed to minimize the

impact of human error on operation. Several well-known accidents caused by human error were noted: the jumpered safety interlocks of *Vermont Yankee*, the *Dresden-2* blowdown in 1971, the *Millstone* seawater intrusion in 1972, the *Browns Ferry* fire in 1975, inversion of control rods, and the *Rancho Seco* control rod drive failure in 1975. Three specified recommendations were made: (1) that control room design be standardized to minimize the possibility for human error, (2) that control room simulators be updated to increase operator proficiency, and (3) that operating/maintenance procedures be more closely controlled.

The operation, maintenance and decontamination of nuclear power plants requires ever-increasing exposure of personnel to ever-increasing levels of radiation. Even normal operation and maintenance requires increased personnel radiation exposure, having increased from an average 188 man-rem per year in 1969 to an average 544 man-rem per year in 1973. This increased exposure to personnel in order to operate and maintain a plant is further complicated by a number of technical retrofits, and more radiation exposure is required for safety improvements.

Little thought has been given to the decontamination of nuclear power plants, which poses additional risk of radiation exposure to personnel and will further aggravate already critical radioactive waste storage problems.

Finally, the engineers recommend that the NRC practice of "grandfathering," the waiver of current safety requirements for older plants, should cease and that older plants should be required to meet current safety requirements. Examples are improved cable separation, cable insulation modification, improved remote shutdown capability and redesign of cable spreading rooms, installation and improvement of emergency core cooling systems, and continued evaluation based on the most current seismic data.

The men made several recommendations and suggested areas of further inquiry by the Committee.

Nuclear News, a leading industry periodical published by the American Nuclear Society, lamented that the month of February, 1976, was "probably the most traumatic that the U.S. nuclear energy community has experienced."[4] If the resignations and public statements of Bridenbaugh, Hubbard, and Minor were the cause of this trauma, it was aggravated by another significant resignation the same month. Robert D. Pollard, an electrical engineer and project manager of the NRC's Division of Project Management, charged that U.S. nuclear plants are constructed and operated without meeting government regulatory standards. (His testimony appears in Appendix B.) Pollard referred especially to Units 2 and 3 of the Indian Point, New York, station.

According to Pollard, "It will be just a matter of luck if Indian Point doesn't sometime during its life have a major accident." Consider that this statement came from a man with management-level responsibility *within NRC*, one who also had the authority for licensing a number of plants. In a report delivered to NRC on February 9, 1976, he described four specific safety problems which allegedly exist at Indian Point.[5]

Pollard also gave an illuminating description of the NRC's practice of suppressing critical views. Unsuccessful in his attempts to present his case within NRC, Pollard reported that "at each turn I'm told [that] I shouldn't rock the boat. I shouldn't keep raising these [safety] concerns. This is going to reflect adversely on my employee situation. . . . Keep it from the public, keep it low key, and that's why I'm resigning."[6]

An instructive example of the "low key" approach was the public statement of William Anders, NRC chairman, after a meeting between agency officials and the three former General Electric engineers. Anders stated that that the purpose of the meeting was to determine if Bridenbaugh, Hubbard, and Minor had

> any specific information which might require immediate regulatory action. Their major concerns appear to be philosophical in nature, over the long-term use of nuclear power. They raised several general safety issues, all of which are under active consideration by the NRC staff. They said they had no additional detailed facts or data that, in their view, required immediate regulatory action. They suggested a review of the safety of plants which presently are operating. In the normal conduct of its regulatory responsibilities, the NRC has licensed nuclear facilities under constant surveillance, and this type of review will continue. Thus the NRC finds no basis for taking immediate regulatory action as a result of the discussions today.[7]

Anders's statement makes it clear that the NRC reacted to the direct accusations (dereliction of regulatory duty, collusion, and failure to protect the safety of the public) by ignoring them. Its evasiveness and narrow interpretation of the testimony make one wonder if there isn't something to these accusations after all. However lame and inadequate the NRC response may have been, from a practical standpoint its range of possible responses was limited, for the testimony was extraordinary in that it implied that *the entire agency* was derelict.[8] In turn, the NRC implied that the three engineers were forgivably unaware that the federal government had everything under control.

The explicit language of the NRC statement is disquieting, nonetheless. It is astonishing that Anders, who had demonstrated superior technical skill earlier as an Apollo astronaut, would dismiss nearly one hundred pages of detailed, technical testimony as "philosophical." One assumes that the ability to distinguish Sputnik from Spinoza was a prerequisite for space flight. Readers of the NRC statement are left with the suspicion that Anders was fully aware of the real nature of the engineers' testimony but phrased his rebuttal as though nothing of consequence had been said—the ultimate "low key" tactic. His co-option of and guarded response to the engineers' charges reminds us that in Washington, D.C., moral considerations are often smothered by technical objectives, and that technical problems must often give way to political imperatives even when matters critical to public safety are at issue.

Hope remained that public opinion would consolidate behind the four engineers who had sacrificed their jobs in order to focus national attention on nuclear industry safety problems. But it didn't take long for the hope to die. Rather than address the issues in public and promote some sort of dialogue, the industry closed ranks and released at least one red herring: loyal GE employees convened a press conference of their own where they implied that Bridenbaugh, Hubbard, and Minor had been manipulated by the Creative Initiative Foundation, a California educational-religious group to which all three belong.[9] The press obliged by limelighting this furor for a while.[10] Meanwhile, public attention was diverted from the safety issues.

The unequivocal stand taken by the four men is permanently on the public record, however, and their words continue to haunt NRC and the industry. Now that their testimony is in print, plant safety inadequacies are more a matter of fact than of opinion. A reasonable doubt, expressed in the national media by engineers with unimpeachable credentials and extensive industry or government experience, exists today about nuclear plant safety and about NRC's ability and willingness to rigorously enforce safety standards. Pollard's statements are especially significant because they confirm the charges of the three former GE engineers by an independent source. The next move, or non-move, is up to the NRC. Since that agency is at least as politically sensitive as its predecessor, the AEC, it is unlikely to levy major reforms as the result of testimony by only four men.

The basis of the deadlock may be the way that NRC and its critics view safety imperatives. Nuclear opponents tend to insist on "safety at any cost" and open themselves to criticisms that they are impractical, have ulterior motives, or are crusading for safety as an end in itself. The NRC appears to take a more "realistic" view: since absolute safety is impossible to achieve without shutting down all plants for good, the only question remaining is how far *relative* safety will be traded off against political-economic considerations such as industry profits, continuous electric production, maintenance, and construction schedule deadlines. As long as no major accident occurs, the burden of proof that plants are unsafe rests with the critics, and tradeoffs will continue.*

But this is so because government and industry together have controlled, until recently, the rules by which the game is played,[13] and have won public acceptance of their premises that nuclear power is inevitable because "we must have it to survive," that plant shutdowns are bad for business, and that risk is inherent in anything we undertake.

The public is beginning to wonder if nuclear power entails, instead, uncommon risks.[14] If a state nuclear initiative passes, voters may turn the tables on the manipulators and require industry to bear

---

* The conflict of expert opinions on the matter of safety suggests that debate on this topic is unproductive and that a deeper issue is involved. Such has been suggested by Amory Lovins, a physicist and British representative of Friends of the Earth, who comments on nuclear fission technology as follows:[11]

> [U]ltimately its safety is limited not by our care, ingenuity, dedication or wealth (as has been true of all previous technologies[12]), but by our inescapable human fallibility; limited not by our good intentions, but by gaps between intention and performance; limited not by our ability to solve problems on paper, but by our inability to translate paper solutions into real events. If this view were correct, it would follow that nuclear safety is not a mere engineering problem that can be solved by sufficient care, but rather a wholly new type of problem that can be solved only by infallible people. Infallible people are not now observable in the nuclear industry or in any other industry.

the burden of proof that the plants are safe as they claim. It may be difficult for the industry to carry this burden[15] and to avoid being placed in a position where it has to marshal evidence to show that plants are safe. The industry spent between $6 and $8 million in 1976 alone in media campaigns aimed at defeating state nuclear initiatives.

Having outspent nuclear opponents by a ratio of four to one, the industry uses voter rejection of the initiatives as proof of citizen support of nuclear power. No doubt it is, but this tactic sidesteps the fundamental question of nuclear plant safety. Eventually, either a major accident or passage of a state initiative will put the controversy in a much different light. Once the battle is joined in either case, the agencies promoting nuclear power will be at several disadvantages: they will be on the defensive, and anything they say in their defense may be construed as self-serving. Consequently, the public may be skeptical of any efforts on their part to bear or shift the burden of proof of plant safety and increasingly unwilling to accept "objective" safety studies published by them. This could lead to polarization of public opinion and further erosion of citizen confidence in nuclear power.

Perhaps a less divisive path can be found to resolve the doubts raised by Pollard and the three engineers from General Electric. The political realities are inescapable, however: a great deal of public pressure for reform will have to build up before NRC bucks the industry. By then it may be too late, for pressures of this magnitude will be created only if a major accident calls attention to the long history of safety compromises and to the rationalizations that sustained them.

Meanwhile, industry and government have evidently decided in favor of risking public health and loss of confidence in government that would certainly follow such an accident. They have arrived at this after considering that even if things do not get much better, they probably will not get substantially worse very quickly if the present course is pursued, and that this course is preferable to the discomfort and dislocation that must accompany substantial reform. Bureaucratic dynamics that require parity in the magnitudes of pretext and consequence where reforms are involved are likely to find themselves reduced to an absurdity if we must have an accident to prevent one.

16
The Atomic Industrial Complex

Several citizen, research, and environmental groups have successfully alerted a growing segment of the American public to the problems of nuclear technology. Despite tiny budgets, limited staffing, and a host of procedural obstacles, these groups have challenged the nuclear industry. This challenge is expressed in recent legislative and court actions, pressure on regulatory agencies, and occasionally in citizen confrontations, sit-ins, and other direct actions here and abroad.[1] By 1976, these efforts had achieved enough momentum to put, in the ballots of several states, initiatives that allowed the public to decide under what conditions nuclear plants could be constructed and operated.

Until recently, the industry sought to limit the debate to technical questions best resolved by specialists. When the debate expanded to include social and moral issues, nuclear promoters optimistically assumed that federal agencies and elected representatives subject to effective checks and balances would decide these issues for all U.S.

citizens. In 1976, a General Electric attorney-spokesman, Fritz Heimann, pointed out that "the acceptability of nuclear power must be
settled by the political process. Thus there is no alternative to opening public debate of the issues raised by the opponents of nuclear
power."[2] This "public debate" has, of course, been going on for
twenty years in Congress during the consideration of much nuclear
legislation. Unfortunately, some legislators may not have been altogether objective when voting on nuclear issues.

Recent investigations have unearthed evidence that many key
congressmen and senators, some even sitting on the Joint Committee on Atomic Energy, have received direct cash payments from at
least one company which may have been interested in fostering
favorable legislation. During the 1973 Watergate investigations,
Special Prosecutor Archibald Cox persuaded Gulf Oil Company and
several other prominent U.S. corporations to make a voluntary public disclosure of their illegal political contributions. More than a
hundred corporations obliged. Shortly after, pressure from court
actions and the Securities and Exchange Commission investigations
led Gulf to disclose illegal payments *exceeding $10 million* in and
outside of the United States. A Special Review Committee led by
John J. McCloy, former president of the World Bank, was formed
to look into Gulf's illegal activities, an investigation that eventually
led to the resignation of Gulf's chief executive officer, Bob Dorsey.

McCloy's committee identified a former Gulf executive and
chief lobbyist, Claude Wild, Jr., as the chief distributor of funds.
Wild's testimony during 1973–75 confirmed that a covert system of
dispersing money under the guise of "political contributions" had
been operating for more than ten years. Since the "contributions"
accompanied no evident transactions other than the implied expectation of congressional favors, one may more accurately regard them
as bribes.

A January 9, 1976, article by John Hall, Washington corre-
spondent for the *San Francisco Examiner,* published specifics:[3]

> [Wild] had so much money to distribute that he decided it
> was physically impossible to do it all himself. So he had to assign
> three staff underlings to hand-carry the cash in sealed envelopes
> to political figures here and across the country. According to one
> of Gulf's Washington executives, Frederick A. Myers, envelopes
> were passed behind a barn in New Mexico to Edwin Mechem

(R-N.M.), who was a senator; in an Indianapolis motel washroom to Richard Roudebush (R-Ind.), who was a congressman (and subsequently appointed to head the Veteran's Administration); and in a Las Vegas office to Howard Cannon (D-Nev.), who heads the Senate Ethics Committee, which regulates the standards and conduct of members. A lawyer retained by Gulf, Thomas Wright, said that Wild had told him [that] Senate Republican Leader Hugh Scott received $5,000 twice a year—spring and fall—from Gulf "for personal matters—never as a political contribution."

Myers said [that] he passed envelopes between March, 1961, and late 1972 to staff members of Sen. Wallace Bennett (R-Utah); Rep. James Burke (D-Mass.); Rep. Herman Schneebeli (R-Pa.); Sen. Howard Baker (R-Tenn.); Sen. Marlow Cook (R-Ky.); and Rep. Joe Evins (D-Tenn.). Norval Carey, a vice president for Gulf's subsidiary, General Atomic Company, said [that] he delivered envelopes between 1970 and 1974 to Sen. Baker and Reps. Melvin Price (D-Ill.); Craig Hosmer (R-Calif); and Chet Holifield (R-Calif.). All were members of the Joint Committee on Atomic Energy at the time.[4]

In all, according to the McCloy report, more than $10 million was secreted overseas, laundered and distributed to politicians at home and abroad in the 14 years that ended in mid-1973. The beneficiaries of Gulf's largesse and that of the other corporations, in response to SEC and press inquiries, explain the payments variously;

• They simply remember nothing. Roudebush said [that] he couldn't recall the envelope nor the washroom.

• They say staff or campaign aides handle such transactions, and there are no records to verify them.

• They say [that] they thought the donations were legal, personal contributions—either from the men who brought around the envelopes or from voluntary employment groups like Gulf's "Good Government Fund."

Though Gulf's contributions are by far the most extensive, systematic and covert yet exposed, it is clear that Gulf was by no means unique. An audit of Ashland Oil Corporation identified $724,000 in illegal political donations, ranging from $5,000 to a candidate for township supervisor in Niles, Michigan,

to $50,000 to Rep. Wilbur Mills (D-Ark.) while he was Chair-
man of the House Ways and Means Committee. Likewise, The
Northrop Corp., a California aerospace firm, made $250,000 in
illegal corporate political contributions during a ten-year period.

The recipients did not ask where the money came from. The
donors did not ask how it would be used. And the institutions
that normally expose and investigate corruption did not ask
much of anything. Not a single investigative arm, either of
Congress or the executive branch, and not a single judge tried
to get to the bottom of the scandal when it became clear [in
1974] that the payments to the 1972 Nixon campaign were only
the tip of an iceberg.

Until the SEC's late-blooming probe was forced by stockholders' suits, no one bothered to demand a complete accounting of
corporate donations. The Senate Watergate Committee rushed
through its investigation of campaign contributions and closed
its books when the evidence began pointing at powerful members of Congress.

Wild, well-known on Capitol Hill as Gulf's man in Washington, was finally called as a witness, but the questions were
confined almost solely to his contributions to the Nixon campaign. When Chief Counsel Samuel Dash finally asked him
about Gulf's contribution of $10,000 to Senator Henry Jackson
and $15,000 to Mills [for] their fleeting 1972 Presidential bids,
the cigar-chomping Wild wryly remarked, "I thought you'd
never ask."

Wild was whisked off the witness stand without further questions of him or other Gulf executives. The committee steadfastly
refused to call Jackson or Mills as witnesses or to explore the
extent of Gulf's political outlays to other members.

One explanation is that the Watergate Committee's mission
was confined to the 1972 Presidential campaign. Ervin steadfastly refused to allow the committee to get off the track. But
another explanation is that the investigators of the coverup
themselves perpetrated a coverup. According to Wright's testimony, based on notes he kept of his meetings with Wild, the
Gulf lobbyist told him that he "had assisted all the senators in
Watergate except Senator Ervin."

The SEC's McCloy Committee said such contributions, if
Wild did make them, could not have been from Gulf's "Good

Government Fund," except in two instances. It said the only two members of the panel who received checks from the fund were Baker and Sen. Lowell Weicker (R-Conn.). Any cash delivered to members of the Committee "must be presumed to have been corporate funds." Former staff members of the committee who wanted to pursue the Gulf contributions said they were firmly told to lay off. One staff members recalls "a lot of yelling and screaming" over the question. The Watergate Special Prosecutor's office, under Archibald Cox, Leon Jaworski and Henry Ruth, developed only information sufficient to wring guilty pleas out of the executives of more than twenty corporations charged with making illegal contributions.

With two exceptions, not a single political recipient of illegal corporate donations was brought to justice. Only Jack Chestnut, manager of Hubert Humphrey's 1970 Senate race, and Maurice Stans, finance manager of the 1972 Nixon race, were convicted of accepting corporate donations. With the Gulf, Ashland and Northrop studies now demonstrating that the practice was widespread, some observers have begun questioning why either the special prosecutor or the courts did not probe more deeply and demand complete accounting of donations by every corporation in exchange for the light fines handed out to the donors. One explanation, offered by the Prosecutor's staff, was that a thorough investigation was more or less preempted by an obscure law pushed through in the waning days of the 1974 session. That measure, part of the Campaign Reform Act that required strict accountability for all contributions and limitations on how much could be raised and spent, prohibited prosecutions for illegal corporate campaign donations after three years. The statute of limitations had a five-year limitation before the law was enacted.

General Atomic Company, until 1973 a wholly-owned subsidiary of Gulf Oil, was a leading manufacturer of high-temperature gas-cooled reactors and reactor supplier of the Fort St. Vrain nuclear plant in Platteville, Colorado. It was in a position to benefit from Wild's handouts. Gulf Oil's political influence in Washington may well have been organized in part to protect its subsidiaries' nuclear commitments. Other corporations may also have made similar, hitherto undisclosed, contributions to assure congressional support of nuclear power.[5] If so, citizens have a reasonable explanation how such an uneconomic technology could gain the fiscal and political momentum that characterize the nuclear juggernaut today. Perhaps the most disquieting implication of all is that this practice of illegal payments may be common throughout federal and state legislatures.

The extent and duration of congressional bribery may never be known and the McCloy report gives us only a random sampling of this behavior. Industry interlocks[6] with other governmental agencies which are in a position to promote and safeguard nuclear investments are much better documented, thanks to a 1976 Common Cause study. After almost a year of work and contested Freedom of Information Act requests, Common Cause's investigation of ERDA and NRC concluded that "the potential for serious conflicts of interest exists in both agencies."

ERDA, which is charged with "implementing a balanced and practical program that will anticipate the energy needs of our nation," draws 52 percent of its top employees (72 of 139 GS-17 and above) from corporations with an especial interest in energy; 39 percent of these employees from the private sector (54 of 139) formerly worked for ERDA contractors.[7]

At the executive level, the conflicts are even more shocking. Six of nine executive-level positions surveyed are filled by former employees of ERDA contractors. The three other positions are filled by former employees of firms regulated by, or contracted with, the NRC or the Federal Energy Administration.

The Common Cause study states, "Represented in the employment background of executive-level personnel are giants in the energy industry and the nuclear industry in particular." These include Union Carbide, which operates Oak Ridge National Laboratory and other ERDA facilities, and General Electric, the only constructor of boiling water reactors.

In December, 1975, ERDA explained that it makes a deliberate effort to recruit personnel from the private sector and to establish "partnerships" with private industry. While tapping expertise and having good working relations with industry are noble pursuits, hiring ex-contractors for a "balanced and practical" energy program is questionable, especially since the nuclear industry is heavily represented in top management.

Conflicts of interest between the regulator and the regulated are even more serious. The study found that 67 percent of top NRC employees (292 of 434 GS-15 and above) came from private firms

involved in energy activities. More than half—58 percent (253)—
of those NRC officials were from commercial enterprises, laboratories, or universities holding NRC licenses, permits, or contracts.

The "Big Five" reactor suppliers—Babcock & Wilcox, Combustion Engineering, General Atomic, General Electric, and Westinghouse—have seventy former employees in the NRC. Seventy-one NRC employees came from Rockwell International, DuPont, Bechtel, Allis-Chalmers, and General Dynamics, all of which are heavily involved in the nuclear industry.

The substantial recruitment of employees from the nuclear industry raises doubts about the NRC's objectivity in its regulatory decisions. In addition, nuclear industry influence through congressional political contributions and incestuous appointments to federal agencies are only two *domestic* examples of established practices for *internationally* augmenting industry profits. In 1976, Friends of the Earth obtained confidential documents belonging to an Australian corporation, Mary Kathleen Uranium Ltd., of Melbourne. These documents provide strong evidence that representatives of Australia, Canada, France, South Africa, and Rio Tinto-Zinc Corporation met in 1972, and thereafter, to fix the international price of uranium.[8] After agreement on uranium prices was reportedly reached, the world price jumped from $6 to $41 a pound.

According to the hundreds of pages of leaked confidential files and charts, the five world uranium producers met to set uranium prices, rig uranium bids, and establish a market quota system. They also established a policy of price discrimination against U.S. uranium "middlemen," such as the Westinghouse Corporation.

Mary Kathleen Uranium Ltd., a subsidiary of Conzinc Riotinto, is controlled by Rio Tinto-Zinc. Rio Tinto-Zinc is a member of the uranium producers' group known informally as "The Club."

A spokesman for Mary Kathleen Uranium has confirmed that the confidential documents delivered to Friends of the Earth were from his company's files. He also acknowledged the existence of the uranium cartel but denied that his company belongs to it. Cartels are legal in many foreign countries but are a violation of U.S. antitrust laws.

Operation of a foreign uranium cartel could cost Americans billions of dollars in exorbitant charges for imported uranium to fuel U.S. nuclear power reactors. In response to the impending shortage of domestic reactor-grade uranium, the U.S. government has lifted the current ban on uranium imports as of 1977.

Collusion by world uranium producers may already have cost Americans enormous sums because U.S. producers have used international price escalations to justify rapid increases of domestic uranium prices. These price increases are passed on to consumers in the form of higher utility bills.

The confidential uranium documents indicate that the uranium producers' club was formed in 1972 to make uranium production more profitable for its members by restricting price competition among uranium suppliers. At that time, the world uranium supply far exceeded demand and uranium prices were between $6 and $8 a pound. The five large uranium producers assembled in Paris during February, 1972, for a meeting of the "Uranium Marketing Research Organization." They reportedly analyzed the world uranium supply situation and discussed uranium marketing and price stabilization. The Mary Kathleen files show that these discussions led to setting world uranium prices for 1974 to 1978. During the talks, specific percentages of the world uranium market were allocated to each club member. These files show that the French delegation proposed frequent meetings on uranium prices and a specific market-sharing agreement. The records also indicate an intention to obtain governmental support for price-fixing activities. As one memo indicated:

> It was recognized that producers cannot come to an acceptable agreement without collaboration of the respective governments since major policy issues are involved. Every effort will be made to maintain contacts as appropriate between producers and their respective governments.

The files contain confirmation of such collaboration on the part of the Australian government, whose representative attended a producers' meeting in Australia at which he reportedly "stressed the need for extreme secrecy" in the uranium talks.

Subsequent to the initial Paris meeting, a series of club talks took place in Paris during 1972. These talks were supplemented by exchanges of correspondence between members, during which agreements and policy were discussed openly. In one exchange, H. F. Melouney, the former manager of Rio Tinto-Zinc Services, chided Louis C. Mazel, another executive, as follows:

In your letter, you use a word which we would not even like to mention, as some of the members of the club are rather worried about informal price agreements. I would like to stress very strongly that under all circumstances there can be only an unofficial agreement and whatever agreement is struck, it should be on a strictly confidential basis.

For the outside world, all Paris and subsequent meetings will be in connection with the exchange of marketing information.

In August, 1972, another club meeting discussed restricting sales of uranium "to middlemen such as reactor manufacturers, e.g., Westinghouse, General Electric" and pricing uranium to them at higher discriminatory rates, to discourage them from selling uranium.

Further talks took place in Johannesburg, London, Paris, and Las Palmas during 1973, at which new minimum prices and market shares were set. Consultations continued in 1974 when the club discussed uranium marketing policies for the 1981–1983 period and decided to give the group a new facade by converting it from a "Uranium Marketing Research Organization" to the "Uranium Institute," ostensibly a foundation set up to conduct uranium-related research. Despite its new trappings, however, the basic purpose of the new group appears to have remained identical to its predecessor's.

Nuclear activists in the U.S. have long believed in the existence of a uranium cartel, although the reports of the stolen files did not surface until mid-August, 1976. But the U.S. press scarcely paid any attention to the story until the end of the same month, at which time three members of the California Energy Commission and a member of the California Public Utilities Commission sent news of the Australian disclosures to U.S. Attorney General Edward Levi and Senator Frank Church, chairman of the Senate Subcommittee on Multinational Corporations. The California officials requested both Justice Department and Senate-level investigations, and the story was suddenly catapulted into the headlines.

In their letter to Levi and Church, the commissioners said:

A foreign uranium cartel would defeat America's current policy of energy independence. This policy proposes an accelerating commitment to nuclear power. Inherent in this commitment is a growing requirement for foreign uranium.

Already, U.S. utilities have more foreign than domestic uranium under contract for delivery from 1980 to 1984. These import commitments amount to over $3 billion worth of uranium at current prices.

If a foreign cartel continues and America's nuclear commitment increases as planned, we will be at the mercy of a uranium OPEC (Organization of Petroleum Exporting Countries).

The Justice Department empaneled a federal grand jury in October of 1976 to investigate evidence of price-fixing derived from the Mary Kathleen papers. If it finds grounds for prosecution of domestic collusion or illegal links between U.S. uranium suppliers and the international cartel, criminal complaints may be filed.

If prosecutions lead to a drop in uranium prices, uranium producers' profits would probably also fall, along with the incentive to develop supplies. Reductions in output, however, could eventually worsen the long-run uranium shortage that is likely to occur as more reactors are built.

Another probable outcome of criminal action against U.S. uranium suppliers would be the filing of civil damage suits against uranium suppliers by purchasers, who might contend that they had been forced to buy their uranium supplies at illegally inflated prices. A substantial drop in uranium prices resulting from antitrust action would tend to make nuclear power economically more competitive with alternative generating sources. It would also have the effect of making nuclear fuel reprocessing less attractive.

Even if none of these repercussions occurs, the spectacle of an international cartel rigging uranium prices at the expense of American citizens is not likely to improve the public image of nuclear power as an inexpensive source of energy. On the other hand, the public has become so resigned to price-fixing as a permanent fixture in the economic sector that sustained demands for reform never seem to be asserted; in addition, no way of preventing price-fixing has ever been devised. The Mary Kathleen case is the first instance in recent years where this practice tainted the nuclear industry, however; and the existence of price-fixing on an international scale suggests that the Atomic Industrial Complex is no more honest in its dealings, internally and with the public, than any other business organization. The important difference, of course, is that the AIC is responsible for developing a technology with an unprecedented potential for damage. Nuclear industry executives are therefore understandably sensitive that public confidence in their professional and technical competence may be seriously eroded by evidence of any shady business practices.

One may assume that the nuclear industry did not welcome the publication of the Mary Kathleen papers and hoped for an early end to the controversy. These hopes must have been dashed two months later when, in October, 1976, Westinghouse Corporation filed suit in the U.S. District Court for Illinois[9]* charging that twenty-nine uranium suppliers illegally rigged U.S. uranium prices. This action increases the intensity of legal cross-fire, which now involves Westinghouse, as well as dozens of electric utilities and uranium suppliers, and which is likely to add further economic uncertainties to future uranium price and supply forecasts. Westinghouse earlier was sued for breach of contract by the utilities whose nuclear fuel supply agreements it had abrogated. The company admitted in September of 1975 that it did not have 65 million of the 85 million pounds of uranium yellowcake that it agreed to sell to its reactor customers over a period of years.

A judgment in Westinghouse's favor on the uranium price-fixing complaint would greatly bolster its defense against the twenty-seven utilities. In that defense, Westinghouse contends that it had to terminate its uranium supply agreements because "unforeseen and unexpected" uranium price increases made fulfillment of the con-

* The defendants named by Westinghouse: Gulf Oil Corporation, Kerr-McGee Corporation, Anaconda Company, Getty Oil Company, Phelps-Dodge Corporation, Utah International, United Nuclear Corporation, Pioneer Nuclear, Reserve Oil and Minerals Corporation, Federal Resources Corporation, Homestake Mining Company, Rio Tinto Zinc Corporation, Engelhard Minerals and Chemical Corporation, Denison Mines, Atlas Corporation, Western Nuclear, Rio Algom, Ltd., of Canada, Rio Algom Corp. of Canada, Rio Tinto Zinc of Britain, RTZ Services of England, Conzinc Rio Tinto of Australia, Mary Kathleen Uranium of Australia, Pancontinental Mining of Australia, Queensland Mines of Australia, Nuclear Fuels of South Africa, Denison Mines of Canada, Noranda Mines of Canada, Anglo-American of South Africa, Englehard Minerals and Chemicals Corporation of South Africa, Gulf Minerals of Canada. The suit charges that these companies, assisted by a steering committee organized through the Atomic Industrial Forum (a trade association with offices in New York and Washington, D.C.), conspired to set prices, to manipulate the international uranium market with rigged bids and contrived market divisions, to keep uranium supplies off the market unless buyers accepted exorbitant prices, and to force Westinghouse out of the market via collective boycott.[10]

tracts "commercially impracticable" under the terms of the Uniform Commercial Code.

At stake for Westinghouse, one of the largest nuclear equipment suppliers, are potential losses of more than $2 billion which the company will have to pay if it is required to meet its outstanding uranium supply commitments to utility customers at the current price of $40 per pound. That price is more than $30 above that specified in Westinghouse's long-range uranium supply contracts, which were signed before the price began rising.

Although the company's legal counsel claimed losses may exceed $2.5 billion "with crippling potential impact for Westinghouse," some experts doubt that the plaintiff-utilities that own Westinghouse nuclear equipment would want to force the giant firm into bankruptcy, even if they could.

For householders using nuclear-generated electricity, the legal battle between the big energy firms may mean higher utility rates. The initial announcement of the Westinghouse fuel shortage caused a sharp rise in uranium prices, and market supplies were immediately restricted.

Should Westinghouse be relieved by the courts of some or all of its obligation to provide low-cost uranium at its long-term contract price of $9.50 per pound, its utility customers will have to buy alternate supplies at much higher prices, causing nuclear power costs to rise.

Westinghouse's complaint against the uranium suppliers and their agents contends that the companies conspired to raise and fix uranium prices since February of 1972. During that time, yellowcake increased in price by more than 500 percent. According to Westinghouse, uranium sellers also have "curtailed and manipulated" supplies of the fuel. Alleging that the defendants thereby violated the Sherman Antitrust Act and the Wilson Tariff Act, the company demanded treble damages. The exact sum being sought was not specified. Citing information already made public by Friends of the Earth relating to operations of an alleged foreign uranium cartel,[11] the Westinghouse brief also charged that U.S. uranium suppliers conspired to impose on their U.S. uranium customers prices set by the international consortium of uranium producers. Westinghouse alleged that a series of meetings between the defendants and major U.S. uranium producers began "at least as early as the fall of 1972," and continued to the present.

The lawsuit specifically charged that the defendants conspired to raise prices during a meeting at Oak Brook, Illinois, in March, 1973, at which producers demanded prearranged uranium price increases and threatened supply curtailments unless its price conditions were met. Westinghouse asserted that members of the Atomic Industrial Forum's Mining and Milling Committee were active in helping uranium suppliers set up this meeting. Westinghouse further stated that it had been seriously injured by the alleged refusal of foreign and domestic uranium producers to sell uranium to Westinghouse. The company said it had been "boycotted" by suppliers because it had previously resold some of its uranium purchases to utilities "below cartel prices."

Issues and Comments

From the absence of substantive reforms of these practices, we can only conclude that most citizens believe that bribery, conflict of interest, and price-fixing practices are common and, for the most part, impossible to prevent. Once established, these practices tend to become part of the environment of business and politics, accepted as necessary evils by participants and, in the end, by the voting public as well. Occasional scandals remove executives and politicians from office but seldom result in reforms that go to the heart of the problem. Replacements for these people are subject to the same forces as their predecessors, and the practices tend to resume before long.

Two important books may be of interest to the general reader who wants additional information on congressional bribery and industry-government collusion. A. Ernest Fitzgerald, a former undersecretary of the air force, describes how the military-industrial complex operates in *The High Priests of Waste,** which focuses on Pentagon attempts to cover up cost overruns on the Lockheed C-5A giant transport. *The Washington Payoff,*† by Robert Winter-Berger, a former Washington lobbyist, is an unconfirmed, detailed report of Capitol Hill bribery and suggests that practices evident in the Gulf Oil scandal are indeed widespread throughout Congress and the federal and state courts. Both books imply that corporate power in our society has become a form of ultimate power and that its misuse has

* See Bibliography.

† See Bibliography.

gone far toward corrupting all levels of our government.

This topic deserves extensive treatment by other researchers, and we mention it here because industry-government collusion may have contributed to many of the nuclear industry's problems. If we extrapolate from the specific examples presented in this article, the possibility emerges that corporate irresponsibility may be the common denominator to many difficulties which our society faces and for which particular solutions have often been prescribed without a prior understanding of the universal problem. Specifically, the pursuit of corporate profits has reached the point where human lives are being shortened involuntarily by all forms of pollution. Citizen attempts to stop this pollution have been fended off until recently by legal maneuvering, "political contributions," and media manipulation funded mostly by corporate profits that were derived from activities that contribute to pollution.[12]

Instead of investigating the causes of this pollution in the first place and significantly curtailing corporate activities that contribute to it, our government—with enthusiastic industry support—attacks the symptoms by underwriting technical fixes. Recent panaceas, some of which involve extensive government funding of industry research and development, include ongoing investment in a cancer cure (medication and treatment, rather than prevention), and building more nuclear plants. Rarely, if ever, does a company voluntarily restrain its production or marketing activities because its processes or products are polluting the environment.

Ralph Nader commented recently on this reluctance of corporations to sacrifice profits and/or significantly reduce production levels as a first step in reducing pollution:

> If you were to send 10 pickpockets to New York and each of them were to do 50 jobs a day, that would, I suppose, make front-page news. But set up a system of acquisition that is derivative and obscure, like the oil-depletion allowance or the oil-import quotas, and nobody pays much attention. You can get away with it on that scale. That's the genius of the corporate system—the genius of being able to achieve the same thievery three stages removed. The genius of developing an abstraction of burglary!
>
> There are a variety of conditions that pose both imminent and long-range threats to human life and health, conditions that represent new styles of violence exerted on a huge scale but somehow kept, by the workings of the prevailing system of power in our society, from reaching a popularly perceived threshold that triggers irresistible public pressure for reform. These new styles

of violence are different from the violence of actual automobile accidents, which as a cause of trauma and death have no challenger in this country, and of which millions of citizens are only too keenly aware. What we must get across to people is that endemic forms of violence are going on daily in this country on a scale that completely dwarfs the common concept of violence as meaning primarily "crimes in the street." Air pollution (whether emanating from a car or a nuclear plant accident) is a devastating form of violence. It takes far more lives and maims far more victims each year than street crime, and destroys more property each year than all the bank robbers' hauls a thousand times over. According to government figures, bank robbers have gotten away with more than $20 million a year in recent years—their biggest ever—but the cost of air pollution alone in a recent year ran to *more than $16 billion. . . .*

The complexity and seriousness of the impact that pollutants and contaminants have on us seem to escape people's sensory ability to comprehend. They don't respond in any active fashion. The threats that people *do* tend to respond to are mostly of the immediate, physical kind. People don't react to the concept of infiltration, and infiltration is an essential process of the corporate system. Industry infiltrates the whole person of the citizen. It invades his privacy as an individual. It infiltrates the bodies of millions of people in this country with its contaminants. Its violence is all-pervasive, whatever its various motivations may be.

The fact is that those who define what is permissible and impermissible violence in our society have been responsible for committing a very large part of it. This is perhaps why so much of the industrially caused violence that is taking place either is excluded from the embrace of the law or has forged about it a protective system where the law has virtually no application. The law has to be made to develop definitions of violence that correspond to the conditions we face. In terms of the violence it is currently committing, industry is largely beyond the law. One problem is that the law in this country has developed in such a way as to require positive duties of citizens. The law has developed to punish certain positive acts, not certain kinds of *inaction.* As matters stand, institutions are structured in such a way that corporations can achieve their economic objectives by inaction —for example, by reducing their costs through *not* taking steps to reduce the pollution they create. And by the same inaction they escape the law. When you couple this deficiency of the law with the new style of violence that doesn't provoke an image of a criminal act, such as a common street crime, you have a combination of the power to devastate increasingly large sections of humanity and a decreasing degree of personal accountability. This is a prescription for social suicide. The president of a chemical company knows that no matter what his company does, *he* won't go to jail as long as he *personally* does not act, or instruct his subordinates to act in violation of the law. Our law is basically a legacy of ancient ethics in which A can harm B only by positive action against him. But increasingly, as people have become members of large organizations . . . they have arrived at a position of being able to harm people more by doing nothing.[13]

Further on, Nader pointed out that, because "people don't know [the names of executives] heading the biggest corporations in the country" our corporate system is able to cultivate "anonymity . . . abstraction and physical detachment."[14] Insulated from devastating court actions and unresponsive to demands for significant internal reform, the mega-corporations of the Atomic Industrial Complex have concentrated unprecedented power and wealth in the last century.

To maintain that power and to insulate themselves from citizen-initiated reforms, the AIC invests considerable amounts of that wealth toward:[15]

- Defeating state initiatives designed to temporarily halt nuclear power development and to permit state governments to reassess this source of energy before it is exploited further.
- Extensive federal and state lobbying, most likely bolstered by the kinds of "political contributions" identified by Winter-Berger and in the Gulf Oil case.
- Subsidizing trade, scientific, and public relations organizations that generate thousands of news releases, advertisements, periodicals, and other publications each year for the media, the technical community, and the general public.

Whether the nuclear industry is merely following conventional business practices in these cases is beside the point. The technology being promoted here is not in the least conventional. Furthermore, at the same time that it has precipitated debates which our technical and political institutions have not yet even developed adequate dialectic to cope with,[16] a number of organizations that stand to reap substantial profits seem to be in an inordinate hurry to foist this technology on the public.

Even more serious are the aggregate efforts by *all* industries to subvert the democratic process to the extent that long-term public considerations appear to be compromised in favor of corporate interests. Examples of the impending consequences are international nuclear weapons proliferation, gradual isotopic pollution of the earth, and the risk of genetic damage for millennia. Citizens ought now to reconsider these delayed costs of nuclear development and address the question of whether the public interest is being served by industries which aggressively market this technology.

The unprecedented affluence that is enjoyed by a majority of American citizens, and that was made possible to a certain extent by corporate-government interlocks in the first place, tends to discourage widespread citizen demands for substantial reforms leading to less influence by corporations over government decisions. As a result, corporate interests will continue to receive disproportionate representation in federal and state governments so long as the public remains uninterested in tampering with the lobbying apparatus which U.S. industry insists serves a useful purpose. As we have seen, however, such lobbies seek to promote their own interests in ways that appear to equate profits with human welfare when, in fact, public health is being compromised.

Corporate efforts to maintain the level of public dissatisfaction well below the point where reforms are imminent may be the most insidious of all. Such efforts, whether consciously directed or not, operate through a network involving primarily the medium of television, which equates the good life with affluence and which identifies both with corporate stability.[17] If the majority of U.S. citizens prefer the status quo, it may be because they perceive that the dovetailing of government and industry satisfies their fundamental needs—which may be created in part by industry advertising—including the need to have their attention diverted from news and thoughtful analysis of unpleasant side effects of that dovetailing.

A NEW VIEW
OF THE NUCLEAR
CONTROVERSY

17
The Dawning of
the Nuclear Age
Philip Herrera

On September 6, 1954, ground-breaking ceremonies for the United States' first nuclear power plant took place in Shippingport, Pennsylvania. President Dwight D. Eisenhower, though not present, commented that he saw, in the development of the peaceful atom, mankind coming "closer to fulfillment on a new and better earth." Most Americans tended to agree. Enthralled by the technological fervor of the times, they took the Shippingport nuclear reactor as a wondrous symbol of implacable progress.

On the other hand, the word "nuclear" also evoked a vision of August 6, 1945, and a solitary U.S. warplane called the *Enola Gay* droning over Hiroshima with a cargo of one bomb. Dropped from 32,000 feet, that bomb created a fireball 110 yards in diameter which briefly reached 300,000 degrees centigrade, sufficient to melt surfaces off granite 1,000 yards away. Then the mushroom cloud billowed up 50,000 feet. Many other people remembered the Japanese fishing vessel *Lucky Dragon*. Unlucky enough to be seventy-two miles off a Marshall Island A-bomb testing site in March, 1954, the boat was dusted with white radioactive ash. Soon thereafter, the crew was stricken ill and made for home and hospitalization. When one fisherman died the following September, a point about radiation had been made. As Harry Slater, an executive of Niagara Mohawk Power Corporation was to say, the peaceful atom's first problem was that it was "conceived in secrecy, born in warfare, developed in fear."

The problem was not adroitly handled. In 1957, just before the Shippingport reactor went into operation, the AEC[1] released a document entitled "Theoretical Possibilities and Consequences of Major Accidents in Large Nuclear Power Plants."[2] Researched by AEC scientists at the Brookhaven National Laboratory, the report attempted to answer two questions: (1) what is the likelihood of a major reactor accident; and (2) what would be the consequences of such an accident? The scientists took as a hypothetical case a 100- to 200-megawatt reactor in a suburban location near a large body of water about thirty miles from a city of one million people. They then assumed that everything went wrong at the precise moment when the plant had built up the maximum amount of radioactive products. "It is as if someone were to try to figure what would happen if every nut, bolt and other structural support suddenly broke in New York's Pan Am Building and that the building fell into Grand Central Station at rush hour," says Paul Turner of the Atomic Industrial Forum, an industry group. "The situation is obviously impossible." Indeed, scientists pointed out that the possibility of the hypothetical accident was so remote as to be near the vanishing point. One reason: the radioactive products would somehow have to escape a succession of physical barriers within the plant. Another: every additional safety feature—e.g., core spray and containment spray systems—would also have to fail.

Despite such qualifications, the report's conclusions were all too simple: if 50 percent of the hypothetical plant's fission products were released to the atmosphere, 3,400 people would be killed and 43,000 injured within a radius of forty-five miles. In addition, property damage within an area of up to 150,000 square miles would exceed $7 billion. The figures, if not their context, were chilling—very bad public relations. Nuclear power plants were to grow much larger than 200 megawatts and be sited closer to cities than thirty miles.

The reason the study was made in the first place was to help to determine the limits of insurance for nuclear plants. Yet by itself, the insurance industry was unwilling to enter the area. In 1956, Herbert W. Yount, vice-president of Liberty Mutual Insurance Company, told the Joint Congressional Committee on Atomic Energy:

> The catastrophe hazard is apparently many times as great as anything previously known in the industry. . . . We have heard estimates of catastrophe potential under the worst possible circumstances running not merely into millions or tens of millions but into hundreds of millions and billions of dollars. It is a reasonable question of public policy as to whether a hazard of this dimension should be permitted, if it actually exists. Obviously there is no principle of insurance which can be applied to a single location where the potential loss approaches such astronomical proportions. Even if insurance could be found, there is serious question whether the amount of damages to persons and property would be worth the possible benefit accruing from atomic development.[3]

Still, the electric utilities were not going to develop atomic power without adequate insurance. To resolve the dilemma in 1957, the federal government stepped in with an act named after its sponsors, Congressman Melvin Price and Senator Clinton Anderson, both members of the Joint Committee. As originally enacted, the Price-Anderson Act forces every reactor operator to buy all available maximum insurance from private companies ($60 million in 1957, subsequently raised to $82 million), above which ceiling the federal government would supply another $500 million of coverage (subsequently lowered to $478 million).

Inevitably, critics compared the Brookhaven report's theoretical damages of $7 billion with the Price-Anderson Act's limit of $560 million. It required little imagination to conclude that in its insurance the private utilities were being subsidized by taxpayers and that, no matter how horrendous an accident might be, the utilities would not suffer any financial loss. The fact that the electric utilities have had an excellent record—there have been only sixteen minor incidents involving insurance to date—does not in any way vitiate the critics' argument. The act provides an umbrella for the utilities' operations, inasmuch as they would be unwilling to build nuclear reactors without it.

From 1957 to 1962, small nuclear plants were planned, built and put into operation at such places as Morris, Illinois; Rowe, Massachusetts; and Buchanan, New York. Though each plant was clearly experimental, each community warmly welcomed the nuclear neighbor. The only significant exception was a peculiar project in northern California. Like the rumble of distant thunder, the happenings at Bodega Bay gave warning that the nuclear program was in for trouble.

To millions of Americans, Bodega Bay, California, is familiar as the setting for Alfred Hitchcock's movie, *The Birds*. Hitchcock himself was enchanted by the little fishing village of Bodega and its protected bay—the safest harbor of refuge between San Francisco, 50 miles to the south, and Coos Bay, Oregon, 250 miles north along a rugged, often forbidding coastline. When the film maker first saw the area in the early 1960s he pronounced it "a rare find, remote and unspoiled by man." In a word, it was perfect for his purposes.

Hitchcock was not the first to reach that conclusion. Ever since 1944, California's Division of Beaches and Parks had wanted land around Bodega for the state parks system. In particular, the division wanted Bodega Head, a peninsula which hooks sharply from the mainland of Sonoma County two miles out into the Pacific Ocean. In 1956, the University of California began negotiating with Beaches and Parks to get some land on Bodega Head for a marine laboratory. Indeed, it was a unique site. Two unpolluted ecosystems were set on each side of the peninsula. On the bay, tidal mud flats teemed with cockle, clam, shrimp and myriad other aquatic creatures. On the ocean frontage were tide pools alive with anemones, starfish, abalone, mussels and snails. But in 1957, both the University and Beaches and Parks dropped their plans. Pacific Gas & Electric (PG&E), the nation's second largest power company, had also found Bodega.[4]

What did PG&E see in the area that took precedence over a state park or research facility? Although company officials did not spell out the answer until 1961, Bodega Head seemed to be the perfect place for a nuclear power plant. Close to the company's largest industrial customers around San Francisco, the peninsula was also safely isolated from centers of population. Moreover, the harbor would allow easy shipment of heavy generating equipment by sea. Once built, the $61 million, 325-megawatt plant could gulp great draughts of bay water for cooling and then discharge the heated water into the ocean, where natural turbulence would quickly disperse the heat. Inland, dairy ranches covered the rolling hills—just the kind of open countryside that is ideal for transmission lines. Even geology seemed to be on PG&E's side. A company engineer testified at a public hearing: "The natural granite rock foundation for the plant provides an ideally stable platform." He was to be proven wrong.

Between 1957 and 1962, PG&E quietly developed its plans. Few citizens protested. One who spoke out was oceanographer Joel Hedgpeth; he liked nothing about PG&E's plans and said so in public, disregarding vague threats to his career at the University of the Pacific. Another was Rose Gaffney, a major landowner on Bodega Head who did not want to lose her property to a power company. "I am not asking for your sympathy," she later said speaking for all landowners threatened with eminent domain. "I am asking for justice."

But in general, people were quiet. Perhaps they were entranced by the prospect of the high taxes that PG&E would pay in the area. More likely, they were silent because they had precious little opportunity to object in a meaningful way. PG&E seldom gave any advance warning of its plans. Worse, the Sonoma County Board of Supervisors, which had to pass on all land-use proposals connected with the project, routinely made public decisions in private and then simply announced the results. For example, the residents of Bodega Bay suddenly learned that their airstrip was to be removed from nearby Doran Park because planes might be endangered by tall transmission towers and lines that were not yet approved or built. So complete was the early exclusion of citizen participation that it took U.S. Congressman Clem Miller's active intercession in 1962 to get the first public hearings in the area—and those hearings did not concern the power plant but an access road to it.

Still, when citizens did have the chance to attend three days of public hearings in March, 1962, before California's Public Utilities Commission [PUC], hardly any objectors turned out. But then the PUC started receiving letters of protest urging that the lovely peninsula not be taken for a nuclear plant. In response, the PUC announced it would reopen hearings.

All at once, the opposition found both its voice and the vital assent of an indefatigable leader. David Pesonen, a forester by training, established a 2,000-member citizens group with the awesome name of the Northern California Association to Preserve Bodega Head and Harbor, Inc. For the next two years, it fought the nuclear plant through each step of the licensing process. When that failed, Pesonen's group turned to the courts[5] and to the press. While he succeeded in starting a steady drumbeat of opposition, PG&E pressed ahead with the project. It dug what became known as "the hole in the Head," a pit 140 feet in diameter and 70 feet deep which was designed to hold the nuclear reactor. Excavated rock went into construction of an access road across the bay's tidal flats. The road destroyed much of the mud flats' wildlife values.

In retrospect, it is easy to separate the basic arguments against the power plant into distinct categories, each of them an important harbinger of controversies yet to come.[6] At first, aesthetics were the main issue: should Bodega Head ever be developed? "The proposed reactor will take a priceless scenic resource for its site, but will produce nothing but common kilowatts," the Sierra Club argued in a "friend of the court" brief. Dave Pesonen compared the PG&E project to using Yosemite Falls for a hydroelectric project. In 1963, California's Lieutenant Governor Glenn Anderson said that the company's "massive facility" would "permanently scar one of the nation's truly unique, and particularly appealing coastal regions."

But if such an argument rests squarely on the side of the angels —who can object to beauty?—it also depends on the eye of the beholder. PG&E gleefully collected newspapermen's on-the-spot descriptions of Bodega and its surroundings: "a seaside slum," "ugly," "desolate," "inaccessible." The company also noted that California ranked Bodega ninety-third of 123 possible park sites—a low priority which presumably spoke volumes about the area's real aesthetic appeal, but may equally have reflected practical factors.

A related and somewhat less subjective debate took the Bodega reactor as a symptom of California's slapdash preparations for the future. Conservation writer Karl Kortum testified: "There is only one population curve that bears on the situation wholly—and that is 17 million people in our state in 1963 and 45 million in our state 37 years later. Forty-five million desperate people—not desperate for more kilowatts but desperate for space, for nature, for the sea, for room for the soul."[7]

Dr. Daniel Luten, a geographer at the University of California, also looked ahead as he dealt specifically with the power company's enticements of new tax revenues, new growth, new progress for Bodega:

> You are being told that this power plant will increase the tax base, and it will, of course. And you are either told or left to infer that it will therefore lower your taxes. There was a time when all of us were inclined to believe that if we could increase our tax base, God would provide that no increase in costs would accompany it. Now we know better. All of the rapidly growing parts of the country have increased in tax bases. Are these the areas with the lowest taxes? No, they aren't.[8]

Luten then pinpointed the real problem, one that was to serve as a basis for a whole series of new environmental laws:

> Let me ask you the general question, "Does this society exist merely to serve its economy?" No one will answer "yes" to this question, yet by and large our decisions on public policy are made primarily or wholly in terms of economic considerations. This is attractive because it submits to quantitative analysis. There is a strong and recurring tendency, when both economic and social considerations are involved, to dismiss the social considerations because they cannot be expressed quantitatively and cannot be measured up side by side with the economic terms. But if a society does not exist to serve its economy, such procedure is improper and intolerable.[9]

What lent such arguments special cogency was the fact that the public agencies with some authority over the project considered only limited issues. The Sonoma Board of Supervisors, the state Public Utilities, the AEC—each considered a few of the problems involved and dismissed the others with a shrug as if to say, "Sorry, it's not within our jurisdiction." In this process, the environment was completely neglected. Nor was the utility about to provide the missing perspective. At a remarkable public meeting in 1962, physicist Richard Sill pointed out: "The Pacific Gas & Electric Company, one must remember, is not a philanthropy, and while service is their business, profit is correctly their motive."

Then how could ordinary citizens gain technical competence to contest the reactor proposal? Dr. Sill knew the answer: *"You cannot find out for yourself."*

When the project's opponents searched for information on the risks of having a reactor in their backyard, what they discovered frightened them. Oceanographer Joel Hedgpeth conceded no one could predict with certainty the effects of the plant's discharge of at least 250,000 gallons of hot water per minute into the Pacific Ocean. Under certain conditions, like a period of calm water and no circulation, he warned, "A tiny thing like this could kill off organic life along several miles of seacoast." More significant, the reactor—biggest to date—was "experimental." Looking over the history of nuclear plants did not reassure them that the Bodega reactor would be safe. They quickly discovered that there had been 300 reactor accidents in the world. Though most were minor, an incident in 1961 at the Arco reactor in Idaho had killed three operating personnel. In another accident at Canada's Chalk River installation, 900 safety devices did not prevent a combination of mechanical defects and operating errors resulting in severe damage to the reactor.

Adding to the understandable fears was a certain amount of equally understandable anger. The critics noted that PG&E's announcement to build a nuclear plant closely followed the Atomic Energy Commission's decision to cut the costs of enriched uranium fuel by 34 percent—in effect, a subsidy of nuclear power. Similarly, the Price-Anderson Act of 1957 seemed an unfair indemnification of the hazards of nuclear reactors because the burden of potential losses would be borne by the taxpayer. "We have been launched headlong into the peaceful atom program," David Pesonen argued a little helplessly, "not by reason but by insurance policy."

Fortunately for opponents of the Bodega reactor, there was one issue that was wide open to specific debate. The plant would be built near the San Andreas Fault, the most active earthquake area in the United States. Just how near was important. PG&E engineers reckoned that it would be 1¼ miles from the fault itself and 1,000 feet from the western edge of the fault zone. They insisted that the reactor would be designed to withstand any earthquake, even one as severe as the 1906 tremor that ruined parts of San Francisco and caused Bodega Head itself, an old-time resident remembered, to "roll like the ocean." Moreover, PG&E geologists maintained that the land had withstood countless numbers of earthquakes with virtually no movement. . . .

But the company soon got some bad scientific news. The Scripps Institute of Oceanography found a new complex of bedrock faults ninety feet seaward of Bodega Head. Another geologist reported that what the company had thought to be a thick pad of granite under the reactor was really sixty feet of silt, clay and sand.[10]

In October, 1964, the Atomic Energy Commission's Division of Reactor Licensing released its official report:

> In our view, the proposal to rely on unproven and perhaps unprovable design measures to cope with forces as great as would be several feet of ground movement under a larger reactor building in a severe earthquake raises substantial safety questions.
> . . . It is our conclusion that Bodega Head is not a suitable location for the proposed nuclear power plant at the present state of our knowledge.[11]

"We have repeatedly stated," responded PG&E President Robert H. Gerdes, "that if any reasonable doubt exists about the safety of the proposed Bodega plant, we would not consider going forward with it." He therefore canceled the project,[12] reaping abundant praise for acting in the public interest. The company had spent between $3 to $4 million at Bodega Head.

The electric utility industry chose to dismiss the whole Bodega incident because it was so "specialized." The AEC, industry leaders were to rationalize, would never, ever, have licensed such a poorly sited plant. But if that were true, why did the AEC allow the situation to snowball for so many months? Whatever the answer may be, the industry was much more interested in other, minor facets of the case. *Nuclear News* magazine implied that a relatively small number of people—most of them "nonlocal"—clearly did not have the real facts (i.e., did not believe PG&E's assurances of safety.) What really interested the magazine was not the merits of the critics' case but where they got their funds. "It is impossible to impute motives to others," the magazine saw fit to comment, "but those close to the situation feel that most of the deep-rooted opposition was from ultraliberal groups who fervently champion federally owned power as opposed to privately financed power."

The nuclear industry bloomed as if nothing had happened at Bodega. Economics was the main reason. Nukes seemed cheaper to operate and fuel than conventional fossil-fired plants. In a euphoria, power companies ordered so many new reactors between 1964 and 1966 that *Electric World* magazine predicted that 56 percent of all generating equipment to come on line between 1971 and 1975 would be nuclear. The coal companies, beginning to worry about the future of their best market, put pressure on railroads to reduce rates for hauling coal to the electric utilities. Even the supercautious AEC was moved to make a prediction. By 1970, it said, there would be 11,500 megawatts of installed nuclear generating capacity. Of course, that forecast depended on everything going well, from the national economy to public acceptance of nukes.

Yet problems have plagued the industry. Equipment manufacturers fell behind schedule in making deliveries. Construction of nuclear plants took longer than expected. Inflation sent all costs spiraling. Sales of new reactors slowed. And environmentalists started to revive the doubts raised first at Bodega Bay.

18
How Safe Is
Diablo Canyon?
David Pesonen

On January 16, 1976, in an excellent piece of investigative journalism by David Perlman of the *San Francisco Chronicle,* Californians learned that, for the fourth time, serious questions of public safety connected with reactor operation in their seismically active state were again revealed at PG&E's proposed Diablo Canyon site at San Luis Obispo.

In view of PG&E's history in nuclear power development along the coast and its chronic problems with elusive earthquake faults, the new revelations at Diablo Canyon are astonishing.

We find that after some six years of construction and the investment of close to a billion dollars of the ratepayers' money, a major earthquake fault between 80 and 150 miles in length lies virtually at the threshold of the Diablo Canyon plant. The history of this elusive fault is interesting for what it reveals about the role both of the atomic industry in California to safeguard the public and the role of the federal regulatory agencies charged with the responsibility of overseeing the safety of nuclear facilities. PG&E tells us that it knew nothing of the notorious Hosgri fault at the time it applied for a construction license. Throughout the entire licensing procedure for this facility, the company made no mention of the fault even though a full geological survey is the utilities' responsibility. We learned further that, according to Carl J. Stepp, chief of seismology for the Nuclear Regulatory Commission, the recent findings of the U.S. Geological Survey respecting the Hosgri fault have been known to the AEC and the NRC for at least a year and a half. Dr. Clarence A. Hall, Jr., chairman of the UCLA geology department, published a report in *Science*[1] which emphasized the proximity of the Hosgri fault to the Diablo Canyon plant. His report, the result of two years of study, only came to public attention through leaks after higher officials in the federal government suggested that references to the Diablo Canyon plant be deleted from his report.

Throughout this entire history no suggestion was made that construction be halted or even slowed at the Diablo Canyon plant. Rather, PG&E was permitted to deepen its commitment to construction of the facility at the ratepayers' expense. The result is that, despite grave questions of public safety posed by this facility, the enormous momentum generated by nearly a billion dollars of investment in the facility leads the Nuclear Regulatory Commission to conclude that "the likelihood of this new data killing the plant is very small."[2]

We must face the fact that either PG&E is unable to find an 80-mile-long earthquake fault at the doorstep of its Diablo Canyon plant, or it has deliberately covered up its discoveries. Neither conclusion inspires any confidence in the industry's ability to safeguard the public.

Several years after PG&E's withdrawal from the Bodega Bay site, the company initiated plans, in 1964, to build another plant on the Mendocino County coast, again a seismically active area. The same assurances that were heard early in the Bodega controversy[3] were again repeated at Mendicino: the site was on solid rock, it was far from known faults, the site had been thoroughly evaluated by experts of the highest qualifications retained by the company, and so forth. Following the filing of its preliminary safety analysis report, one geologist at the U.S. Geological Survey who had done his doctoral thesis at the site used aerial photos from his files to establish the probability of severe seismic activity at the site. He was joined by another expert of the U.S. Geological Survey who discovered that critical data from PG&E's consultants' reports had been deleted from the application filed with the Atomic Energy Commission. Only after this situation was revealed again in the public press—not by the Atomic Energy Commission—did PG&E reluctantly and grudgingly withdraw from the site in 1972.

Finally, at the Humboldt Bay facility, we know now that there is strong evidence of faulting directly under the site of this operating reactor. This information has been known to the NRC for at least the last two years, and no steps have been taken to restrict operations at this facility or to shut it down, despite the fact that it is clearly not in compliance with existing regulations.

In 1973, after the U.S. Geological Survey had been alerted to the evidence of active faulting beneath the site by oil company drillers working in the area, the Survey reviewed the data submitted in connection with PG&E's original operating license[4] for the Humbolt plant and concluded that: (1) "There are inconsistencies and inadequacies of information" in the original application, (2) PG&E's consultants "have not supplied sufficient data to substantiate their case" which asserted that three well-known faults in the area are "inactive," and (3) the structural design of the facility based on PG&E's consultants' data "does not appear to be justified."[5] After a meeting with PG&E over these new findings, the AEC's site analysis branch recommended to L. Manning Muntzing, the AEC director of regulation, that the Humboldt plant be "shut down immediately." The regulatory section took no action. The only response of the NRC and PG&E has been a two-year, foot-dragging, leisurely program of reevaluation of the site, and each progress report undertaken during this period points to confirmation of the Geological Survey revelations. Yet this facility, described by the Advisory Committee on Reactor Safeguards at the time of the original license review as one involving a "high population density relatively close to the site," has continued operation.

Diablo Canyon is not a mere accident, an isolated but costly oversight—it is the fourth step in an unbroken record of conduct by this utility that cries out for a full legislative investigation by a committee endowed with the power of subpoena and the power to compel testimony under oath . . .

Issues and Comments

Three months after Pesonen entered this testimony into the public record, Diablo Canyon suffered another setback. The former QA manager in charge of testing all electrical cable manufactured by Raychem Corporation, Menlo Park, California, destined for installation in Diablo Canyon's control systems, announced that his company's product did not meet several important industry specifications during tests between May, 1973, and February, 1975.

Frederick A. Slautterback, an engineer specializing in quality assurance and employed by Raychem until he resigned in November, 1975, released his report[6] to the Nuclear Regulatory Commission, PG&E, and *Nucleonics Week,* an industry newsletter, in April and May of 1976. Slautterback identified the material in question as Flamtrol, a type of insulation developed by Raychem for its flame resistance and other characteristics, and reported that, from late 1971 until late 1975, Raychem shipped hundreds of miles of Flamtrol-insulated cable to several nuclear projects,* and that some or all of this cable may not meet key specifications based on widely recognized industry standards (IPCEA S-66-524, IEEE Standards 383 and 323, and Raychem Spec 60). This cable is now installed in Diablo Canyon Units 1 and 2.

According to Slautterback, one of the standard requirements that Flamtrol did not meet was an "accelerated water absorption test," where the amount of water absorbed over two weeks is measured by observing changes in insulator electric characteristics. The test requires that the cable be subjected to a continuous voltage of 80 volts per mil (.001 inch) of insulation thickness. For example, 45 mils of Flamtrol is tested at 3600 VAC (volts alternating current). Much of the cable tested failed, according to Slautterback, when this minimal 3,600 volts were applied. A good insulation should hold voltages of at least 20,000 VAC without breakdown.

Both Slautterback and Raychem agree that this failure occurred because electron beam treatment of the Flamtrol polymer during

---

* Arkansas Power & Light Company, Arkansas Nuclear One (Unit 2); Georgia Power Company, Edwin I. Hatch Nuclear Plant (Unit 2); Pacific Gas & Electric Company, Diablo Canyon (Units 1 and 2); Control Wire and Cable. Carolina Power & Light Company, Brunswick (Units 1 and 2); Southern California Edison, San Onofre Nuclear Generating Stations (Unit 1: Rewire Instrumentation and Control Systems; Units 2 and 3: Original Instrumentation and Control Systems); Washington Public Power Supply System (WPPSS Nuclear Project No. 2).

manufacture backfired.[7] Apparently, some of the electrons aimed at the insulation material to cross-link the polymer during the manufacturing process stalled on their way to the conductor in the cable's core. These electrons accumulated and built up millions of small electric charges throughout the length of the cable. After a while, these charges released and passed through the insulation material, burrowing holes that allowed water to enter and short out the conductor during subsequent testing.

Slautterback strenuously objected to shipping cable that did not conform to water absorption specifications and to Raychem's unwillingness to discuss the alleged nonconformance with purchasers of the cable as a first step in taking corrective action.[8] Raychem management ordered him to release the cable for shipment, arguing that the outer waterproof jacket surrounding the Flamtrol material would prevent any water from reaching the Flamtrol underneath. After skirmishes on this issue, Slautterback was asked to leave the QA manager position on February 25, 1975, to work full-time on the water absorption problem. In early November, 1975, he left the company.

On July 1, 1976, *Nucleonics Week* reported that Slautterback's complaints had been "dismissed by NRC, the utilities involved, and Raychem . . . after various tests and inspections." Although Slautterback agrees that there didn't appear to be any danger of immediate, catastrophic failure, he was concerned about long-run cable deterioration. During subsequent interviews he pointed out that:[9]

- If it were necessary for cable to meet industry standards during manufacture and if the cable were not able to pass identical tests immediately after manufacture, then the entire batch of cable should have been rejected.
- NRC's unwillingness in this case to enforce industry standards in the critically important area of cable integrity for Class IE applications* amounts to technical and legalistic wriggling. If NRC upholds the standards on which the tests were based, then Diablo Canyon Units 1 and 2 are wired extensively with degraded cable, and reactor systems safety is open to question. This should yield persuasive grounds for denying Diablo an operating license. It's much easier, however, to throw out the standards than bankrupt PG&E.
- Either the industry standards and QA requirements are precise specifications to assure quality and safety (and they must be enforced to be taken seriously) or they are merely guidelines. If

* Class IE systems are used for the safe and orderly shutdown of the reactor in case of emergency.

government and industry aren't willing to enforce these standards, then they are admitting that the standards are only "guidelines" that can be interpreted or compromised to fit any situation.

If NRC licenses Diablo Canyon, it will be a clear signal to other manufacturers and utilities that the government will permit operation of two 1,000-MWe reactors even though they happen to be near violently seismic areas, and even though the basic components of the plant (instrumentation and control cables in this case) may not meet basic industry specifications.

Despite Slautterback's lone, forthright stand, it is unlikely that the NRC will deny Diablo Canyon an operating license on the strength of only one engineer's objections. As of February, 1977, both the NRC staff and a subcommittee of the Advisory Committee on Reactor Safeguards were reviewing the thorny seismic questions posed by intervenors: How long is the Hosgri fault? What is likely to be the magnitude of an earthquake in the area? and What design basis should be used for Diablo Canyon, now that the Hosgri fault must be reckoned with?

The *length* of the Hosgri fault affects both the severity of an earthquake originating in the fault and the likelihood that earthquakes centered in other, nearby faults may be picked up by the Hosgri fault. Both the AEC and the NRC received reports that the Hosgri fault may link with a nearly 200-mile-long fault to the north, the San Gregorio fault, which in turn connects with the San Andreas fault. According to Dale Bridenbaugh, former General Electric engineer who has been an active intervenor in the Diablo Canyon controversy, the NRC has apparently ignored these reports and consistently understated the consequences of an earthquake on the Hosgri fault for the safe operation of the Diablo Canyon plant.[10]

An earthquake that registered 7.3 on the Richter Scale* shook the entire California coastline from San Francisco to Point Conception in 1927 and tentatively has been attributed to the Hosgri fault. Diablo Canyon is designed for an earthquake force substantially less than a force this large, active fault could deliver to the plant. Consequently, a major earthquake near the plant could cause a catastrophic nuclear accident.

Questions of fault length and earthquake magnitudes may occupy NRC for some time. It is possible that intervenor suits may moot the issue of revised design basis, and that PG&E may simply have

* The Richter Scale is logarithmic: and earthquake measuring 7.3 is *ten times* more severe than one measuring 6.3.

to admit publicly, as it did in the Bodega Bay case, that seismic problems and public safety take precedence over industry profits. A great deal more money is involved in Diablo Canyon, however, and whereas the Bodega plant was cancelled after only the excavation phase was completed, the entire Diablo Canyon facility is nearly finished. Observers of the controversy are betting that the NRC will issue an operating license after additional reinforcements are installed at the plant. If they are proved correct, then the public will be left to ponder this latest variation of nuclear industry management: in the case of the General Electric Mark III containment concept, the system was sold before it was designed. Diablo Canyon may be the first plant on record that had to be significantly redesigned after it was built.

19
Isotopes
and Hogwash
Philip Herrera

Even though a certain amount of background radiation has always been present in nature, the basic fact is that any radiation is dangerous. In large quantities it kills swiftly; in smaller doses it causes cancer in present populations and genetic defects in future ones. The Federal Radiation Council has reported: "There are insignificant data to provide a firm basis for evaluating radiation effects for all types and levels of radiation. There is particular uncertainty with respect to the biological effects of very low doses and dose rates. It is not prudent, therefore, to assume that there is absolute certainty that no effect may occur." The Council laid the cornerstone for establishing U.S. radiation standards in this statement: "There should not be any man-made radiation without the expectation of benefit resulting from such exposure."[1]

Many environmentalists have argued cogently that, since such a concept of risk and benefit ends up involving subjective choice rather than scientific certainty, the affected public should at least be included in the decision of whether to allow a nuclear plant into its community. However, the AEC sets the national standards for maximum permissible doses of radioactivity. These limits are based on the Federal Radiation Council's national guidelines, which in turn reflect the recommendations of both the National Council of Radiation Protection and Measurements, and the International Commission on Radiation Protection.[2] As the nuclear age entered the late 1960s, the limit for exposure to ionizing radiation was established at 170 millirems per year (a millirem is 1/1,000 of a rem) over and above natural background radiation—roughly the equivalent of four chest x-rays a year.[3] It is worth remembering that all nuclear plants operating at that time threw off less radiation than this maximum.

Was the AEC standard adequate? The industry insisted it was. Pressed for more specific information, industry spokesmen responded with such vague comparisons as "the average American is annually exposed to more radiation from sitting in front of his TV set than from a nuclear plant." Most antinuke critics have refused to be convinced by bland assurances. After all, if there was no "safe" level of exposure to radiation, then even 170 millirems was sure to do damage. In 1969, two scientists at the AEC's Lawrence Livermore Laboratory quantified this fear. Dr. John Gofman and Dr. Arthur Tamplin announced that if all Americans were annually exposed to 170 millirems of radiation, there would be an increase of 32,000 deaths from cancer each year. They asked that the AEC reduce the limits by a factor of ten.[4]

The calculations and the logic were soon disputed by other scientists—how could *all* Americans be so exposed?—but Gofman and Tamplin took their thesis before a public avid for intelligible, objective information. Their conclusions, while possibly incorrect, and their continuous criticism of the AEC were important factors in moving the United States toward a needed reduction in radiation standards. Even more significant was the Monticello case.

It is ironic that one of the most important nuclear controversies grew directly out of a power company's enforced awareness of environmental problems. During 1964, Northern States Power (NSP), Minnesota's biggest electric utility, had been roasted both locally and in the national press for its most recent big coal-fired facility on the St. Croix River. Part of the problem was that private boat owners also liked the stretch of river that they would have to share with the plant; another part had to do with thermal pollution; and finally there was a fiercer than usual debate over transmission lines. The whole incident was extremely embarrassing to the company and NSP wanted to avoid a repetition.

When the company found in 1965 that it needed additional generating capacity, it decided to go nuclear. Nukes were then being touted as environmentally "clean" because they do not cause the familiar sorts of air pollution. Better yet, they seemed especially economic. The two major equipment manufacturers, Westinghouse and General Electric, were contracting to build not only the reactor itself but the entire plant at favorable—and fixed—costs. After carefully picking a remote, nonrecreational site at Monticello, thirty-four miles up the Mississippi River from Minneapolis and St. Paul, NSP in 1966 signed an $82-million contract with GE for a 545-megawatt boiling water reactor. With the scribble of executive signatures, the company effectively slipped from the fossil frying pan into the nuclear fire.[5]

"It is difficult to determine now exactly when it was that organized opposition developed to the plant," an NSP official named Bjorn Bjornson was later to recall. "Maybe it stemmed from the precocious plumber who first spoke out against the plant at an early AEC hearing. Or it may have started with the bearded professor who took pot shots at us from his ivory tower." Dr. Dean E. Abrahamson, luxuriously bearded and an associate professor of anatomy and laboratory medicine at the University of Minnesota, might never have gotten even vaguely interested in the Monticello plant had it not been for some questions first raised by the Minnesota Pollution Control Agency (MPCA), responsible for judging Northern States Power's application to discharge industrial wastes from the Monticello facility.

"One day," Abrahamson recalls, "a member of the Pollution Control Agency came to the university and asked if anyone knew anything about reactors. Apparently the agency had its dander up. Some member had asked NSP about radioactive wastes and the company engineer had replied that it was none of MPCA's business —without bothering to explain that radioactivity was the AEC's responsibility. Well, I had worked on reactor design in the late 1950s at Babcock & Wilcox. One thing led to another. I had to read a lot to catch up with the technology."

Among the documents Abrahamson read were some internal GE memos about the kind of reactor NSP would get at Monticello.

Thus, when a GE expert testified—quoting optimistic figures—at a MPCA open hearing in early 1968, Abrahamson rebuked him with GE's own information. "All hell broke loose," Abrahamson says. "No one had ever doubted the veracity of the experts before." Moreover, the incident got his dander up, too.

Together with another University of Minnesota scientist, Abrahamson went on to raise some even more troubling issues at the hearing.[6] Would the discharge of low-level radioactive wastes constitute a threat to marine life in the Mississippi River? The scientists noted that most of the 1.6 million residents of the Twin Cities draw their drinking water from the river a few miles downstream from the Monticello plant's discharge pipes. How much radioactivity would they be taking in with every glass of water? In effect, the focus of discontent with nuclear power plants suddenly shifted from the possibility of major accidents to the reactor's day-to-day operations. Abrahamson's earnest pot shots were shrewdly aimed, and anxious citizens began to organize against the plant.

Northern States Power tried to reassure the people by taking full-page ads stating that its nukes "will meet every applicable safety standard established by the AEC" and that the company stood ready to "spend whatever is necessary to insure the protection of public health and safety." But how safe were the AEC's standards on radioactivity? Not having the expertise to answer such a question, the Pollution Control Agency did precisely what any organization in a quandary would do. In July, 1968, it engaged a consultant, Dr. Ernest C. Tsivoglou of the Georgia Institute of Technology, who soon got the glib nickname of "the rambling Czech from Georgia Tech."

It is true that Tsivoglou took his time about reaching any conclusion. As befits a prudent scientist, Tsivoglou accepted help from members of both sides of the controversy. GE and NSP furnished him with complete data on the Monticello plant, then helped to formulate a series of environmental goals and objectives to cut radioactive wastes as far as technically possible (down to 2 or 3 percent of AEC limits). Meantime, Abrahamson and other university scientists had formed the Minnesota Committee for Environmental Information to provide the public with the continuous technical information which PG&E's opponents at Bodega Bay had so much trouble in finding. Tsivoglou asked the scientists to evaluate the risks and benefits of nuclear power plants and report back to the

MPCA. While their resulting report did not say that the risks outweighed the benefits in so many words, it raised significant doubts. The report's conclusion says it all: "The clearest fact that emerges from this discussion is the uncertainty, the experimental nature of the nuclear program."

That was an understatement. In a letter to *Science* magazine, nuclear critic Sheldon Novick reported that as of 1968, seventeen civilian nuclear power plants had been completed. Of them, five had been shut down, Novick wrote, "as impractical or unsafe; a sixth, the Fermi reactor, was never made to operate properly . . . ; a seventh, the Humbolt plant has operated within allowable radiation release limits only by reducing power output. The remainder have had various degrees of difficulty. . . ." Once doubts were planted about such a fertile subject as radioactivity, they took root and grew. Citizens jumped to obvious conclusions, and the cry went out that low-level radioactivity would disrupt biological rhythms, cause birth defects and increase leukemia deaths in children.

Dr. Tsivoglou then suggested an eminently logical way out of the Minnesota Pollution Control Agency's troubles. In his own final report, released in March, 1969, he changed a key definition. Instead of regarding a standard as a limit below which NSP could dump its low-level radioactive wastes, he urged that the waste discharged from the Monticello plant be held to the lowest possible level. What might that level be? Tsivoglou had NSP's and GE's data to give him the answer. Using the information, he suggested setting new standards—not goals—about fifty times tougher than the AEC's. The MPCA adopted the recommendations in May.[7]

Northern States Power was aghast. "Stated simply," said Bjorn Bjornson, "what was planned as a floor, the Tsivoglou permit made a ceiling." The problem was not really the extra expense that the permit's restrictions would entail, the company insisted. It was that in looking exclusively at the waste aspects of the plant, the new rules neglected other practical considerations. To meet the permit's gaseous release levels, for example, the company would have to shut down the plant with some frequency for fuel changes or modifications. During the shut-downs and start-ups, radioactive emissions would be greater than normal. Moreover, said NSP's president, Robert H. Engels, those fuel changes could expose the plant's personnel to "real and significant increases in radiation doses." As a result, the company criticized the permit as "arbitrary" and "unworkable." Board Chairman Earl Ewald insisted that the chances of anyone's health being damaged by the plant were in the vicinity of one in 100 million. "If this plant would harm anyone," he said, "I would be the first to order it shut down."

Environmentalists, on the other hand, had good reason to praise the strict new standards. Being so much tougher than the AEC's limits, Minnesota's at least began to ease public fears. Moreover, several nuclear critics noted that the AEC had been licensing nuclear power plants as if each were the only one in operation. But it was clear that there were going to be many plants on such major bodies of water as the Mississippi River; Minnesota's standards would lessen the accumulation of low-level wastes flowing downstream. The scientists also warned that radioactivity concentrates in food chains, first in small organisms and then in the fish that eat those organisms. Dr. Donald I. Mount of the National Water Quality Laboratory in Duluth, Minnesota, put it well: "We've got to think of New Orleans."

The AEC kept supplying reassurances. *Our standards take the effect on food chains into account,* AEC scientists said. *The important thing to consider is the concentration and the type of isotope. We don't let any dangerous ones out into the atmosphere. You get more radioactivity in a dentist's office or an airplane than you do from living next door to a nuclear plant.*

But the critics were not easily mollified. Tritium, the long-lived radioactive isotope of hydrogen, would unquestionably leak into the water supply. While the AEC insisted that tritium would not injure human health, independent scientists were not so sure. At worst, tritium might conceivably enter all living organisms with deleterious effects, they said. Krypton, a radioactive gas, might also be distributed throughout the biosphere.

Local citizens sensibly preferred to take the conservative view of minimizing risks. On Mother's Day, 1969, women and children dramatically marched in protest against the plant. "This is getting to be a damned hot political issue," said the citizens group's attorney, Lawrence Cohen. "It's getting to the point where people are going to say, 'We don't want facts; we just don't want you screwing up the environment. All we know is that you're doing it, and we want the right to say you can't.'" *"You can't,"* repeated the politicians, from Governor Harold LeVander to his state legislators, who then praised the state standards and roundly condemned the AEC and NSP. The AEC replied: "Hogwash."

Hogwash? AEC Chairman Glenn Seaborg wrote Governor Le-Vander, saying Minnesota could not legally set its own standards for nuclear plants because Congress gave the agency sole jurisdiction over atomic development. LeVander retorted that "Minnesota will not be content to follow a minimum standard." That state, after all, was only telling the nuclear industry that it had to live up to a technology that was better than the standards controlling it. But the AEC feared a possible precedent, contending that states so lack the expertise to cope with atomic matters that leaving the standards to them would lead to "total and utter chaos." In addition, NSP's Earl Ewald said, "It is impossible for us to function under two divergent regulatory authorities."

At about this time, Dean Abrahamson recalls, "Things started slipping away from the real issue of low-level emissions. Even the question of the AEC's credibility was neglected. The Monticello dispute turned into a battle over states' rights." That might have been true on the national scene, but in Minneapolis–St. Paul, citizens did everything they could to delay granting of all permits to the plant. NSP began to feel desperate about meeting its commitments to supply a burgeoning demand for power. In a sure sign that it was willing to compromise, it voluntarily offered to install extra environmental safeguards.[8]

Meantime Northern States Power took Minnesota and its Pollution Control Agency to court to answer just one question: Does the authority to regulate a nuclear-powered electric generating plant lie solely with the AEC or with states as well?[9] Minnesota argued that under the Tenth Amendment to the U.S. Constitution it had "the power and duty to regulate and to prevent pollution of its lands, waters and air above it." The suit was emphatically supported by several other states, including Vermont, Illinois, Maryland, Texas, Missouri, and Michigan. Indeed, there is something faintly ridiculous about the notion that a state cannot set safer rules than the federal government. The basic theory of federalism is that the states are sovereign (except in clearly defined areas like foreign policy) and history has proved that they can serve extremely well as social laboratories for the entire nation. It was California, for example, not the federal government, that pioneered new laws governing auto-exhaust emissions.

In December, 1970, nevertheless, U.S. District Court Judge Edward J. Devitt ruled that Congress had expressed "an unambiguous mandate to pre-empt the field" for the AEC in a 1959 amendment to the Atomic Energy Act of 1954.[10] Devitt took pains to make clear that he was ruling on a point of law, not on the adequacy or inadequacy of any regulations before him. He did say, however, that "prudence dictates" stiffer standards than the AEC's. Governor LeVander's successor, Wendell Anderson, a Democrat, promptly appealed the decision.

Over the next four months, the last technical and legal objections were resolved at the Monticello plant and the big facility started generating electricity—ten months after it was scheduled to go into commercial operation. The cost of the delay: an estimated $18.9 million in purchases of power from other utilities and in use of NSP's older, inefficient plants.

Aside from the fact that interested environmentalists could quite easily delay the operation, if not the actual construction, of a nuke, the Monticello case proved other points. Back in the fall of 1969, the prominence of the case may well have been a contributing factor in the AEC's decision to confront for the first time its public critics. The agency sent a group of its members and scientists to Vermont to take part in three open forums. It was as surprising—and inconclusive—a move as if the Department of Defense had debated Dr. Benjamin Spock. Perhaps, too, the Monticello case helped to spur the extensive public hearings[11] on "The Environmental Effects of Producing Electric Power" in late 1969 and early 1970. Held before the Joint Congressional Committee on Atomic Energy, the proceedings fill three thick volumes, which indicates, if nothing else, that the environmental problems of nukes are extensive.

Directly, the Monticello case called national attention to questions about radiation standards. Three other states—Illinois, Vermont, and Maryland—eventually followed Minnesota's lead in setting their own tough restrictions on routine radioactive releases from power plants. In total, the attorneys-general or governors of twenty-four states have indicated support of Minnesota's contention before the courts that a state can establish tougher standards than the federal government. Many electric utilities, perhaps anticipating trouble from state governments and environmentalists, have started to submit plans for new reactors complete with added devices to cut radioactive emissions.

In June, 1971, the Atomic Energy Commission took a belated step. It proposed to lower its long-standing and staunchly defended limit on permissible releases of radiation from conventional (light water) reactors by 99 percent.[12] The new rules, the AEC said, would be expressed in reactor design and operating criteria that would hold radioactive releases beyond the plant's site to above five millirems per year above what a nearby resident would get from natural background radiation. This reduction is precisely what Gofman and Tamplin had been urging since 1969, though the AEC insists that it was not responding to their criticism. Rather, the crucial point was whether large new reactors could comply with the lower standards. Apparently, they can. Dr. Gofman hailed the AEC's decision as "a tremendous victory for our position," quickly adding that the nuclear program's problems are still far from solved.

20
Poisoned
Power

John W. Gofman
and Arthur R. Tamplin

Suppose that we have just developed a new, "wondrous" technology with a by-product that is very poisonous. Let's call this poison "Q." If this poison is released into the environment, it may affect hundreds of millions of human beings. How much escaping "Q" are you willing to tolerate in exchange for the benefit of the "wondrous" technology? Of course, you must be concerned not only about whether people will drop dead immediately from exposure to "Q"

Editor's note: Drs. Gofman and Tamplin were among the first to openly criticize the AEC's radiation standards in a public forum. A short list of their internationally recognized publications on this topic appears in the notes to this article. Originally published in 1971, Poisoned Power is a forthright analysis for the layman of the medical aspects of the nuclear controversy and of how our government and the nuclear industry lull the public into accepting less than adequate standards for radiation protection. Dr. Gofman holds two doctorates from the University of California: a Ph.D. in nuclear-physical chemistry and an M.D. Dr. Tamplin graduated from the University of California with a Ph.D. in biophysics. Both were research associates at Lawrence Livermore Radiation Laboratory, Livermore, California; Dr. Gofman was associate director of the laboratory.

but also about possible long-range effects upon the entire human species. Cancer and leukemia cases that might show up in 5, 10, or 25 years must be our immediate concern. Genetic damage that might take generations to show up had certainly *better* worry us.

The promoters of the new technology tell us (in two-page ads in all national magazines) that life on earth will be miserable unless the technology is developed as soon as possible and made available to everybody. These promoters, naturally, want to spend as little money as possible on protecting the people from exposure to "Q," thereby reducing unprofitable capital investment and operating costs. Therefore, they prefer permissive regulations against releasing "Q" into the environment.

How should society decide on the amount of "Q" that is allowed to reach humans? One view suggests that the promoters of any technology have the burden of proof that releasing any "Q" into the environment will not result in harm to humans.* This view also insists that the promoters prove safety beyond a reasonable doubt *before* they release any "Q."

How does the Atomic Industrial Complex actually manage the problem of radioactivity? The promoters of atomic energy and the

---

* Editor's note: The best statement of this view was made in 1957 by Ralph and Mildred Buchsbaum in their book *Basic Ecology* (Pittsburgh, 1957) as follows:

> The religion of economics promotes an idolatry of rapid change, unaffected by the elementary truism that a change which is not an unquestionable improvement is a doubtful blessing. The burden of proof is placed on those who take the "ecological viewpoint:" unless *they* can produce evidence of marked injury to man, the change will proceed. Common sense, on the contrary, would suggest that the burden of proof should lie on the man who wants to introduce a change; *he* has to demonstrate that there *cannot* be any damaging consequences. But this would take too much time, and would therefore be uneconomic. Ecology, indeed, ought to be a compulsory subject for all economists, whether professionals or laymen, as this might serve to restore at least a modicum of balance. For ecology holds that an environmental setting developed over millions of years must be considered to have some merit. Anything so complicated as a planet, inhabited by more than a million and a half species of plants and animals, all of them living together in a more or less balanced equilibrium in which they continuously use and re-use the same molecules of the soil and air, cannot be improved by aimless and uninformed tinkering. All changes in a complex mechanism involve some risk and should be undertaken only after careful study of all the facts available. Changes should be made on a small scale first so as to provide a test before they are widely applied. When information is incomplete, changes should stay close to the natural processes which have in their favour the indisputable evidence of having supported life for a very long time.[1]

agencies setting the standards require, in effect, that the public must prove that it is being harmed by radioactivity before they will stop radioactive pollution. Where environmental poisons are concerned, it has always been up to the public to show harm, rather than up to the polluter to prove safety.

Let's assume that private citizens supporting the conservative, cautious standard prevail. They insist, with excellent reasons, that no "Q" be released to the environment until its safety is established. They are certain to be faced with two of atomic energy's favorite clichés:

"Do you want to stop technological progress?"

"Don't you realize the benefits outweigh the risks?"

Society answers, "Of course we wish to receive all the benefits technological progress can give us, but we insist on knowing the hazards involved. After all, since we are the potential victims, you must convince us that what we stand to gain is greater than what we stand to lose. And if there is a risk, prove to us that we cannot receive the same benefits through less hazardous means."

If the promoters of "Q" technology follow the pattern of the Atomic Industrial Complex, they will answer,

"We just know the benefits are marvelous. The benefits just *must* outweigh the hazards. And furthermore, we have seen no evidence that the amount of "Q" we plan to release will cause cancer, leukemia, and genetic damage to humans."

"But you are not saying that "Q" has been proved safe," the public responds. "Your statement of 'no effects observed' simply reflects your ignorance concerning "Q." If you have made inadequate observations with 'Q,' or none at all, how can you possibly know the answers?"

In answer, the "Q" promoters appoint a body of expert scientists who will hold a long, serious conference and emerge from it with a magic number—plucked out of thin air—a permissible standard for the safe release of "Q." And the public will be told it need have no fears, that the expert standard-setters will be watching the situation carefully. If too many corpses appear, they will confer once more, and set the safe standards for "Q" lower.

The public will certainly denounce the plan: "What utter nonsense it is to release the poison 'Q' into the environment and wait to see what happens! Surely there must be a more rational approach."

The "Q" technologists propose next that they be permitted to release "Q" in *some* amount. Presumably some accidental exposures will occur to sizable groups of humans. The experts plan careful studies of how many cancers, leukemias, and genetic mutations are occurring in exposed humans. "Then we will know precisely how poisonous 'Q' is. If the numbers turn out to be too high, we'll reduce the 'permissible' levels of 'Q.' " (This is precisely what happened with atomic energy. The standard-setters waited for the corpses to appear in Hiroshima survivors *before* they would believe increased cancer occurs in humans exposed to radiation.) Meanwhile, of course, all 230-million people in the country might have been irreversibly injured by the "Q" already released.

Obviously, *disaster* is the fate of a people willing to accept a poison in their environment, hoping that an accident will show them how dangerous the poison is. Worse yet, they will come to realize too late that the technology spawns many "Q" poisons, not one, and that all of them together might mean the end of the human species.

This entire scenario about the new poison "Q" may sound far-fetched. It is a precise description of how the radiation hazard question has been handled in the course of developing atomic energy. Far worse, both the nuclear electricity promoters and the standard-setting bodies still insist vehemently that they must be allowed to proceed with the same idiocy in the future.

Atomic technology was pushed hard by two government agencies: the Atomic Energy Commission and the Joint Committee on Atomic Energy. Accredited biological experts were assembled, on one committee or another, to consider radiation and radioactivity and decide how much people could be exposed to. Obviously, the pressure was on. These expert bodies must have realized that they should burden atomic technology with the fewest possible restrictions.

Did these experts tell the technologists, "The burden of proof of safety is upon *you*"? Did they say, "We refuse to allow you to expose *anyone* to man-made radiation because we don't know how much physical damage it will cause"?

They did not.

Instead, they pulled some numbers out of a hat and declared that the numbers represent "acceptable" standards for human radiation tolerances. And atomic technology proceeded under the blanket of respectability of these "allowable" doses.

By now it is obvious that this level was too high, since these "acceptable" doses have had to be lowered 100-fold in the past two decades. They *have* been lowered, and something certainly was wrong with the original standards. Perhaps the experts *did* know that people wouldn't drop dead immediately from "acceptable" doses they set at first. But for such late effects as leukemia, cancer, and genetic diseases, the "experts" could hardly have been further off-base than they were.

If there had been no information available to the "experts" about the potential danger of cancer and genetic injury to humans, it might be argued that the men who set the standards had no way of knowing that such radiation effects were possible. But the knowledge *was* available. These scientists *knew* that radiation causes cancer and genetic damage. And still, they set totally unacceptable standards! It is impossible to believe anything but that these agencies responded to pressure from atomic technology promoters for "standards we can live with." The technologists were presented with a set of numbers for human exposure that presumably wouldn't make the promoters too unhappy, while those who set them probably prayed the disaster to the human species wouldn't be too severe.

In our discussion of "Q," we expressed dismay that anyone might even *suggest* waiting for some catastrophic consequence of accidental human exposure to evaluate late-effect hazards of the poison. Yet, this is *precisely* what the various standard-setting bodies for radiation are doing and have been doing for many years.

A number of groups of humans actually have been exposed to ionizing radiation in high doses (Hiroshima-Nagasaki bomb survivors, 14,000 British subjects with arthritis of the spine treated with x-rays). The experts have seized upon these groups with enthusiasm. They have announced that as the cancer and leukemia corpses appear in the human groups, they will be counted.

Only when a sufficient number of corpses have appeared, say, from lung cancer, will the experts accept that lung cancer is produced by radiation. If these men determine that too many cancers and leukemias are occurring, the allowable radiation dose to the public will *then* be lowered. Incredible as it may seem, this barbarian approach to public health practice is truly occurring!

The evidence is indisputable. Even though cancer of the lung, the breast, the thyroid, the pharynx, the stomach, the lymph glands and bone have been unequivocally proved to occur in human subjects as a result of radiation, the standard-setting agencies are just beginning to consider some of these concerns in their calculations.

VIEWS
OF THE
FUTURE

21
Clean Energy Alternatives
John Berger

Contrary to the claims of nuclear advocates that the United States is suffering from an energy shortage, the energy available from the sun, wind, oceans, and from the nation's large coal reserves far exceeds our present needs—now or in the foreseeable future. The virtually inexhaustible energy flux that surrounds us is clear evidence that our energy supply is ample. Yet we *do* have a real crisis—a crisis in energy policy and a closely related environmental crisis.

This crisis is evident in the government's failure to invest adequate resources in either clean energy systems or in energy conservation. Solar power and conservation have not been neglected through accident, oversight or simple ignorance. The U.S. energy bureaucracy is unresponsive to citizens' needs for safe, monopoly-free energy because it is overly responsive to the pervasive pressures of giant energy corporations that wield enormous influence. Our current energy policy reflects the desire of these large energy corporations not to disturb the current economic-political *status quo,* which de-

pends on rapid energy growth rates and wasteful energy use. Implicit
in the possibility that energy consumption might be curtailed is the
threat that the more affluent sectors of society, which now use a
disproportionately large share of energy, will be required to limit
their consumption. Also implicit is the idea that corporate profits
will fall as more individuals and small firms build safe, decentralized
energy systems. For this reason, major energy policy reform is a likely
step toward a more pervasive *political* change that could diminish
the influence of large corporations upon our lives.

Before a U.S. energy policy that emphasizes conservation and
clean technologies can gain acceptance, the U.S. must renounce the
ill-advised expansion of nuclear fission. When this extravagantly
costly option is foreclosed, then adequate resources can be devoted
to the development of safe alternatives. Public awareness of these
alternatives must be vigorously nurtured to create the intense public
pressure that will be required to bring about a basic change in U.S.
energy policy. Widespread knowledge of the alternatives can only
hasten the accelerating erosion of public tolerance for nuclear power.
Fortunately, energy conservation facts are easier to grasp than arguments about long-term nuclear hazards.

The U.S. has a massive, hidden energy reserve equal in value to
billions of barrels of oil, trillions of cubic feet of natural gas, or
hundreds of billions of tons of coal. This energy has lain untapped
for eons. Yet to benefit from it, we don't need to invent any fundamentally new energy technology; we don't need to construct vast
numbers of power plants; we don't need to ravage the environment;
we don't need to extract energy from expensive marginal sources
such as oil shale and tar sands.

The energy bonanza is conservation. Its potential is so enormous
that, if employed extensively, it makes the building of uranium
fission plants completely unnecessary.

Dr. George Kistiakowsky, a former science advisor to President
John F. Kennedy, attested recently that "a thirty to forty percent
reduction in energy use is entirely feasible."[1] Many other knowledgeable scientists support this view. Dr. Marc H. Ross, director of
the Center for Environmental Studies at Princeton University, and
Dr. Robert H. Williams, senior scientist for the Ford Energy Project, indicate that forty-five percent of U.S. energy could be saved
by a comprehensive energy conservation effort.[2] If this energy saved

by conservation was all converted to electricity at normal power plant efficiencies, we would save as much electricity as could be produced by 680 nuclear reactors. This would destroy the justification for creating those plants.

To determine how much energy our society truly needs (as opposed to the quantity demanded), it may be helpful to compare the amounts of energy we use with energy consumption data from foreign countries. On the average, each American consumes six times as much energy per person as people in other countries.[3] Although we comprise only six percent of the world's population, we use nearly a third of the total energy consumed worldwide. Citizens in Sweden and Switzerland use less energy per person and have living standards similar to those of people in the United States as measured by their Gross National Products (GNP).[4] Denmark, using roughly half our per capita energy, has a per capita GNP equal to that of the United States. The prosperous Scandinavians are certainly not suffering economically because of their careful energy use. They are merely using energy more efficiently than we are, and their lifestyles are somewhat more energy-conserving as well. Conservation based on energy efficiency can *increase* rather than decrease a country's Gross National Product.

But because Americans have become so dependent on electricity and labor-saving appliances, they tend to believe electric utility advertising that traditionally has equated conservation with deprivation. Energy conservation does not cause valid energy needs to be ignored or denied. It simply means increased energy efficiency instead of shortages; it also means *optimized* energy use so that we can enjoy greater well-being for whatever energy is supplied.

One way to conserve is to reduce energy demand by improving the design of energy-use systems. For example, adding insulation to a home so that less heat is required alters the end-use system by a "leak-plugging" strategy.

A distinctly different conservation approach reduces energy need by improving the efficiency of mechanical devices used for converting energy from less to more useful forms. The Franklin stove, in which the energy in wood is changed to heat, needs less wood than a fireplace to warm a house. An air conditioner with a high performance rating provides more cooling for less power. Both devices enable us to do more with less, not to do without. Neither of these two conservation approaches necessitates changes in lifestyles. However, potentially large energy savings often can be had in exchange for relatively slight changes in the way we do things. Recycling of used newspapers and metals, opting to use large appliances early in the morning or on weekends when power demands are lower, and the use of feet, bicycles or public transport all entail energy-saving lifestyle changes.

A third approach, one based on alternative energy sources, relies on the generation of methane (the main constituent of natural gas) from agricultural wastes and the conversion of urban refuse into heat and power. One writer on agricultural wastes, Farno Green, has even suggested that utilizing two-thirds of U.S. agricultural residues would provide as much energy as one-third of U.S. coal production.

Obviously, none of the three conservation tactics entails direct economic cutbacks to any but the energy industry itself, and cutting back there—by producing less unneeded energy—will have a stimulating effect on the rest of the economy. This follows because, for each dollar invested in the highly mechanized energy sector, relatively few jobs and only a relatively small improvement in GNP result.[5] Alternatively, a dollar invested in manufacturing, commerce or service industries tends to produce far more jobs.

These facts may be obscured by recent industry efforts to convince the public that conservation efforts will impose substantial hardships and cannot possibly yield enough energy savings to meet the nation's voracious demand.[6] But these assertions are based on two assumptions: that large-scale energy conservation will not be employed, and that demand for energy will continue to grow at or near the excessively high rates of the 1965–1973 period. Neither assumption is necessarily true.

As energy prices started to climb in 1974, people cut back on energy consumption and the demand for more electricity began leveling off. In response to the oil embargo and higher oil prices, many companies voluntarily instituted limited energy conservation programs. The results were encouraging: IBM Corporation saved 30 percent of its fuel in 33 major plant, laboratory and headquarters facilities.[7] J.C. Penney Company and Sears, Roebuck & Company's distribution center in Atlanta achieved similar results. Other firms reported greater savings. Yet by no means did these measures involve more than "belt-tightening" measures. The companies merely eliminated waste by turning off unneeded lights and equipment. Thermostat settings were lowered to 68 degrees. No major investments were required. No impairments of plant operations resulted. And no significant lifestyle changes were required of employees, except perhaps to put on a sweater or turn off a light occasionally.

Currently, U.S. industry consumes about 17 percent of the nation's fuels to make steam. A recent National Science Foundation Study coordinated by the Dow Chemical Company found that, if industry would use some of that steam to make electricity, this cogeneration of steam and electricity would save the equivalent of 680,000 barrels of oil a day by 1985, four percent of our current daily oil consumption.[8] That is roughly equivalent to the electrical output of thirty large (1,000 MWe) nuclear power plants.

Improvements in the efficiency of industrial fuel, proper adjustment of combustion equipment and the use of heat recovery systems do not necessarily require long time-lags for implementation. Engineer Charles A. Berg, a Federal Power Commission expert on industrial conservation, pointed out that, "far from requiring 15 years or more for significant changes to be brought about, industrial production can be revolutionized in less than a decade if the incentives are sufficiently strong."[9]

Enormous savings are also possible in the transportation sector, which accounts for about a quarter of all our energy consumption. According to Dr. Edward Teller, a prominent physicist, ". . . except for the mental inertia in Detroit, we could have a fifty-mile-per-gallon car in perhaps as short a time as three years." Whether the cause of Detroit's sluggishness is mental inertia or reluctance to retool because of profit considerations, it is clear that improvements in car mileage would result in substantial energy savings at no inconvenience to the public. And doubling the mileage of all cars would result in saving six percent of our energy (the energy equivalent of 90 nuclear power plants). Additional savings can be achieved by encouraging more people to shift from reliance on one-occupant passenger cars to car pools, buses and trains. At the same time that they reduce ambient air pollution, all these modes of transport require far less gas and oil per passenger.

Much energy can also be conserved in homes, where 19 percent of the nation's energy is consumed. A major study of energy efficiency in buildings by the American Institute of Architects [AIA] concluded that, by thorough insulation and other efficiency measures in both old and new structures, more energy could be saved by 1990 than could be produced either from domestic oil resources (including oil shale) or by a great expansion in nuclear power. More than 12 million barrels of oil a day could be saved, said the AIA, compared with our daily domestic oil production of 11 million barrels.[10]

Simple calculations show that it would take about 500 large nuclear power plants to save those 12 million barrels of oil if there was enough high-grade uranium to fuel these nuclear plants. Construction of these plants would require at least $500 billion, plus additional tens of billions for fuel enrichment and waste management plants. Reprocessing, if federally approved, would require the construction of more facilities. The AIA found that, although the new building standards would be expensive, they would be cheaper over the buildings' lifetimes than expanding fuel supplies; and whereas the initial cost would be higher, the long-run environmental costs of the conservation approach would be decisively less than the costs of expanding the energy supply.

Since conservation would mean that less fuel would be needed, less prospecting, less oil drilling and less strip-mining would be necessary. This would mean less offshore oil drilling, fewer oil spills, fewer supertanker terminals, fewer pipelines, fewer refineries, fewer dusty carloads of coal on the rails, and fewer coal- and oil-storage areas. Dispensing with these developments would create "second-order" environmental benefits. For example, reducing the amount of mining would save scarce water needed for western agriculture, and would avoid disruptive water-diversion projects. Less mining and drilling would slow the industrialization and commercialization of rural resource-extraction areas.

With less resource extraction taking place and less industrial development, fewer roads, factories and trucks would be needed. Naturally, less fuel would be combusted on the road and in power plants, so there would be less air pollution and less ash for disposal from plant sites. Perhaps, also, less land would be needed for power transmission lines (more than 7 million acres have already been set aside for this purpose).[11]

Another major environmental benefit would be that further additions to the heat burden of the earth would be unnecessary. Scientists have found that the total heat from fossil-fuel burning or nuclear fission, in conjunction with air pollution, may trigger major world climatic changes.

Finally, extracting less energy from the earth husbands its limited stock of fossil-fuel deposits, which took hundreds of millions of years to form and accumulate. The complex hydrocarbon chains in petroleum are necessary for many medicines, plastics, fertilizers, and other petrochemicals. One day, future generations may regard our current energy extravagance as not only thoughtless, but antisocial and criminal.

In addition to the environmental benefits, conservation offers some economic advantages. First, it saves money, Secondly, it tends to exert a downward pressure on fuel prices and, to the extent that this reduces consumers' energy expenses, it weakens the energy combines. The reduction in fuel prices could create a dilemma for conservationists, however, in that lower fuel prices will tend to raise energy demand. However, the new equilibrium demand should be less than the preconservation demand level.

As fuel demand falls due to conservation, the average cost of fuels tends to drop, or at least to rise less rapidly. This tends to hold down the rate of inflation throughout the economy because energy is a basic resource essential to nearly all forms of production. Energy conservation also leads to the creation of more jobs. This occurs because the energy industry is highly mechanized with high capital requirements and low labor requirements. As energy conservation begins taking effect, consumers find they have to pay less of their incomes directly to the energy industry. The money saved may be directed toward service or manufacturing industries that produce far more jobs than energy industries.

These energy conservation benefits may be multiplied by alternative energy systems available in a few years. Thermal batteries, heat pumps and *external* combustion engines, such as the Stirling engine, may be the essential components of decentralized energy-use systems of the 1980s. These systems could be linked to solar, wind, geothermal or conventional generating devices with resulting overall fuel efficiencies which are fantastic by conventional standards. Therefore, the fuel requirements of such self-contained energy systems could be spectacularly low, or absent (the Stirling engine can be run on stored solar heat or heat from a windmill-powered heat pump). Although these comprehensive energy systems are still not commercially available, various safe and clean energy sources are useable now. The largest and most obvious of these sources is solar energy.

The sun is actually a gigantic, lustrous fusion reactor, far bigger, hotter and, at a distance of 93 million miles from earth, far safer than man can ever hope to build. It deluges the earth with an enormous energy field, far in excess of our energy needs. While the energy companies talk of energy crises, the surface area of the United States alone receives about nine thousand trillion kilowatt-hours per year in solar energy, an amount roughly 600 times the current energy use. An average of 0.7 kilowatts falls on every square meter of land during daylight—about 64 watts per square foot. Even in the rainy northwest, the equivalent of three times the average household electricity use can be collected from residence roofs.[12] Viewed from a global perspective, the solar energy reaching the Saudi Arabian desert each year equals the world's entire reserves of oil, gas and coal.[13]

The earth's daily solar energy bath not only heats our planet but also provides the basis for plant growth, weather phenomena, and the oceans' currents. Thus wind energy, hydroelectric power and fossil-fuel energy are all actually manifestations of solar energy. The winds occur as a result of atmospheric expansion and contraction caused by solar heating. Hydroelectric power is made possible by solar-initiated water evaporation and subsequent rain. Fossil fuels are but the remains of prehistoric organic compounds transformed by physical and chemical processes into the familiar substances we know as oil, coal and gas. Direct solar energy, in the form of heat and incident sunlight, supplies us with fuel, warmth, cooling, food and fresh water. The versatile sun also can supply us directly with enough heat to propel vapor-driven power plants, including refrigeration devices such as heat pumps.

The technology for using the sun is safe and can be used with minimal damage to the environment and the public. No energy technology is without *any* adverse effects, however. Solar electric power plants will add to the earth's heat burden by increasing the absorption of solar energy and then converting it at relatively low efficiency to electricity with resultant releases of heat.

Unlike fossil or fissile energy, the sun's energy is virtually inexhaustible and can be tapped without seriously depleting limited energy sources, save for those used in constructing and maintaining the solar plant. Since no fossil fuels are burned in the course of plant operation, solar power plants will not foul the air with particulate or radioactive discharges.

Solar power thus has a simplicity and elegance that no other technology can match. Another advantage is that energy is available from the sun in large amounts during the late afternoon when electric demand is greatest. Moreover, a large proportion of the solar energy needed is distributed without cost by nature to the point of use. There is no need to pay an energy conglomerate to bring the sunshine to our homes.

The effective use of solar energy is especially challenging because the sun is a diffuse and intermittent energy source that varies in intensity daily, seasonally, and by geographic region. These characteristics add to the complexity and cost of solar systems. Nevertheless, the problems are being successfully overcome. One reason for this success is that, in contrast to nuclear fission or fusion, the technology required to use most forms of solar energy is quite simple. In fact, solar technology for home heating and cooling has been available for at least 20 years. Today, that technology could "easily contribute 15–30% of the nation's energy requirements," according to the NSF/NASA Solar Energy Panel report published in 1973. The panel reported that "the total electric energy consumed in the United States [in 1969] could have been supplied by the solar energy incident on 0.14 percent of the U. S. land areas . . . under the . . . assumption of a 10 percent conversion efficiency and U. S. average solar incidence." The panel was confident that "solar energy can be developed to meet sizable portions of the nation's future energy needs."[14]

The development of solar power in the United States is closely linked to the nation's nuclear commitment. Because the giant energy corporations know that solar power development could make further nuclear fission plants unnecessary, they are deeply opposed to it—until their gargantuan investments in the nuclear boondoggle can be recovered. Only after trying to exact this economic tribute from society will major corporations begin large-scale commitments to solar power. Meanwhile, the orgy of profiteering on nuclear fission power has robbed taxpayers of the legitimate benefits their energy research and development dollars could have provided. The United States *could* have made a real contribution to world welfare by pursuing a vigorous solar energy development program in the early '50s, instead of chasing a will-o-the-wisp "safe nuclear fission."

In a statement before the Federal Energy Administration's Consumer Affairs/Special Advisory Committee, Edwin Rothschild, representing the Consumers Solar Electric Power Corporation asserted that:

The fossil/nuclear industry is fighting the rapid development of solar electric power which poses such a serious threat to the continued economic and political power of that industry. This explains, in part, why our company and another small company, International Solarthermics, have suffered intimidation and harassment. This explains, in part, why Exxon, Mobil and Shell have bought out small solar electric power companies. This explains, in part, the government's long time-frame for the development on a commercial scale of solar electric power. This explains, in part, why the larger, more powerful companies, especially those which have the most to lose, are getting many of the federal grants in solar energy research and development. . . . I do not, however, believe that the government can continue to cover up the sun.[15]

Indeed, present token solar efforts by several large corporations may be designed more to capture lucrative government grants than to pioneer the solar industry or promote the use of cheap solar energy devices. If so, ERDA's 1977 fiscal year budget for solar energy research, which totaled an unprecedented $278 million, may be spent by these large corporations in non-productive and self-serving ways. As recently as 1974, some of the largest corporations in the United States continued to demonstrate their bias against solar energy. General Electric Corporation, Westinghouse Electric Corporation, and TRW Systems Group made estimates of solar energy potential under grants from the NSF and asserted that solar could make paltry contributions to energy supplies by 2000 A.D. (1.6 percent, 3.0 percent, and 5.7 percent respectively).[16] These estimates conflict with the MITRE Corporation's report to the National Science Foundation during the same year in which MITRE claimed that solar power could contribute as much as 35 percent to our energy budget by 2000 A.D.[17]

Many other industries allied with the nuclear complex, including the electric utilities, deluge the public with "scientific information" that portrays solar energy as an exotic, pie-in-the-sky energy source that is so expensive that it should be dismissed as an impractical energy option for the present. The fears of the big corporations and their research-group acolytes are understandable. For if the bulk of the nation's power can be derived from the sun directly, bypassing the big utilities and the nuclear conglomerates behind them, then a wide range of Atomic Industrial Complex investments may be threatened.

It is still not too late for solar energy to supply most of our energy needs and free us from dependence on nuclear power and foreign energy sources, provided only that the nation makes the *political* decisions to develop it rapidly. The major obstacles are emphatically *not* scientific ones. For ERDA and the nation's largest energy corporations are laying the groundwork so that they can sell us the sun at maximum profit to the energy combine.[18] To oppose this ploy, diligent, sophisticated citizen and activist efforts are needed in order that solar energy contributes to the process of liberating people from the dominance of large corporations.

22
The Scarcity Society
William Ophuls

Historians may see 1973 as a year dividing one age from another. The nature of the changes in store for us is symbolized by the Shah of Iran's announcement in December, 1973, that the price of his country's oil would thenceforth be $11.87 per barrel, a rise of 100 percent over the previous price. Other oil-producing countries quickly followed suit. The Shah accompanied his announcement with a blunt warning to the industrialized nations that the cheap and abundant energy "party" was over. From now on, the resource on which our whole civilization depends would be scarce, and the affluent world would have to live with the fact.

Our first attempts to do so were rather pitiful. In Europe, the effect was to reduce once-proud nation-states to behavior that managed, as one observer put it, to combine the characteristics of an ostrich and a flock of hens. In America, which then lacked almost any observable leadership, the reaction to the statement was merely a general astonishment, followed by measures even more inappropriate than those adopted by the Europeans (except for Kissinger's efforts to promote international cooperation).

In one sense, Iran's move marked a dramatic geopolitical "return of the repressed," as the long-ignored Third World for the first time acted out its demand for a fair share of the planet's wealth. And the powerful new Organization of Petroleum Exporting Countries (OPEC) is only the first such group; resource cartels in copper, tin, bauxite, and other primary products may soon follow OPEC's example. But in another, more important sense, the Shah laid down a clear challenge to the most basic assumptions and procedures that have guided the industrialized democracies for at least 250 years. That challenge is the inevitable coming of scarcity to societies predicated on abundance. Its consequences, almost equally inevitable, will be the end of political democracy and a drastic restriction of personal liberty.

For the past three centuries, we have been living in an age of abnormal abundance. The bonanza of the New World and other founts of virgin resources, the dazzling achievements of science and technology, the availability of "free" ecological resources such as air and water to absorb the waste products of industrial activities, and other lesser factors allowed our ancestors to dream of endless material growth. Infinite abundance, men reasoned, would result in the elevation of the common man to economic nobility. And with poverty abolished, inequality, injustice, and fear—all those flowers of evil alleged to have their roots in scarcity—would wither away. Apart from William Blake and a few other disgruntled romantics, or the occasional pessimist like Thomas Malthus, the Enlightenment ideology of progress was shared by all in the west.* The works of John Locke and Adam Smith, the two men who gave bourgeois political economy its fundamental direction, are shot through with the assumption that there is always going to be more—more land in the colonies, more wealth to be dug from the ground, and so on. Virtually all the philosophies, values, and institutions typical of modern capitalist society—the legitimacy of self-interest, the primacy of the individual and his inalienable rights, economic laissez-faire, and democracy as we know it—are the luxuriant fruit of an era of apparently endless abundance. They cannot continue to exist in their

---

* Marxists tended to be more extreme optimists than non-Marxists, differing only on how the drive to Utopia was to be organized.

current form once we return to the more normal condition of scarcity.

Worse, the historic responses to scarcity have been conflict—wars fought to control resources—and oppression—great inequality of wealth and the political measures needed to maintain it. The link between scarcity and oppression is well understood by spokesmen for underprivileged groups and nations, who react violently to any suggested restraint in economic growth.

Our awakening from the pleasant dream of infinite progress and the abolition of scarcity will be extremely painful. Institutionally, scarcity demands that we sooner or later achieve a full-fledged "steady-state" or "spaceship" economy. Thereafter, we shall have to live off the annual income the earth receives from the sun, and this means a forced end to our kind of abnormal affluence and an abrupt return to frugality. This will require the strictest sort of economic and technological husbandry, as well as the strictest sort of political control.

The necessity for political control should be obvious from the use of the spaceship metaphor: political ships embarked on dangerous voyages need philosopher-king captains. However, another metaphor—the tragedy of the commons—comes even closer to depicting the essence of the ecopolitical dilemma. The tragedy of the commons has to do with the uncontrolled self-seeking in a limited environment that eventually results in competitive overexploitation of a common resource, whether it is a commonly owned field on which any villager may graze his sheep, or the earth's atmosphere into which producers dump their effluents.

Francis Carney's powerful analysis of the Los Angeles smog problem indicates how deeply all our daily acts enmesh us in the tragic logic of the commons:

> Every person who lives in this basin knows that for twenty-five years he has been living through a disaster. We have all watched it happen, have participated in it with full knowledge . . . The smog is the result of ten million individual pursuits of private gratification. But there is absolutely nothing that any individual can do to stop its spread . . . An individual act of renunciation is now nearly impossible, and, in any case, would be meaningless unless everyone else did the same thing. But he has no way of getting everyone else to do it.

If this inexorable process is not controlled by prudent and, above all, timely political restraints on the behavior that causes it, then we must resign ourselves to ecological self-destruction. And the new political strictures that seem required to cope with the tragedy of the commons (as well as the imperatives of technology) are going to violate our most cherished ideals, for they will be neither democratic nor libertarian. At worst, the new era could be an anti-Utopia in which we are conditioned to behave according to the exigencies of ecological scarcity.

Ecological scarcity is a new concept, embracing more than the shortage of any particular resource. It has to do primarily with pollution limits, complex trade-offs between present and future needs, and a variety of other physical constraints, rather than with simple Malthusian overpopulation. The case for the coming of ecological scarcity was most forcefully argued in the Club of Rome study, *The Limits to Growth*. That study says, in essence, that man lives on a finite planet containing limited resources and that we appear to be approaching some of these major limits with great speed. To use ecological jargon, we are about to overtax the "carrying capacity" of the planet.

Critical reaction to this Jeremiad was predictably reassuring. Those wise in the ways of computers were largely content to assert that the Club of Rome people had fed the machines false or slanted information. "Garbage in, garbage out," they soothed. Other critics sought solace in less empirical directions, but everyone who recoiled from the book's apocalyptic vision took his stand on grounds of social or technological optimism. Justified or not, the optimism is worth examining to see where it leads us politically.

The social optimists, to put their case briefly, believe that various "negative feedback mechanisms" allegedly built into society will (if left alone) automatically check the trends toward ever more population, consumption, and pollution, and that this feedback will function smoothly and gradually so as to bring us up against the limits to growth, if any, with scarcely a bump. The market-price system is the feedback mechanism usually relied upon. Shortages of one resource—oil, for example—simply make it economical to substitute another in more abundant supply (coal or shale oil). A few of these critics of the limits-to-growth thesis believe that this process can go on indefinitely.

Technological optimism is founded on the belief that it makes little difference whether exponential growth is pushing us up against limits, for technology is simultaneously expanding the limits. To use the metaphor popularized during the debate, ecologists see us as fish in a pond where all life is rapidly being suffocated by a water lily that doubles in size every day (covering the whole pond in thirty days). The technological optimists do not deny that the lily grows very quickly, but they believe that the pond itself can be made to grow even faster. Technology made a liar out of Malthus, say the optimists, and the same fate awaits the neo-Malthusians. In sum, the optimists assert that we can never run out of resources, for economics and technology, like modern genii, will always keep finding new ones for us to exploit or will enable us to use the present supply with ever-greater efficiency.

The point most overlooked in this debate, however, is that politically it matters little who is right: the neo-Malthusians *or* either type of optimist. If the "doomsdayers" are right, then of course we crash into the ceiling of physical limits and relapse into a Hobbesian universe of the war of all against all, followed, as anarchy always has been, by dictatorship of one form or another. If, on the other hand, the optimists are right in supposing that we can adjust to ecological scarcity with economics and technology, this effort will have, as we say, "side effects." For the collision with physical limits can be forestalled only by moving toward some kind of steady-state economy—characterized by the most scrupulous husbanding of resources, by extreme vigilance against the ever-present possibility of a disastrous breakdown, and therefore, by tight controls on human behavior. However we get there, "Spaceship Earth" will be an all-powerful Leviathan—perhaps benign, perhaps not.

The scarcity problem thus poses a classic dilemma. It may be possible to avoid crashing into the physical limits, but only by adopting radical and unpalatable measures that, paradoxically, are little different in their ultimate political and social implications from the future predicted by the doomsdayers.

Why this is so becomes clear enough when one realizes that the optimistic critics of the doomsdayers, whom I have artificially grouped into "social" and "technological" tendencies, finally have to rest their different cases on a theory of politics, that is, on assumptions about the adaptability of leaders, their constituencies, and the institutions that hold them together. Looked at closely, these assumptions also appear unrealistic.

Even on a technical level, for example, the market-price mechanism does not coexist easily with environmental imperatives. In a market system, a bird in the hand is always worth two in the bush.* This means that resources critically needed in the future will be discounted—that is, assessed at a fraction of their future value—by today's economic decision-makers. Thus decisions that are economically "rational," like mine-the-soil farming and forestry, may be ecologically catastrophic. Moreover, charging industries—and therefore consumers—for pollution and other environmental harms that are caused by mining and manufacturing (the technical solution favored by most economists to bring market prices into line with ecological realities) is not politically palatable. It clearly requires political decisions that do not accord with current values or the present distribution of political power; and the same goes for other obvious and necessary measures, like energy conservation. No consumer wants to pay more for the same product simply because it is produced in a cleaner way; no developer wants to be confronted with an environmental impact statement that lets the world know his gain is the community's loss; no trucker is likely to agree with any energy-conservation program that cuts his income.

We all have a vested interest in continuing to abuse the environment as we have in the past. And even if we should find the political will to take these kinds of steps before we collide with the physical limits, then we will have adopted the essential features of a spaceship economy on a piecemeal basis—and will have simply exchanged one horn of the dilemma for the other.

Technological solutions are more roundabout, but the outcome —greater social control in a planned society—is equally certain. Even assuming that necessity always proves to be the mother of invention, the management burden thrown on our leaders and institutions by continued technological expansion of that famous fishpond will be enormous. Prevailing rates of growth require us to double our capital stock, our capacity to control pollution, our agricultural productivity, and so forth every fifteen to thirty years. Since we already start from a very high absolute level, the increment of required new construction and new invention will be staggering. For example, to accommodate world population growth, we must, in roughly the next thirty years, build houses, hospitals, ports, factories,

---

* Of course, noneconomic factors may temporarily override market forces, as the current Arab oil boycott illustrates.

bridges, and every other kind of facility in numbers that almost equal all the construction work done by the human race up to now.

The task in every area of our lives is essentially similar, so that the management problem extends across the board, item by item. Moreover, the complexity of the overall problems grows faster than any of the sectors that comprise it, requiring the work of innovation, construction, and environmental management to be orchestrated into a reasonably integrated, harmonious whole. Since delays, planning failures, and general incapacity to deal effectively with even our current level of problems are all too obvious today, the technological response further assumes that our ability to cope with large scale complexity will improve substantially in the next few decades. Technology, in short, cannot be implemented in a political and social vacuum. The factor in least supply governs, and technological solutions cannot run ahead of our ability to plan, construct, fund, and man them.

Planning will be especially difficult. For one thing, time may be our scarcest resource. Problems now develop so rapidly that they must be foreseen well in advance. Otherwise, our "solutions" will be too little and too late. The automobile is a pertinent example. By the time we recognized the dangers, it was too late for anything but a mishmash of stopgap measures that may have provoked worse symptoms than they alleviated and that will not even enable us to meet health standards without painful additional measures like rationing. But at this point we are almost helpless to do better, for we have ignored the problem until it is too big to handle by any means that are politically, economically, and technically feasible. The energy crisis offers another example of time lag, for even with an immediate laboratory demonstration of feasibility, nuclear fusion cannot possibly provide any substantial amount of power until well into the next century.

Another planning difficulty: the growing vulnerability of a highly technological society to accident and error. The main cause for concern is, of course, some of the especially dangerous technologies we have begun to employ. One accident involving a breeder reactor would be one too many: the most minuscule dose of plutonium is deadly, and any we release now will be around to poison us for a quarter of a million years. Thus, while we know that counting on perfection in any human enterprise is folly, we seem headed for a society in which nothing less than perfect planning and control will do.

At the very least, it should be clear that ecological scarcity makes "muddling through" in a basically laissez-faire socioeconomic system no longer tolerable or even possible. In a crowded world where only the most exquisite care will prevent the collapse of the technological society on which we all depend, the grip of planning and social control will, of necessity, become more and more complete. Accidents, much less the random behavior of individuals, cannot be permitted; the expert pilots will run the ship in accordance with technological imperatives. Industrial man's Faustian bargain with technology therefore appears to lead inexorably to total domination by technique in a setting of clockwork institutions. C. S. Lewis once said that "what we call Man's power over Nature turns out to be a power exercised by some men over other men with Nature as its instrument," and it appears that the greater our technological power over nature, the more absolute the political power that must be yielded up to some men by others.

These developments will be especially painful for Americans because, from the beginning, we adopted the doctrines of Locke and Smith in their most libertarian form. Given the cornucopia of the frontier, an unpolluted environment, and a rapidly developing technology, American politics could afford to be a more or less amicable squabble over the division of the spoils, with the government stepping in only when the free-for-all pursuit of wealth got out of hand. In the new era of scarcity, laissez-faire and the inalienable right of the individual to get as much as he can are prescriptions for disaster. It follows that the political system inherited from our forefathers is moribund. We have come to the final act of the tragedy of the commons.

The answer to the tragedy is political. Historically, the use of the commons was closely regulated to prevent overgrazing, and we need similar controls—"mutual coercion, mutually agreed upon by the majority of the people affected," in the words of the biologist Garrett Hardin—to prevent the individual acts that are destroying the commons today. Ecological scarcity imposes certain political measures on us if we wish to survive. Whatever these measures may turn out to be—if we act soon, we may have a significant range of responses—it is evident that our political future will inevitably be much less libertarian and much more authoritarian, much less individualistic and much more communalistic than our present. The likely result of the reemergence of scarcity appears to be the resurrection in modern form of the preindustrial polity, in which the few govern the many and in which government is no longer of or by the people. Such forms of government may or may not be benevolent. At worst, they will be totalitarian, in every evil sense of that word. At best, government seems likely to rest on engineered consent, as we are manipulated by Platonic guardians in any one of several versions of Brave New World. The alternative will be the destruction, perhaps consciously, of "Spaceship Earth."

There is, however, a way out of this depressing scenario. To use the language of ancient philosophers, it is the restoration of the civic virtue of a corrupt people. By their standards, or by the standards of many of the men who founded our nation (and whose moral capital we have just about squandered), we are indeed a corrupt people. We understand liberty as a license for self-indulgence, so that we exploit our rights to the full while scanting our duties. We understand democracy as a political means of gratifying our desires rather than as a system of government that gives us the precious freedom to impose laws on ourselves—instead of having some remote sovereign impose them on us without our participation or consent. Moreover, the desires we express through our political system are primarily for material gain: the pursuit of happiness has been degraded into a mass quest for what wise men have always said would injure our souls. We have yet to learn the truth of Burke's political syllogism, which expresses the essential wisdom of political philosophy: man is a passionate being, and there must therefore be checks on his will and appetite. If these checks are not self-imposed, they must be applied externally as fetters by a sovereign power. The way out of our difficulties, then, is through the abandonment of our political corruption.

The crisis of ecological scarcity poses basic value questions about man's place in nature and the meaning of human life. It is possible that we may learn from this challenge what Lao-tzu taught two-and-one-half millennia ago:

> Nature sustains itself through three precious principles, which one does well to embrace and follow.
> These are gentleness, frugality, and humility.

A very good life—in fact, an affluent life by historic standards —can be lived without the profligate use of resources that characterizes our civilization. A sophisticated and ecologically sound technology, using solar power and other renewable resources, could bring us a life of simple sufficiency that would yet allow the full expression of the human potential. Having chosen such a life, rather than having had it forced on us, we might find it had its own richness.

Such a choice may be impossible, however. The root of our problem lies deep. The real shortage with which we are afflicted is that of moral resources. Assuming that we wish to survive in dignity and not as ciphers in some antheap society, we are obliged to reassume our full moral responsibility. The earth is not just a banquet at which we are free to gorge. The ideal in Buddhism of compassion for all sentient beings, the concern for the harmony of man and nature so evident among American Indians, and the almost forgotten ideal of stewardship in Christianity point us in the direction of a true ethics of human survival—and it is toward such an ideal that the best among the young are groping. We must realize that there is no real scarcity in nature. It is our numbers and, above all, our wants that have outrun nature's bounty. We become rich precisely in proportion to the degree in which we eliminate violence, greed, and pride from our lives. As several thousands of years of history show, this is not something easily learned by humanity, and we seem no readier to choose the simple virtuous life now than we have been in the past. Nevertheless, if we wish to avoid either a crash into the ecological ceiling or a tyrannical Leviathan, we must choose it. There is no other way to defeat the gathering forces of scarcity.

23
A Conservation Ethic
David Brower

The Third Planet: Operating Instructions

This planet has been delivered wholly assembled and in perfect working condition, and is intended for fully automatic and trouble-free operation in orbit around its star, the sun. However, to insure proper functioning, all passengers are requested to familiarize themselves fully with the following instructions.

Loss or even temporary misplacement of these instructions may result in calamity. Passengers who must proceed without the benefit of these rules are likely to cause considerable damage before they can learn the proper operating procedures for themselves.

A. Components

It is recommended that passengers become completely familiar with the following planetary components:

1) Air

The air accompanying this planet is not replaceable. Enough has been supplied to cover the land and the water, but not very deeply. In fact, if the atmosphere were reduced to the density of water, then it would be merely 33 feet deep. In normal use, the air is self-cleaning. It may be cleaned, in part, if excessively soiled. The passengers' lungs will be of help—up to a point. However, they will discover that anything they throw, spew, or dump into the air will return to them in due course. Since passengers will need to use the air, on the average, every five seconds, they should treat it accordingly.

2) Water

The water supplied with this planet isn't replaceable either. The operating water supply is very limited: If the earth were the size of an egg, all the water on it would fit in a single drop. The water contains many creatures, almost all of which eat and may be eaten; some of these creatures may be eaten by human passengers. If disagreeable things are dispersed in the planet's water, however, caution should be observed, since the water creatures concentrate the disagreeable things in their tissues. If human passengers then eat the water creatures, they will add disagreeable things to their diet. Passengers are advised especially not to disdain water, because that is what they mostly are.

3) Land

Although the surface of this planet is varied and seems abundant, only a small amount of land is suited to growing things. That portion is essential and should not be misused. It is also recommended that no attempt be made to penetrate the surface too deeply inasmuch as the land is supported by a molten and very hot underlayer that will grow little but volcanoes.

4) Life

The above components help make life possible. There is only one life per passenger and it should be treated with dignity. Instructions covering the birth, operation and maintenance, and disposal for each living entity have been thoughtfully provided. These instructions are contained in a complex language, called the DNA code, that is not easily understood. However, this does not matter, as the instructions are fully automatic. Passengers are cautioned, however, that radiation and many dangerous chemicals can damage the instructions severely. If, in this way, living species are destroyed, or rendered unable to reproduce, the filling of reorders is subject to long delays.

5) Fire

This planet has been designed and fully tested at the factory for totally safe operation with fuel constantly transmitted from a remote source, the sun, provided at absolutely no charge. *The following must be observed with greatest care:* The planet comes with a limited reserve fuel supply, contained in fossil deposits, which should be used only in emergencies. Use of this reserve fuel supply entails hazards, including the release of certain toxic materials which must be kept out of the air and the food supplies of living things. The risk will not be appreciable if the use of the emergency fuel is extended over the operating life of the planet. Rapid use, if sustained only for a brief period, may produce unfortunate results.

B. Maintenance

The kinds of maintenance will depend upon the number and constituency of the passengers. If only a few million human passengers wish to travel at a given time, no maintenance will be required, and no reservations will be necessary. The planet is self-maintaining, and the external fuel source will provide exactly as much energy as is needed or can be safely used. However, if a very large number of people insist on boarding at one time, serious problems will result, requiring costly solutions.

C. Operation

Barring extraordinary circumstances, it is necessary only to observe the mechanism periodically and to report any irregularities to the Smithsonian Institution. However, if, owing to misuse of the planet's mechanism, observations show a substantial change in the patterns of sunrise and sunset, passengers should prepare to leave the vehicle.

D. Emergency Repairs

If, through no responsibility of the current passengers, damage to the planet's operating mechanism has been caused by ignorant or careless action of previous travelers, it may be appropriate to request the Manufacturer's assistance (best obtained through prayer).

Upon close examination, this planet will be found to consist of complex and fascinating detail in design and structure. Some passengers, upon discovering these details in the past, have attempted to replicate or improve the design and structure, or have even claimed to have invented them. The Manufacturer, having among other things invented the opposable thumb, may be amused by this. However, it is reliably reported that, at this point, it appears to the Manufacturer that the full panoply of consequences of this thumb idea of His will not be without an element of unwelcome surprise.

Appendix A

Testimony of Dale G. Bridenbaugh, Richard B. Hubbard, and Gregory C. Minor before the Joint Committee on Atomic Energy February 18, 1976

I. INTRODUCTION

When we first joined the General Electric Nuclear Division, we were very excited about the idea of this new technology—atomic power—and the promise of a virtually limitless source of safe, clean and economic energy for this and future generations. There was a sense of excitement in the industry that approached a missionary zeal in those early days. Like many of our colleagues, we felt that the results of our work would provide major benefits to mankind.

But now, after a combined total of more than 40 years of experience in all facets of the sale, design, manufacture, construction, and operation of nuclear power plants, we see that the vision has faded. The promise is still unfulfilled.

The nuclear industry has developed to become an industry of narrow specialists, each promoting and refining a fragment of the technology, with little comprehension of the total impact on our world system.

On February 2, 1976, we simultaneously resigned our management positions from the General Electric Company.

We did so because we could no longer justify devoting our life energies to the continued development and expansion of nuclear fission power—a system we believe to be so dangerous that it now threatens the very existence of life on this planet.

We could no longer rationalize away the fact that our daily labor would result in a radioactive legacy for our children and grandchildren for hundreds of thousands of years. We could no longer resolve our continued participation in an industry which will depend upon the production of vast amounts of plutonium, a material known to cause cancer and produce genetic effects, and which facilitates the continued proliferation of atomic weapons throughout the world.

We know that this Committee has heard abundant testimony over the past 30 years on these aspects of nuclear power, but we feel it is important to express our deep concern about the entire technology before turning to the specifics of our experience.

We resigned our jobs to commit ourselves totally to the education of the public on all aspects and dangers of nuclear power as we have learned them over our many years of experience in the industry.

We hope that this Committee, in exercise of its legislative oversight function, will provide the vigorous safety evaluation this program so desperately needs.

II. NUCLEAR POWER—DEFICIENCIES IN MANY AREAS COMBINE TO MAKE IT UNSAFE

The nation is continually assured by the industry, the power plant owners, and the NRC that nuclear power plants are designed to be very safe. Overlapping emergency systems and redundant components are incorporated into the plant design to reduce the consequences of a single malfunction, and hence, to increase plant reliability. Complex mathematical models of the plants are then promoted to the public as predicting a very low probability and impact of a major accident. The most recent example, of course, is WASH-1400, the "Rasmussen Report."

But "actual" performance does not meet the "theoretical" projections. The unplanned anomaly is the "human factor"; for example, the designers who had not properly considered all the relevant design parameters. The result is an incomplete or uncoordinated design.

The industry, with the concurrence of the NRC, has overemphasized the theoretical approach in design verification with insufficient prototype, laboratory, or field-test verification. The result is inadequate and unsafe design.

In the implementation of today's regulatory process, there is a tendency to review each safety concern separately and to then conclude that the specific concern, by itself, does not present an undue safety hazard to the public. But what attention is given to the summations of all the individual safety hazards? How many specific safety concerns does it take to conclude that the whole system is unsafe?

A. Design Defects

Through our experiences we are aware of many design deficiencies and unresolved regulatory issues that have a serious impact on the safety of nuclear power plants. We have listed herein some examples of safety hazards and NRC regulation inadequacies in the design, manufacture, construction, operation, and maintenance of a nuclear facility. The examples of design deficiencies in nuclear plants in operation and under construction are grouped as follows: core, control rods, reactor vessel, primary containment, materials, and supporting systems.

1. FLOW-INDUCED VIBRATION IN THE CORE

Light water reactors have been plagued by numerous flow-induced vibration problems in both BWR's and PWR's. Such problems result from the continual, rapid flow of cooling water through the complex core geometry. Figure 1 will assist in understanding the problem.

The first instance where this problem occurred involved the core shroud and thermal shield seal at the Big Rock Point Plant in Northern Michigan in 1964. The repair of this flow-induced vibration problem required approximately one year of outage time and millions of dollars.

Flow-induced vibrations have been experienced at PWR's as well, and, as a matter of fact, PWR's are somewhat more susceptible, as they are designed with higher flow velocities in the reactor vessel than occur in boiling water reactors. Flow-induced vibration problems at PWR's were experienced at, among others, Duke Power's Oconee Plant in 1972, where failure of an in-vessel component resulted in severe damage to the tube sheet of a steam

General Electric Reactor Pressure Vessel

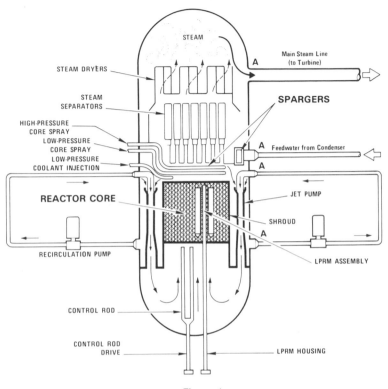

Figure 1

generator, at the Rochester G & E Ginna Plant in 1971, and at the Chooz PWR and Sena PWR Plants in Europe in the late 1960s.

a. Sparger Failure

The fatigue failure of the in-vessel spargers is a continuing flow-induced vibration problem. A sparger is generally a large heavy-walled pipe designed to distribute fluid uniformly within the vessel. (See Figure 1.)

An early example was the fatigue failure of the liquid poison sparger at the Garigliano Plant in Italy in 1964. The failure was first discovered when leakage was noted coming from an in-core housing beneath the reactor vessel. Investigation revealed that the poison sparger had failed, and pieces of the sparger had dropped to the bottom of the vessel. Flow-induced vibration caused the sparger piece to rub a hole through the in-core housing, resulting in the leakage of water through the housing. The "fix" involved was extensive. But, of greater significance, was the fact that because the poison sparger could not readily be replaced, an analysis was performed which indicated that the sparger was not really necessary to distribute the boron poison throughout the core should emergency shutdown require it. It was decided that continued operation was permissible without the sparger.

This is but one example of the subtle undermining of the safety margins that occur during the operation of a nuclear plant. The plant becomes so radioactively hot and inaccessible that it is very difficult to make even normal repairs. As a result, shortcuts are often taken that, in the long run, jeopardize the safety of the plant.

The failure of the liquid poison sparger was the beginning of a series of flow-induced vibration sparger problems. Within the last year, feed-water sparger failures have been observed at a number of boiling water reactors: Northeast Utilities' Millstone Point Plant, Northern State Power's Monticello Plant, Boston Edison's Pilgrim Plant and Commonwealth Edison's Dresden and Quad Cities Plants. In each case, vibration of the sparger, induced by the flow of the feed-water through it, resulted in early failure of the sparger and required redesign and replacement of the sparger under extremely difficult field conditions. Such replacements required significant personnel radiation exposure and outage time.

The cracking and breaking of the sparger creates a very unsafe condition. No way has been developed to provide on-line detection of this failure. How many existing plants have this defect? How will we discover the defects before it is too late?

b. Local Power Range Monitor (LPRM) Failure

Another recent problem has been the local power range monitor (LPRM) vibration. The LPRM is an instrument assembly which is spaced periodically throughout the reactor core to measure local power distribution in the core. (See Figure 2.) It is the key neutron sensor of the reactor safety system. During the past year, 11 BWR-4's have experienced failures of

LPRM's and the associated fuel bundle channels resulting from vibration of the LPRM assemblies against the channels. LPRM impact upon the channel caused cracking of the channel and degradation of the signals from the LPRM assembly itself.

The impacts of this problem are manifold. First of all, when the signal from the LPRM is degraded, the effectiveness of the entire power monitoring system is degraded. It also has a significant impact on the loss-of-coolant accident (LOCA). The vibration of the LPRM assembly on the channel has resulted in failure of channels in at least two plants: the Vermont Yankee and NPPD-Cooper Plants. Cracks occurred in

Reactor Core Assembly

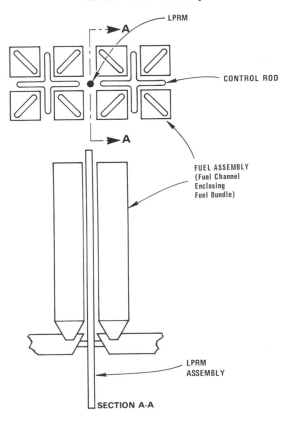

Figure 2

the channels and large pieces have actually broken out of them. This failure impacts on the loss-of-coolant accident since the channels are required to insure the uniform distribution of the emergency core cooling water over the fuel rods. The source of the flow-induced vibration has been traced to the cooling flow passages for the LPRM's and field plugging of these holes has been initiated. Plugging of these passageways changes the ability of the core to reflood following an accident, and has required derating of the plants to various levels to ensure that the calculated accident consequences are not exceeded. A "fix" has been proposed where holes will be drilled in the lower portion of the fuel bundles to provide the necessary cooling flow. This is a very difficult modification since holes must be drilled in the highly irradiated fuel bundles; damage to the fuel is possible, and substantial amounts of outage time are required for the operation. Meanwhile, all eleven BWR-4 plants which have this condition continue to run in a derated condition.

During the testing involved in the development of this LPRM "fix," another surprising fact was discovered. The pressure drop across the core plate and the fuel bundles was sufficient to cause more than the calculated rounding of the square fuel channels. This results in increased bypass-leakage flow and subsequent deterioration of the operating thermal margins of the reactor fuel. The concern here is that all operating limits are based on the theoretically-calculated quantity of cooling water through the core. Discovery of the leakage caused consternation within the affected design groups and created yet another unsafe condition in the eleven BWR-4 plants.

c. Effectiveness of Core Spray

Another area of concern in all BWR's is the effectiveness of core spray in actually cooling the fuel rods following a loss-of-coolant accident. A number of cold tests have been conducted by General Electric in measuring core spray distribution in terms of the geometry of the water patterns above the mocked-up fuel core. But no actual thermal tests have been performed to determine that sufficient quantities of cooling water will be delivered to the hot fuel rods to cool them in seconds—as the system must do to prevent a core meltdown following a loss-of-coolant accident.

Word has been circulating around the industry that tests on this phenomenon have been performed in Europe which show that the steam blasting up out of the fuel bundle during an accident will prevent the delivery of cooling water to the hot rod. This raises a question as to the effectiveness of the core spray cooling system for the reactor. Have the European tests been evaluated by NRC? With what results?

d. End-of-Cycle-SCRAM Reactivity Effect

Another aspect of core performance that leads to reactor safety questions is the phenomenon called "End-of-Cycle-SCRAM Reactivity Effect." In the BWR, as the core nears the end-of-cycle (approximately annually), the reactor operates with all control rods essentially fully withdrawn from the core. It has recently been discovered that more severe transients can occur under such end-of-cycle conditions with load rejection. This results because the negative reactivity resulting from rod insertion is not linear, as was originally assumed, but is effective only as the rods near the end of their insertion. The result is a surge —a higher power transient following the load rejection, a higher pressure transient to the reactor primary system, and much more severe conditions within the core during the SCRAM than were originally anticipated. "Fixes" have been proposed for this phenomenon, but have yet to be implemented. End-of-Cycle-SCRAM Reactivity Effect is another example of conditions encountered during the course of operation which are much different from those assumed during the original design phase.

When a similar effect was observed at the Garigliano Plant in Italy, the "fix" that was applied was to develop a filter system to trim out the peak-flux spike in the flux monitoring instrumentation measuring the surge. This is a typical example of how the industry rigs a patched-up safety system in order to keep operating. Rather than to cure the defect, the "fix" proposed was to modify the instrumentation which senses phenomenon so as not to SCRAM the reactor.

What "fix" is proposed for the End-of-Cycle-SCRAM Reactivity Effect?

Will the NRC require the retrofit to be performed in all BWR's?

2. CONTROL ROD

The control rod system is a critical component in plant safety. It is the system that is normally used to start up and shut down the reactor, and to vary power levels during operation. It is also used as an emergency system when loss of electrical load occurs and a reactor emergency shutdown is indicated; control rods then must be rapidly inserted to reduce the power to the level where the shutdown cooling systems can handle the energy that continues to be produced.

a. Control Rod Life

Control rods have caused problems in reactors from the earliest days. In 1960, the Dresden-1 reactor control rods were found to have extensive cracking after minimal irradiation, and complete replacement of these rods was required after approximately six months of operation. Control rods similar to the replacement design in Dresden-1 are still being used in today's plants. Despite the critical function of the rods in reactor operation, the life of the control rods has yet to be verified. It has been postulated that control rod life may be anywhere from seven to twenty years, but no one really knows.

The end of control rod life may result from two effects: (1) nuclear depletion, which is the inability of the boron contained in the control rods to absorb enough neutrons to provide the shutdown required, or (2) end of the mechanical life of the control rods resulting from material embrittlement or pressure stresses in the boron-containing tubes.

In some Dresden-1 sample rods, it has been found that the (Type 304) stainless steel rods cracked and that the powdered boron leached out of the rods. There is no way to detect when the cracking occurs, and the only way to detect when the boron has been lost from a rod is to do detailed rod-worth tests on each individual control rod in a test configuration. This is not a normal license requirement or condition of operation for any reactor. Therefore, it is entirely possible that the end of control rod life may be reached without the knowledge of the plant staff until the reactor is called upon to shut down, and the control rod worth is discovered to be insufficient to produce shutdown. Control rod life must be determined and the technical specification requirements established to ensure that shutdown margins are adequately tested to account for these effects.

b. Collet Cylinder Tube Cracking

Material failures in the control rod drives also affect the ability of the plant to perform safely. Material failures were experienced early at Dresden-1. Modifications were made which appear to be performing satisfactorily, but similar problems have recently appeared again. One such problem is the cracking of a collet cylinder tube on the control rod drive which could affect the ability of the drive to perform properly.

This cracking has resulted in the issuance of an order to limit the number of inoperable drives permissible for continued operation, but ultimately will require the removal and rebuilding of all control rod drives in operation. This will result in a large penalty in operating time, personnel radiation exposure, and high costs, affecting the economic viability of continued operation of the reactor. Other material problems can be expected to occur in the future with serious potential impact to reactor safety.

c. Rod-Drop Accident and Patches

There is a classic accident analyzed for BWR's and PWR's. It is called a rod-drop accident for the BWR, and a comparable rod-ejection accident for the PWR. This is a case where the rod becomes disconnected from the drive and remains stuck in an inserted position when the drive is withdrawn. The accident is caused by the rod moving from the stuck position to the withdrawn position in one quick motion. This would cause a rapid change in local flux (for a high-worth rod) and likely produce local fuel damage, a rapid increase in local power and possible release of fission by-products.

Historically, the industry has played down the possibility of this accident, yet has added various system patches to mitigate the consequences of such an event. The approach of adding an electronic patch to overcome a mechanical deficiency which is otherwise impractical to solve is a common occurrence in the industry. In this case, systems were added to control via procedures the pattern of rod withdrawal and block withdrawal of high-worth rods out of sequence.

These somewhat ineffective patches have added to the complexity of operating the reactor and often wind up bypassed, ignored or circumvented. Specifically, this includes the Rod Worth Minimizer (RWM), the Rod Block Monitor (RBM) and the Rod Sequence Control (still being defined).

This accident must now be regarded with increased respect and a higher probability of occurrence in view of the chronic problems plaguing the control rod mechanisms and cracking fuel channels—events which make a stuck, disengaged rod more credible. Thus, it is recommended that either the problem be solved or the mitigating systems be improved and made mandatory for operation.

3. PRESSURE VESSEL INTEGRITY

Many questions concerning pressure vessel integrity have been raised in the past and many more undoubtedly will arise in the future. All have important implications concerning the safety of operating nuclear plants. One important concern is the possibility of a breach of the pressure vessel.

The incredibility of a gross-pressure-vessel-failure accident has not been—indeed, cannot be—proven. Yet incredibility is assumed in plant design. Such a distinguished body as the United Kingdom Atomic Energy Authority has recommended against the use of light water reactors because of this. Pressure vessel integrity is questionable for the following and other reasons:

a. Nozzle Break Between Vessel Wall and Biological Shield

A nozzle break occurring between the vessel wall and the biological shield surrounding the vessel could cause movement and forces on the vessel that may have incalculable results.

From past experience with primary piping systems, cracks are most likely to occur at the vessel safe end (Point A on Figure 1). This is the most susceptible point for an instantaneous pipeline break. Failure could cause an instantaneous pressure wave to build up between the inside of the biological shield and the outside of the reactor pressure vessel. This instantaneous pressure wave could be high enough to cause the vessel to move sideways, or to tip over against the foundation causing severe and unknown damage. Such an accident would certainly cause gross distortion of the reactor internals, disruption of the core, and possibly prevent insertion of control rods to shut down the reactor. The emergency core cooling lines could be severed and result in the "incredible" accident. This accident potential exists in most light water reactors, and should be thoroughly evaluated by the NRC.

b. Pressure Vessel Pedestal Acceleration

Another serious problem in "pressure suppression" plants is the possibility óf vessel pedestal acceleration resulting from the loads on the pedestal created by the pressure wave developed in the supression pool. (See Figures 4 and 5.) These loads on the structure may be transmitted through the soil and through the foundation, causing the vessel support pedestal to vibrate to a degree greatly above the seismic design basis for the equipment. This condition could result in failure of the vessel and internals, failure of the core support structure, possible loss of emergency core cooling capabilities, possible loss of insertion capability of the control rods, and, again, untold effects by the accident that could occur. This condition should be thoroughly reviewed by the NRC.

c. Structural Integrity of Pedestal Concrete

A further problem that could affect continued safe operation of BWR and PWR reactors in all containment designs has to do with the structural integrity of the reactor vessel pedestal concrete. A loss-of-coolant accident in even the small or intermediate range can result in a temperature transient inside the containment of such a magnitude that the thermal shock received by the concrete reactor pedestal could result in cracking of the foundation to the point where its strength would be seriously affected, particularly for future seismic or LOCA transients, thus jeopardizing future public safety. The minimum result would be an extensive shutdown while a way was found to replace or strengthen the pedestal. This would almost certainly be a multimillion-dollar expenditure and require months and months of outage time. This accident may already have occurred in 1971 at the Commonwealth Edison Dresden-2 and -3 Plants, where an accidental pressurization of the drywell created a temperature transient which destroyed most of the core monitoring cables, and may well have damaged the foundations on which these reactor vessels sit.

Dresden-2 and -3 should immediately be evaluated for suitability of continued operation, and regulations should be developed to ensure that appropriate action is taken should similar events occur at other plants. This evaluation should include the safety implication of continued operation of plants that are susceptible to such damage.

4. CONTAINMENT

The primary containment of a nuclear power plant is the pressure-containing structure that is provided to contain the most credible accident that is assumed to occur during the life of the plant. In most cases, this is the guillotine break of the largest pipe in the primary system.

a. Mark I Pressure Suppression Containment

In the early BWR's, the system used was a dry containment which was sized to contain the total energy inventory of the primary system that would be discharged in the event of the maximum break. When the first large-scale commercial plants were being designed, it was soon seen that designing a dry pressure containment would be difficult and expensive, and a new concept was developed for boiling water reactors called the "Mark I Pressure Suppression Containment."

In this concept, two chambers are used—a pressure vessel, called the "drywell," which houses the primary reactor system components, and an auxiliary chamber, called the "wet well," which contains a large quantity of water which is used as a heat sink. (See Figure 3.) Vents are piped into the "wet well" from the "drywell" and discharged beneath the surface of the water.

In the event of an accident, the "drywell" becomes pressurized with the discharged steam and fission products. This combination of steam and gases is discharged beneath the surface of the pressure suppression pool, which condenses the steam and mitigates the pressure transient.

The Niagara-Mohawk Nine Mile Point Plant and the Jersey Central Power and Light Oyster Creek Plant ushered in the new era of large commercial power reactors. These were the first two plants of a series of over 30 plants, domestic and

Containment Concept MARK I

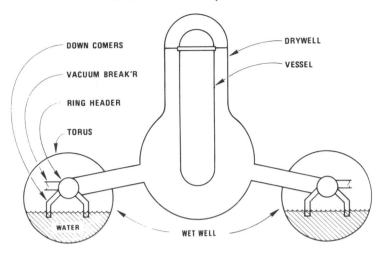

DOWN COMERS

VACUUM BREAK'R

RING HEADER

TORUS

DRYWELL

VESSEL

WATER

WET WELL

Figure 3

overseas, that were built utilizing the Mark I Containment System.

b. Mark II Containment

As experience was gained in the design, construction and startup of the plants, it was soon learned that the rather complex structures involved in the Mark I Containment resulted in severe cost penalties and in lengthened construction schedules. A modification was developed which utilized concrete slip forming techniques and an over/under, drywell/wet well arrangement to simplify construction and to cut costs.

This concept was known as the Mark II Containment (see Figure 4) and a number of plants are currently under construction utilizing this concept. Approximately ten BWR plants are based on the Mark II design; none of these plants has yet gone into service.

c. Mark III Containment

The next iteration in pressure suppression containment was the Mark III type containment. (See Figure 5.) This containment was a further refinement of the concrete construction, but used a suppression pool under the reactor vessel which utilized a

Containment Concept MARK II

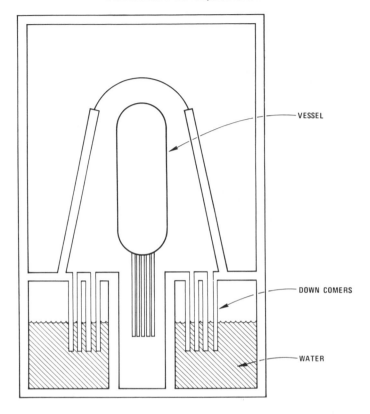

VESSEL

DOWN COMERS

WATER

Figure 4

weir wall and horizontal vents to relieve pressure from the
drywell in place of the vertical downcomers used in both the
Mark I and Mark II containments. It has, incidentally, the
disadvantage of requiring maintenance and operating personnel to routinely enter spaces that will be pressurized with steam
during the loss-of-coolant accident, a questionable arrangement. Because this design was a departure from past technology, General Electric built a 1/3 scale Mark III mock-up to test
this concept. In analyzing the resulting Mark III test data, it
was discovered that the rapid discharge of air through the
horizontal vents into the water resulted in a high vertical swell
of the pool surface creating large hydrodynamic forces on structures and components above the suppression pool.

Containment Concept MARK III

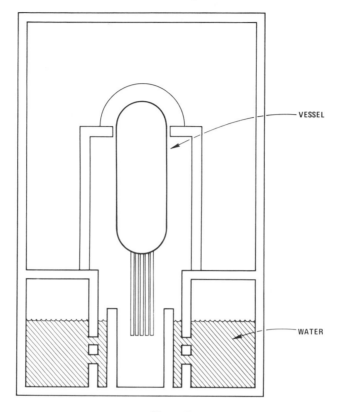

VESSEL

WATER

Figure 5

This phenomenon necessitated structural modification of the Mark III designs, and it was only a matter of time until the question was asked as to the applicability of this phenomenon and associated loads on the Mark II plants and on the 38 Mark I plants in operation or construction all over the world.

This question was initially raised by the NRC in April 1975, in a discussion with General Electric Company. The NRC subsquently requested all licensed Mark I plants in the U.S. to conduct an immediate assessment of the new loads that had been discovered in the Mark III testing. The NRC asked for a thorough review of the design basis of all of the existing plants, and for response within approximately 45 days as to justification for continued operation. Since there were some 20 new loads or phenomena that had been determined and which had not been considered in any way whatsoever during the original design of the Mark I containment, the response time was impossible to meet. To answer the question required development of analytical models, hardware testing, and incredible amounts of structural evaluation. It also required the generation of new acceptance criteria. It was recognized early in the program that the containment could not be found adequate under normal structural design criteria, since that amount of margin is not used for structural design. Accordingly, a concept of "most probably loads" (non-conservative) and "maintenance of function" was adopted.

To address the schedule problem, the utilities proposed to the NRC that a generic evaluation be conducted by the utilities, General Electric and a number of consultants to evaluate all plants simultaneously by categorizing them into typical plant groupings. The plan was to perform representative plant evaluations to determine the probability that the containments, as built, could withstand the forces expected during a design basis, loss-of-coolant accident. The evaluations neither assumed the latest seismic criteria nor load combinations for this "quick and dirty assessment," yet another non-conservatism.

This program was begun in May of 1975, and was to have been completed by the fall of 1975. During the course of the evaluation, a number of results were found to be higher than the permissible maximums established for the screening analysis, so it was necessary to run further tests, refine models and perform further evaluation. Even so, the effect of some of the loads on the torus support system was found to be unacceptable. This resulted in an extension of the short-term program from October, 1975, until January of 1976.

Progress of the program was communicated periodically to the NRC. The utilities planned to review with the NRC, no later than January 28, 1976, the results of the second analysis of the torus support system for all plants. A preliminary review of the results was held by the utilities on January 22, 1976, and on January 23, 1976, the utilities advised NRC by telephone that it appeared all containments would be able to withstand the most probable course of the LOCA event and still maintain function. This conclusion was based on the extrapolation of data from a 1/12 scale test model, and on a substantial amount of engineering judgment regarding the conservatisms inherent in the design basis accident, and in the structural response of the system.

In no way would a competent structural consultant testify that the containments clearly met the requirements for integrity of the system. In many plants, loads on columns, welds, and other critical elements would be expected to exceed the yield strength of the materials, a situation not allowed under normal design codes. In addition, only nominal seismic loads were factored into the evaluation, as all of these plants were designed before new seismic criteria had been developed and, if judged by current day standards, would be found to be inadequate in most cases. Even worse, only the most probable loads were used in the calculations, not bounding case loads, and no coincidental loads due to stuck-open safety relief valves, or other effects were considered to be applicable.

As a result of errors that were found in the analysis over the weekend from January 23 to January 26, the Vermont Nuclear Power Corporation was advised that Vermont Yankee would be subjected to more severe torus uplift loads than had been reported on January 23, 1976. As a consequence of this new information, the Vermont utility made the decision to shut the plant down until further evaluation could be completed. Recent newspaper articles report that temporary repairs have been made and that the plant is returning to service, with extensive changes to be made while the plant is operating.

All other plants except Boston Edison's Pilgrim Plant, which is shut down for a scheduled refueling, have continued to operate.

A complete list of U.S. Mark I plants licensed for operation is shown in Table 1:

TABLE 1
Operating Plants (Mark I Containments)

Utility	Plant Name
Boston Edison Co.	Pilgrim
Jersey Central Power & Light Co.	Oyster Creek

(continued)

Utility	Plant Name
Niagara Mohawk Power Corp.	Nine Mile Point
Commonwealth Edison Co.	Dresden-2 & -3,
	Quad Cities-1 & -2
Northeast Nuclear Energy Co.	Millstone-1
Northern States Power Co.	Monticello
Philadelphia Electric Co.	Peach Bottom-2 & -3
Georgia Power Co.	Hatch-1
Nebraska Public Power District	Cooper
Carolina Power & Light Co.	Brunswick-2
Iowa Electric Light & Power Co.	Duane Arnold Energy Center
Power Authority of the State of N.Y.	Fitzpatrick
Vermont Nuclear Power Corp.	Vermont Yankee
Tennessee Valley Authority	Browns Ferry-1 & -2

The consequences of failure of the primary containment system are frightening. The primary containment system provides the most basic defense to public health and safety by preventing the release of highly radioactive fission products into the biosphere should a loss-of-coolant accident occur. In addition, the torus portion of the primary containment system provides the source of cooling water for the emergency core cooling system, and is in the life-line that prevents the disastrous core meltdown following the loss-of-coolant accident. If the torus support structure fails in the initial phases of the loss-of-coolant accident, it could result in failure of the emergency core cooling system piping systems attached to it, and in loss of the supply of cooling water for the core.

The integrity of this portion of the primary system, then, is an absolute requirement for the protection of the public health and safety and should be an absolute requirement for continued operation of the plant. It is unthinkable that plant operation can be continued on the very tenuous argument that the probability of the accident occurring is low; even the NRC's Rasmussen Report (WASH-1400) postulates that a loss-of-coolant accident will occur within the foreseeable future. It is more probable that such an accident would occur early in the time period considered by WASH-1400, because the techniques, materials, know-how, and design improvement made in later plants have not been incorporated into the early plants.

During the many meetings of the utility representatives involved with this containment assessment program, the responsibility of continued operation of these plants has weighed heavily on the minds of the parties involved. Each individual should be questioned as to his faith in the design, the level of knowledge, and the wisdom of continuing to operate these plants.

It is urgent that this problem be seriously evaluated and the wisdom of continued operation of these plants be reconsidered.

d. Primary Containment Fatigue Life

There are other questions of containment adequacy that require close inspection by the regulatory commission and by the members of the Committee.

One concerns primary containment fatigue life. The same torus support system, which is of questionable integrity for the loss-of-coolant accident, is also subject to substantial load effects when safety-relief valves discharge into the suppression pool. This is not an abnormal event, but occurs on most load transients. At a number of plants, relief valves have stuck open, with the result of continued extensive blow-down into the suppression pool. This problem actually resulted in *rupture* of the primary containment at the Wuergassen Plant in Europe several years ago. Damage was also discovered during 1975 in the pipe restraint systems in the suppression pools at the TVA Browns Ferry Plant, the Iowa Electric Duane Arnold Plant, and the Boston Edison Pilgrim Plant.

Some modifications have been made and the NRC has issued advice to the operators of the other plants, but there is no coherent program underway to ensure the *immediate* assessment of the effects of this phenomenon on total plant safety. In addition, the original containment design estimated that a specific number of cycles would occur, and the containments would be designed to withstand this limited number of cycles over their lives. In fact, valve operation has been more frequent than expected, and the cyclical loading of the containment from periodic valve discharges is expected to far exceed the initial design.

e. Uncertain Pressure Suppression Testing

Other problems that affect the performance of the primary containment are uncertainties with regard to the testing of the pressure suppression phenomenon. A limited number of tests were performed on a segment of the containment that was planned for the Bodega Bay Plant (never built). Depth of downcomer submergence testing was varied, but most of the tests were performed at only one level of submergence: four feet. Waves generated within the pool may uncover the vents; seismic slosh may occur, which makes pressure suppression ineffective; and since the loss-of-coolant accident is almost impossible to mock-up, complete verification of the adequacy of the system to withstand a LOCA has yet to be demonstrated.

f. Erosion of Design Margins

Design margins are continually eroded by modifications resulting from problems that have occurred during plant operation. A good example is the removal of baffles from the torus several years ago. The first Mark I plants were built with anti-slosh baffles installed in the torus. The purpose of these baffles was to ensure that waves would not be built up in the pool resulting in gross disruption to the pool surface and making ineffective the downcomer submergence.

Early in plant operation, it was discovered that the forces produced in the suppression pool by relief valve discharges dislodged a number of these baffles. The "fix" that was implemented was not to strengthen the baffles, but to perform an analysis that indicated that they were not needed in the first place, and subsequently baffles were removed from most existing plants and were eliminated from all future plants. There is a good bit of speculation and concern, particularly by those utilities which removed the baffles, that the containment evaluation program currently underway may prove that baffles are again required for the mitigation of seismic slosh. There will, indeed, be much consternation if hundreds of thousands of dollars (per plant) and a substantial amount of outage time are required to reinstall those baffles.

g. Corrosion Allowance in Material Thickness

Yet another area of weakness in the primary containment of nuclear plants has to do with the fact that no material corrosion allowance has been provided. To ensure that material thickness is maintained for the 40-year life of the plant, it was planned that a chromate corrosion-inhibitor would be added to the suppression pool water. It has since been determined that the chromate materials have a severe environmental impact and they have subsequently been banned for use in most states. The plants, in some cases, now find themselves operating with torus suppression water inhibited with chromate, but with no way to discharge the water, should modifications be needed that would require draining of the torus for maintenance or modification work. This condition seriously jeopardizes the utility's ability to inspect and modify this critical portion of the nuclear plant. If corrosion inhibitor is not used, protective coatings must be applied. Their effectiveness is also questionable under these severe conditions.

h. Containment Electrical Penetration Seals

A major weakness in the containment system is in the areas where nozzles containing electrical power and instrumentation cables breach the containment. Typically, approximately 4000 to 8000 conductors must enter the containment through nearly 40 nozzles, each 12 inches in diameter. Many containment electrical penetrations utilize an epoxy sealant around conductor rods to achieve a pressure barrier. Epoxy adherence to the conductors has proven to be deficient. For example, at the Farley-1 PWR Plant, approximately 5 percent of the installed electrical seals leaked following installation checkout under ambient environmental conditions. During simulated LOCA conditions, the penetration epoxy has in some cases reverted. This resulted in gross leakage paths and inadequate electrical insulation of the epoxy sealed penetrations. This is an example of a common-mode failure which would breach the containment and short out control cables at the time of their most critical need. Containment penetrations are a very weak link in the containment systems and consequently should receive greatly increased scrutiny from the NRC.

i. Wet Well/Drywell Vacuum Breakers

Another area where leakage presents a problem is with the vacuum breakers between the wet well and the drywell of the primary containment system. It was discovered that these vacuum breakers are so difficult to maintain, to the degree that leak tightness is required in the technical specifications, that the Mark I utilities almost unanimously rejected a drywell pressurization scheme as a potential fix for the torus-support problem noted above. Vermont Yankee, the only plant not operating with the containment inerted by nitrogen, felt pressurization was a viable "fix." One plant, Pilgrim, decided to test drywell pressurization and found that their vacuum breakers were leaking so badly that their technical specifications required them to immediately shut down for repair, even though the test was only to determine whether drywell pressurization was a viable alternative.

j. Summary—Primary Containment of BWR's and PWR's

In summary, then, the primary containment system housing almost all of the operating boiling water reactors in the United States contains serious design defects. All of these have been disclosed to the NRC, and programs to some degree are underway to provide for improvement. However, it is of deep concern whether the action taken is rapid enough. Because the potential consequences of failure are great, the advisability of continued operation of these plants under these circumstances is extremely questionable.

While primary containment systems for pressurized water reactors have not been questioned in detail in this testimony, it is certain that similar problems and deficiencies exist in those plants. Pressure suppression is not used in pressurized water reactors; a dry containment with a higher pressure rating is utilized instead.

The same penetration seal problem and the valve leakages will affect the PWR containment's ability to provide the protection required. In addition, the transient assumed to occur during the PWR loss-of-coolant accident assumes that the energy contained in the steam generators is not immediately released to the containment, as it is assumed that the steam generator tubes and tube sheets will remain intact. This assumption is subject to question and should be proof-tested. Integrity of the steam generator under LOCA conditions requires almost instantaneous pressure reversal on the thin tube structure and on the tube-sheet element. This is opposite to the normal direction of loading for which this structure is designed.

It is strongly recommended that full-scale tests be performed on the capability of steam generators to withstand such a transient condition. If this has not been adequately demonstrated, then pressurized water reactor plants also should be shut down until this can be verified.

5. MISCELLANEOUS COMPONENTS

Numerous component reliability and failure problems have been observed that indicate essential safety goals are not being achieved in the light water reactor program. These problems are reported in lengthy summaries prepared by the NRC, but it is not clear that the responsible design organizations are required to take the necessary corrective safety actions on past, present and future plants. Design improvement programs are somewhat haphazardly suggested to the operating utilities, but no program of required modifications has emerged. The following itemization identifies some of the components with poor performance records:

a. Valves—All Types

(1) Main steam isolation valves have not met leakage requirements, and have not met closure times.
(2) Safety and safety relief valves: Set points drift, valves leak, and valves stick open.
(3) Feed water check valves: Leakage requirements are seldom met when tested.

b. Heat Exchangers

All types of heat exchangers have experienced tube failures, tube vibrations, and various other failure mechanisms.

c. Main Condensors

Tube life has been poor at many plants. The "old-time" fix of sawdust injection is prevalent—questionable for use in nuclear technology.

d. Valve and Pump Seals

Length of life is poor; leaks cause many shutdowns.

e. Inspection Techniques

Current non-destructive testing techniques are not effective. Ultrasonic inspection is a "black art." Leaking cracks have been found in lines declared sound by standard inspection.

The list could be endless. What is required is an intelligent program to guide the many organizations in the implementation of mandatory changes in response to field experience. When will the NRC initiate such a program?

6. MATERIAL FAILURES

Material failures in reactor systems are numerous, have been well-documented, and will not be elaborated upon in great length here. However, they represent a significant impact on the continued economic and safe operation of nuclear power plants.

Intergranular stress-corrosion cracking, frequently observed in (Type 304) stainless steel materials, particularly those that have been furnace-sensitized, has been the subject of industry studies in the past.

Piping cracks that were prominent in BWR operations over the past year resulted in the issuance of several industry reports, and the significant question that should be asked is, what is being done to implement material changes in the 59 operating reactors that already exist in the United States? Also, what is being done to preclude utilization of materials and procedures that will result in similar failures on the hundred or so reactors that currently are under construction? The efforts that would be required to update these systems may be impossible for our industrial system to undertake at this time, but inaction leads to seriously jeopardizing the safety of continued reactor operation, and could be economically disastrous for future plants.

Included in this are not only the piping failures, but cracks of the vessel safe-ends (attachment points) of piping systems, failures of vessel internal parts (such as stub tubes, cladding, thermal sleeves, etc.), failures of the feed-water spargers previously mentioned, and cracks that have been observed on the pressure vessel nozzle blend radii. These failures present significant safety problems for continued future operation. A specific evaluation, including recommendations for a detailed program of material replacement, needs to be developed by the NRC to minimize their safety impact.

7. FUEL STORAGE FACILITIES

a. New Fuel

A problem brought about by fragmented responsibility regards criticality-safety in new fuel storage pits.

New fuel storage racks have in the past been designed to be criticality-safe with the fuel stored dry, or also to be safe if the pit happens to be flooded by water. It was recently discovered that the worst criticality condition was neither dry nor flooded, but when a density of less than 0.5 water-to-air mixture was present. This fact became a safety consideration when it was discovered that fog nozzles of the type commonly used for fire protection in power stations could actually achieve a water density of 0.3. This problem results from incomplete review and communication between the nuclear equipment designers and those responsible for fire protection equipment. When this fact was discovered, somewhat by happenstance, a limited survey of fire protection facilities in operating plants was conducted by General Electric. Recommendations were made to plant operators that fog equipment should not be used in the vicinity of the fuel storage pit. Who really is responsible for the coordination of all such technical intricacies? The risk of such an accident is so great that this should not be left to chance.

b. Spent Fuel Storage

Much has been written about the economic penalties to be suffered by the lack of fuel reprocessing facilities. The safety impact of this lack should not be overlooked. Inadequate spent fuel storage facilities may soon exist at many plants. If a defect develops, requiring the rapid discharge of the full core load of fuel, this escalates to an immediate safety problem. What plans has the NRC made to handle such emergencies?

B. Reliability Calculations

Reactor vendors have advertised ambitious goals for plant availability. Yet there are currently no design goals established for subsystem reliability or availability. Instead, the practice is to design the system or subsystem, then analyze its failure modes and, from this data, calculate the failure rate in the unsafe mode. When approached from this viewpoint, the failure rate data (and thence the reliability and availability) obtained are meaningless since there is no rational basis for acceptance or rejection.

Substantial safety risk is incurred by not knowing the acceptance criteria for the reliability of safety and safety-related systems. Any decision affecting the design or modification of the systems must evaluate the effect of reliability. An example is the design of a new safety system for BWR's. In order to determine the need for a test system and the frequency of test, it is necessary to know the desired reliability of the overall system in terms of its impact on the total plant safety. This has not yet been established.

The Rasmussen Report is said to employ a reliability model of an operating BWR (Peach Bottom-2, -3). This is incongruous in light of the lack of reliability data on many of the subsystems. The establishment of such a model is highly recommended. It should be updated regularly as a result of analytical and operating data.

C. Redundancy v. Diversity

Redundancy is the practice of putting equal systems or components in place to back up one another (e.g., two Rod Block Monitors are used; either one can block rod motion). Diversity in system design is the practice of making a portion of the system or back-up system from a different technology, component or design, such that the two portions of the total system would not be subject to a common-mode failure. (A common-mode failure is the failure of more than one redundant system or subsystems due to a single factor or event.)

Redundancy is used in most of the reactor systems. However, the industry has carefully looked at diversity and, to date, has avoided it for all but a few situations (e.g., emergency power and types of core cooling systems). In control and instrumentation systems, the decision to not use diversity has been based mainly on economics.

The Browns Ferry fire demonstrated the vulnerability of redundant and diverse circuits to a common-mode failure when a single event (fire) wiped out the redundant cabling and disabled the diverse core cooling systems. Similar events are both possible and probable in other areas of the control room and cabling areas. These problems would be most likely to appear in older plants (prior to Browns Ferry design) where even less care was taken to provide redundancy and diversity.

It is recommended that the NRC review plants prior to TVA–Browns Ferry and evaluate adequacy of redundancy in the designs as well as the need for diversity in critical systems. The industry, the NRC, and the IEEE should prepare a standard defining diversity requirements.

D. Political, Economic and Technological Pressures on the NRC and Utilities Prevent the NRC from Effectively Regulating Nuclear Power in the Interest of Public Health and Safety

The ability of the NRC to effectively regulate safety within the constraints of the current situation of the diverse commercial nuclear power enterprise is suspect.

The tremendous cost, schedule and political pressures experienced make unbiased decisions, with true evaluations of the consequences, impossible to achieve. This is the ultimate deficiency of our nuclear program.

The Mark I containment assessment program vividly illustrates this. The Mark I safety evaluation covers 19 operating plants representing almost one-half of the United States' nuclear power electrical production. It also involves 6 more units under construction in the U.S. and 13 plants overseas. *THE PRIMARY FOCUS OF THE PROGRAM HAS BEEN TO "PROVE" THE PLANTS ARE SAFE ENOUGH FOR CONTINUED OPERATION—NOT TO OPENLY ASSESS THEIR TRUE SAFETY.* During this program, many statements and concerns were expressed, illustrating the impossibility of true safety evaluation. For example:

(1) The ever-present fear of local intervenors "finding out."
(2) Proposals that responses to NRC questions on containment be made "generically" by the Mark I group. This was considered desirable so that the information would be filed in the generic files of the public document room, making it virtually inaccessible to local intervenors active on specific plant AS&LB hearings.
(3) NRC staff implications that generic filing essentially "buries" the information from the public.
(4) Utility legal representatives expressing serious concern on what is the true definition of an unresolved safety problem—suggesting immediate plant shutdown requirements.

(5) NRC staff suggestion that serious problems should be reviewed directly with the Director of Nuclear Reactor Regulations. The reason for the review was unstated perhaps to avoid unnecessary expression of concern in the public record.

(6) Advice on how best to present abnormally high stress results of the structural analysis. Stress levels are avoided because current codes state limits on stresses. Strain limits avoid such technical debate.

(7) Repeated questions by the NRC staff as to the cost of the evaluation program and on how expensive the "fixes" would be. Repeated questions on the loss-of-power costs to the utility. Such concerns are of general interest, but express an unnecessary or unreasonable concern over issues not the direct responsibility of the NRC. The NRC safety decisions are, by law, to be made on the basis of technical safety and its potential impact on the public health, and should not involve economic evaluations.

(8) NRC public statements that are misleading regarding the status of safety evaluations, particularly on operating plants. The January 30th [1976] NRC press release regarding Mark I containment indicates everything is fine, that the one reactor that was shut down had design and operating characteristics that differ from those of the other reactors. This is technically true, but only marginally so, and is misleading to members of the public and the industry. This was picked up immediately by plants in Japan where the same immediate assessment of "difference" was made on the Mark I plants there to justify continued operation.

(9) The basis of Mark I technical defense for continued operation was low probability of accident and immediately implemented corrective action. Action has been started by most utilities, but no commitment has been made as to when it will be implemented.

It is indeed unfortunate that the commercial and technical proprietary pressures of the business world also work to the detriment of the maximum achievement of safety. An excellent in-depth study of nuclear safety was described by Mr. Reginald Jones, Chief Executive of the General Electric Company, in a December address to New York Security Analysts. Dr. Charles Reed, the top technologist in the Company, amassed a task force of the most knowledgeable people that could be put together in the nuclear business to evaluate the technical and business risks facing General Electric in its nuclear orders. This task force of as many as 70 to 80 people worked for a whole year. It included the finest scientists and engineers in General Electric. The result was a final report that was overwhelming—a five-foot shelf full.

These findings should be shared with the NRC. Nuclear safety affects the welfare of the whole public, and all possible sources of information should be pursued. Has the NRC been presented with a review of these findings and is progress being evaluated? What programs have been implemented by the NRC as a result of this evaluation?

Decisions are forced upon the power plant owners, the vendors, and the NRC to make go, no-go decisions based on technical inadequacies or insufficient data. Compromises are made. There is little disagreement that operating plants are not as safe as we would wish them to be or as safe as they should be. We believe it is imperative that public safety be adequately separated from other considerations such as economics and the degree of dependence of a region on nuclear power to meet the area's energy requirements. This serious deficiency in our total nuclear safety system must somehow be overcome.

III. NRC QUALITY ASSURANCE PROGRAM REQUIREMENTS ARE INADEQUATE

A total Quality Assurance Program has many aspects. The Quality Assurance Program should encompass all activities involved in the design, manufacture, installation and operation of the safety-related equipment.

A. NRC Inspection of Manufacturers Inadequate

The ASME Codes were initially developed in 1911 to protect the public from boiler explosions in public facilities such as schools, auditoriums, and office buildings. In the 1960's, Section III of the Code was developed and applied to nuclear power plant components, such as primary pressure-bearing portions of the reactor system, including the reactor pressure vessel, in-core instrument guide tubes, and primary-loop valves.

These procedures apply only to those items covered by the ASME Code. A problem exists with respect to non-Code safety-related equipment. NRC requirements for safety-related equipment are less stringent than for the ASME Boiler Code items. Because nuclear power plant accidents pose potentially the greatest risk to the health and safety of the public, NRC inspection of all (Code and non-Code) safety system items should be at least as strict as those developed by ASME. The following comparison illustrates the problem:

Quality Assurance Program Requirement	Section III, ASME Code	NRC Regulations Non-Code Safety-Related Items
1. Certification of design specification.	1. Certified by a Registered Professional Engineer retained by the owner or his agent.	1. No requirement for certification by a Professional Engineer.
2. Certification of stress report.	2. Certified by a Registered Professional Engineer retained by the manufacturer.	2. No requirement for certification by a Professional Engineer.
3. Review of stress report.	3. Transmitted to owner or his agent for review.	3. No requirement for review by the owner or his agent.
4. Q.A. Program and Program changes.	4. Detailed implementation Q.A. Program surveyed and approved by the ASME. No changes can be made in the program prior to receiving approval from the Engineering Specialist of the Authorized Inspection Agency.	4. NRC surveys program to the generic program document. For instance, for GE, Section 17 of the GESSAR. No requirement for prior NRC approval of changes to documents that implement the program.

(continued)

Quality Assurance Program Requirement	Section III, ASME Code	NRC Regulations Non-Code Safety-Related Items
5. Material verification.	5. Authorized Inspector reviews all material certifications prior to release of the material for processing.	5. No requirement for third-party review of materials.
6. Review of manufacturing process sheets.	6. Authorized Inspector reviews manufacturing process sheets and places his checks and hold points in the process.	6. No review of manufacturing process sheets. No check or hold points.
7. NDE (non-destructive examinations).	7. Authorized Inspector witnesses or reviews the records of all NDE examinations.	7. No witness or review of NDE examinations.
8. Final acceptance tests—hydrostatic or pneumatic tests.	8. Final acceptance tests witnessed by Authorized Inspector.	8. No requirement for third-party witness of final acceptance tests.
9. Repair of non-conformances.	9. Repair procedures reviewed by the Authorized Inspector.	9. No requirement for third-party review.
10. Release for shipment.	10. Final review and release by Authorized Inspector.	10. No requirement for third-party review and release.
11. Internal audit of Q.A. Program.	11. Results of internal audits are reviewed with the Authorized Inspector.	11. No requirement for review of internal audit results with the NRC.

Regulation and enforcement by the NCR is not adequate. For non-Code safety-related equipment, the NRC should immediately implement a program of detailed third-party inspection similar to the disciplined program presently conducted for ASME Code items.

B. NRC Regulation of Product Qualification Is Inadequate

Underwriters Laboratories, Inc. (UL), founded in 1894, was chartered to establish, maintain, and operate laboratories for the investigation of materials, devices, products, equipment, constructions, methods and systems with respect to hazards affecting life and property. Many household electrical appliances receive the third-party review by the UL.

In contrast, the NRC has no requirement for independent third-party evaluation and product proof-testing of the Class I safety-related electrical equipment that controls and protects a nuclear power plant.

The public has a right to know that an electrical appliance, such as a toaster or hair dryer, has more stringent safety checks than the electrical instruments that control a nuclear plant. This is a clear demonstration of the inadequate attention given by the NRC towards protecting the public safety.

If the manufacturer of a commercial appliance should desire to redesign or change the process requirements for a UL-listed device, the Underwriters Laboratories must review the proposed change to determine if the listing is still valid, or if requalification testing is required. But here again, the NRC regulations contain no similar requirement for third-party review to safeguard the public from the effects of improper product changes.

For all safety-related equipment, the NRC should immediately implement a program of third-party product qualification which is at least as stringent as is required for commercial products listed by UL.

IV. SUMMARY AND RECOMMENDATIONS

For the reasons stated in this testimony, we believe that the continued operation of nuclear power plants creates severe hazards to the public.

Some of the design defects and deficiencies alone create severe safety hazards; for example: the possibility of failure of the Mark I pressure-suppression containments. But the one most important point, and the point we want to emphasize to this Committee, is that the *cumulative* effect of all design defects and deficiencies in the design, construction and operation of nuclear power plants *makes a nuclear power plant accident, in our opin-*

ion, a certain event. The only question is when, and where.

It is for this imperative reason that we submitted our resignations and are now totally committed to educating the public about the truth regarding nuclear power.

The most probable response to many of our concerns from the NRC and from the industry will be that the concerns are not new —and, indeed, many of them are not. Certainly, it is true that intensive efforts are underway to resolve many of the problems. But the level of effort now being expended to resolve the problems has not made the problems go away.

We, therefore, recommend:

1. That, as an urgent matter, the Mark I pressure suppression containment problem be evaluated and the wisdom of continued operation of all 19 domestic nuclear plants with Mark I systems be reconsidered;

2. A thorough review of the total effect on the in-vessel components of flow-induced vibration, particularly with respect to the failure of feed water spargers and the Local Power Range Monitors;

3. A study of the distribution of cooling water over the hot core, which is the basis for the effectiveness of the core spray cooling system, giving particular emphasis to the result of the European studies of this phenomenon;

4. A study of the full effects of the End-of-Cycle-SCRAM Reactivity Effect;

5. That control rod life be determined and the technical specification requirements be established to ensure that shutdown margins are adequately tested to account for these effects;

6. That the "rod-drop" accident be thoroughly reevaluated and satisfactory systems to prevent or mitigate the accident be developed;

7. That the current study of pressure vessel integrity be intensified, with particular emphasis on the possibility of nozzle break between the vessel wall and the biological shield, the problem of excessive loading of the pressure vessel pedestal resulting from pressure waves in the suppression pool, and the structural integrity of the pedestal concrete;

8. That the continuing study of the Mark I containment system give particular emphasis to the effect of seismic slosh and continued erosion of design margins, the corrosion allowance and

material thickness, and difficulties with containment electrical penetration seals, and with the wet well/drywell vacuum breakers. It is recommended that full-scale tests be performed on the adequacy of the steam generators to withstand pressure reversal under LOCA conditions and, if not adequately demonstrated, that pressurized water reactor plants be shut down until it can be verified;

9. That the NRC immediately implement a program of detailed third-party inspections for non-ASME Code safety-related items which is at least as stringent as is required under the ASME Code, and a third-party product qualification procedure for safety-related electrical equipment which is at least as stringent as is required for commercial products listed by the Underwriters Laboratories;

10. That NRC regulations be amended to prevent common-mode failure of safety-related items by requiring that changes in the design or fabrication of subassemblies or safety-related devices can only be accomplished in an NRC-disciplined system requiring appropriate device design reverification and performance requalification;

11. That NRC institute a disciplined NRC-managed field failure identification and analysis system, and replacement parts control system;

12. That the design of control rooms be standardized and that NRC require more frequent retraining on updated control room simulators to increase the proficiency of control room operators;

13. That the NRC begin immediately to study the problem of increased personnel radiation exposure during the operation, maintenance and retrofit of the existing plants, and develop plans and programs for the decontamination of existing plants, and for disposal of radioactive wastes resulting from the decontamination, and for disposal of the decontaminated plant itself;

14. That the NRC practice of waiving current safety requirements with respect to older plants cease, and that older nuclear plants be required to retrofit as is necessary to meet current safety requirements; and

15. That plants currently under construction should not be permitted to start up and become radioactive *until all design issues are resolved.* This is particularly important with regard to the seismic criteria controversy at PG&E's Diablo Canyon Plant.

Appendix B

Testimony of Robert D. Pollard
before the Joint Committee
on Atomic Energy
February 18, 1976

My name is Robert D. Pollard. Until last Friday I was an employee of the Nuclear Regulatory Commission and had been employed there for six and one-half years. During that time I served as Technical Intern, Reactor Engineer, and Project Manager. As a result of my work at the Commission, I believe that the separation of the Atomic Energy Commission into two agencies has not resolved the conflict between promotion and regulation of commercial nuclear power plants. Because I found that the pressures to maintain schedules and to defer resolution of known safety problems frequently prevailed over reactor safety, I decided I had to resign. I could no longer, in conscience, participate in a process which so effectively evades the single legislative mandate given to the NRC —protection of the public health and safety.

The second decision I made was to resign publicly. I chose this course because I believe the American people are being misled into believing that nuclear power holds the promise of a painless solution to the problems called the energy crisis. Not only do I disagree with that but I believe that the unresolved questions of nuclear waste disposal, safeguards, economics and reactor safety need to be resolved before this nation of ours goes any further with commercial nuclear power. I was persuaded to take this public resignation route principally by my own moral and ethical value judgments—no special-interest group or expectation of personal financial gain motivated this action. A secondary reason was the widely held and oft-expressed view by my NRC colleagues that if well-intentioned critics of nuclear power had even just a little expert technical advice, they could be successful.

The official NRC reaction to my resignation was predictable— Chairman Anders requested that I document my concerns but other officials, directly or indirectly, hindered my ability to collect written evidence that unresolved problems are almost routinely ignored. Nevertheless, I managed to prepare two reports regarding the regulatory review process and the inhibition of dissent within the agency. . . .

In my report on the NRC review process, I attempted to show two things: first, that historically, the regulatory policy has been to avoid major problems that are expensive to solve or would delay significantly the schedule for a particular plant. Indian Point Unit 2 was chosen only as one example—other examples exist.

The second point I attempted to prove was that NRC management officials are aware of many unresolved safety problems that apply to whole classes of reactors but continue to press for expediting of individual plants licenses. This NRC policy is so absurd that it bears repeating—the NRC policy appears to be that unresolved problems that are relevant to many reactors need not be resolved and should not be discussed in deciding whether to license a particular reactor.

I think this attitude is wrong even if every member of the NRC staff agrees with it, which is not the case. The NRC staff does not have the final say in whether a plant should be licensed. The Atomic Safety and Licensing Boards, the Commission, and the courts have a role also. Unless the NRC presents all the facts to the public and to the other bodies, then any decision the boards, Commission, and courts make must necessarily be suspect because it would not be based on the whole truth.

The Committee will hear today from three former General

Electric engineers who, like myself, although without my prior knowledge, felt compelled to leave their jobs and also tell the truth about nuclear reactor safety. The Committee will also hear from representatives of the NRC who will deny our charges and assert that all is safe with nuclear reactors and that the NRC is internally open and publicly forthright. In evaluating this conflicting testimony the Committee can use two procedures. First, it can evaluate the testimony itself and, in this evaluation, attempt to determine whether anyone here has a motive to distort or obfuscate the facts. I and the three General Electric employees left responsible, well-paying jobs for substantially less income, or no income, and for no security at all. It is difficult to see why we would incur those disadvantages just to present a false picture. However, I believe such comparisons of motives need not be relied upon exclusively. Instead, the Committee should undertake the second option which is for this Committee to request the General Accounting Office to conduct a thorough investigation of the concerns expressed here today and, without first reporting to any NRC office, report directly to the Congress and simultaneously to the public on their findings.

The issues which I believe should be explored fall in two general categories: the institutional problems of the regulatory process and the safety problems which remain unresolved, in part because of the institutional problems.

The institutional issues which should be investigated are:

1. How does the NRC rationalize licensing and continued operation of reactors in the face of safety problems which have remained unresolved for literally years?
2. To what extent is the existence of these problems and their full safety significance made known to the public and to Atomic Safety and Licensing Boards for their independent evaluation of the propriety and legality of licensing and continued operation of reactors?
3. To what extent do the technical members of the Regulatory Staff believe they are pressured to approve reactor licenses and continued reactor operation when they believe safety problems exist?
4. Does the NRC have a policy of encouraging its technically qualified employees to freely discuss with upper management, the public, and their Senators and Representatives their concerns about safety and about NRC policy? What mechanisms or proce-

dures have been developed and used to encourage such frank discussions?

5. Does the NRC have a policy of revealing to the public the existence of disagreements about safety problems among its employees?

6. How are safety standards, rules and regulations developed? Does the regulated industry or the regulator have more voice in this process? Is that process open to the public early—before the standards are adopted?

7. Have the regulations developed by the old promotional AEC been reviewed to assure they are compatible with the legislative mandate given the NRC?

8. When judgments are made, are they purely subjective? What are the consequences of an error in judgment? What research needs to be done to allow more objective decisions?

The safety issues that should be investigated are:

1. What are the unresolved safety problems that have been identified by the NRC, its Staff, its Advisory Committees and concerned citizens?

2. What is the significance of each unresolved safety problem in terms of its potential effects on the public health and safety?

There are any number of examples of these safety problems to choose from—and they all need to be evaluated. In addition to the generic safety problems affecting many plants, there are specific problems affecting both old and new plants now operating and under construction: vital circuit breakers which cannot be tested adequately anywhere in the world; emergency systems under construction for which not even computer analyses, much less experimental data, are available; unknown earthquake magnitudes that may be greater than plant design limits; construction permits issued without adequate technical bases; a one-of-a-kind reactor with a multitude of unique problems; and several operating reactors which have serious, known safety deficiencies which have been given legal exemptions from the regulations. There certainly is no shortage of problems to be explored.

The concerns expressed today are real and genuine. They will not go away merely because the NRC denies them. If they are not surfaced and resolved now, they will remain and increase in number until the day when one of the unresolved safety problems becomes a catastrophe rather than just an engineer's theory. So far we have been lucky. Unresolved safety problems have merely resulted in unscheduled reactor shutdowns, deratings or expensive backfittings. When the problems become so severe that failure to resolve them can no longer be tolerated by anyone, it could be that our dependence on nuclear power will have grown to the point that the nation will not have the opportunity to pause and reassess the viability of commercial nuclear power.

Bibliography

Anderson, Bruce, and Riordan, Michael. *The Solar Home Book*. Harrisville, N.H.: Cheshire, 1976.

Backus, Charles, ed. *Solar Cells*. New York: IEEE Press, 1976.

Berger, John J. *Nuclear Power: The Unviable Option*. Palo Alto, Calif.: Ramparts Press, 1976.

Bookchin, Murray. *Our Synthetic Environment*. New York: Harper & Row, 1974.

Brodine, Virginia. *Radioactive Contamination*. New York: Harcourt Brace Jovanovich, 1975.

California Assembly Subcommittee on Resources, Land Use and Energy. *Hearings on Proposition 15—The Nuclear Initiative*. Sacramento, Calif. Vols. I–XVI. October 14, 1975, to May 14, 1976.

California Assembly Subcommittee on Resources, Land Use and Energy. *Reassessment of Nuclear Energy in California*. Sacramento, Calif. May 10, 1976.

Carson, Rachel. *Silent Spring*. Boston: Houghton Mifflin, 1962.

Cochran, Thomas B. *The Liquid Metal Fast Breeder Reactor*. Baltimore: Johns Hopkins Press/Resources for the Future, 1974.

Commoner, Barry. *The Closing Circle*. New York: Jonathan Cape, 1972.

———. *The Poverty of Power*. New York: Knopf, 1976.

Curtis, Richard, and Hogan, Elizabeth. *Perils of the Peaceful Atom.* New York: Ballantine, 1969.

Daly, Herman E., ed. *Toward a Steady-State Economy.* San Francisco: Freeman, 1973.

Ehrlich, Paul R., with R. L. Harriman. *How to Be a Survivor.* New York: Ballantine, 1971.

Ehrlich, Paul R., and Ehrlich, Anne H. *The End of Affluence.* New York: Ballantine, 1974.

————. *Population, Resources, Environment.* Rev. ed. 1977. San Francisco: Freeman.

Energy Policy Project of the Ford Foundation. *A Time to Choose: America's Energy Future.* Cambridge: Ballinger, 1974.

Fitzgerald, A. Ernest. *The High Priests of Waste.* New York: Norton, 1970.

Ford, Daniel F., and Kendall, Henry W. *An Assessment of the Emergency Core Cooling Systems Rulemaking Hearings.* Cambridge: Union of Concerned Scientists and San Francisco: FOE, 1974.

Freeman, S. David. *Energy: The New Era.* New York: Vintage, 1974.

Friends of the Earth. *Stockholm Conference ECO,* vols. 1–3. San Francisco: FOE, 1972–1975.

Fuller, John G. *We Almost Lost Detroit.* New York: Readers Digest Press/Crowell, 1975.

Glasstone, Samuel. *Source Book on Atomic Energy.* 3rd ed. New York: Van Nostrand Reinhold, 1968.

Gofman, John W., and Tamplin, Arthur R. *Poisoned Power: The Case Against Nuclear Power Plants.* New York: New American Library, 1974.

Hardin, Garrett. *Exploring New Ethics for Survival: The Voyage of the Spaceship Beagle.* New York: Viking, 1972.

Hayes, Denis. *Rays of Hope: A Global Energy Strategy.* New York: Norton, 1977.

Hearings, Joint Committee on Atomic Energy, "Investigation of Charges Relating to Nuclear Reactor Safety," 94th Congress, 2nd Session. 2 vols. February 18–24 and March 2–4, 1976. Washington, D.C.: U.S. Government Printing Office, 1976.

Holdren, John, and Herrera, Philip. *Energy: A Crisis in Power.* San Francisco: Sierra Club, 1971.

Inglis, David. *Nuclear Energy: Its Physics and Social Challenge.* Reading, Mass.: Addison-Wesley, 1973.

Joint Committee on Atomic Energy. *Atomic Energy Legislation through the 94th Congress, 1st Session.* Washington, D.C.: USGPO, 1976.

Joint Review Committee of the Sierra Club and the Union of Concerned Scientists. *Preliminary Review of the AEC Reactor Safety Study.* San Francisco: Sierra Club and Cambridge: Union of Concerned Scientists, 1974.

Kendall, Henry W. *Nuclear Power Risks: A Review of the Report of the American Physical Society's Study Group on Light Water Reactor Safety.* Cambridge: Union of Concerned Scientists, 1974
———. "Public Safety and Nuclear Power." Testimony prepared for Hearings before the House Committee on Interior and Insular Affairs, Subcommittee on Energy and the Environment. Cambridge: Union of Concerned Scientists, April 29, 1975.

Knelman, Fred. *Nuclear Energy.* Edmonton, Canada: Hurtig, 1976.

Komanoff, Charles. *Power Plant Performance.* New York: Council on Economic Priorites, 1976.
——— et al. *The Price of Power: Electric Utilities and the Environment.* Cambridge: MIT Press, 1974.

Laitner, Skip et al. *A Citizen's Guide to Nuclear Energy.* Washington, D.C: Center for the Study of Responsive Law, 1975.

Lash, Terry R. et al. *Citizen's Guide: The National Debate on Handling Radioactive Wastes from Nuclear Power Plants.* Palo Alto, Calif.: NRDC, 1975.

Lewis, Richard S. *The Nuclear-Power Rebellion: Citizens vs. the Atomic Industrial Establishment.* New York: Viking, 1972.

Lovins, Amory B. *World Energy Strategies: Facts, Issues and Options.* San Francisco: FOE and Cambridge: Ballinger, 1975.

Lovins, Amory B., and Price, John H. *Non-Nuclear Futures: The Case for an Ethical Energy Strategy.* San Francisco: FOE and Cambridge: Ballinger, 1975.

Lowrance, William W. *Of Acceptable Risk: Science and the Determination of Safety.* Los Altos, Calif.: Kaufmann, 1976.

McCloy, John J. et al. *Report of the Special Review Committee of the Board of Directors of Gulf Oil Corporation.* Pittsburgh, Penn.: Gulf Oil Corporation, 1975.

McPhee, John. *The Curve of Binding Energy.* New York: Ballantine, 1975.

Meadows, Donella H. et al. *The Limits to Growth.* New York: New American Library, 1974.

Mesarovic, Mihajlo, and Pestel, Eduard. *Mankind at the Turning Point.* New York: New American Library, 1974.

Metzger, H. Peter. *The Atomic Establishment.* New York: Simon and Schuster, 1972.

Miller, G. Tyler. *Living in the Environment: Concepts, Problems and Alternatives.* Belmont, Calif.: Wadsworth, 1975.

Munson, Richard, ed. *Countdown to a Nuclear Moratorium.* Washington, D.C: Environmental Action Foundation, 1976.

Novick, Sheldon. *The Careless Atom.* New York: Dell, 1969.
———. *The Electric War.* San Francisco: Sierra Club, 1976.

Olson, McKinley C. *Unacceptable Risk: The Nuclear Power Controversy.* New York: Bantam, 1976.

Patterson, Walter C. *Nuclear Power.* Harmondsworth, Middlesex, England: Penguin Books Ltd., 1976.

Potter, Van Rensselaer. *Bioethics: Bridge to the Future.* Englewood Cliffs, N.J.: Prentice-Hall, 1971.

Primack, Joel and Van Hippel, Frank. *Advice and Dissent: Scientists in the Political Arena.* New York: New American Library, 1974.

Research and Education Association. *Modern Energy Technology.* 2 vols. New York: Research and Energy Association, 1975.

Reynolds, William C., ed. *The California Nuclear Initiative: Analysis and Discussion of the Issues.* Stanford University, Institute of Energy Studies, 1976.

Schumacher, E. F. *Small Is Beautiful: Economics as if People Mattered.* New York: Harper & Row, 1973.

Stone, Christopher D. *Should Trees Have Standing? Toward Legal Rights for Natural Objects* Los Altos, Calif.: Kaufmann, 1974.

_____. *Where the Law Ends: The Social Control of Corporate Behavior.* New York: Harper & Row, 1975.

Thompson, Theos, and Beckerley, J.G. *The Technology of Nuclear Reactor Safety.* Cambridge: MIT Press, 1973.

Union of Concerned Scientists. *The Nuclear Power Issue: A Source Book of Basic Information.* Cambridge: UCS, 1974.

_____. *The Nuclear Fuel Cycle: A Survey of the Public Health, Environmental, and National Security Effects of Nuclear Power.* Rev. ed. Cambridge: MIT Press, 1975.

United Nations, Department of Economic and Social Affairs. *World Energy Supplies 1950–1974.* ST/ESA/STAT/SER.J/19. New York: United Nations, 1976.

U. S. Atomic Energy Commission. *The Safety of Nuclear Power Reactors (Light Water-Cooled) and Related Facilities.* WASH-1250. Washington, D.C.: USAEC, 1973.

U. S. Energy Research and Development Administration. *Solar Energy: A Bibliography* vol. 1: Indexes; vol. 2: Citations. TID-3351-RIP 1/2. Springfield, Va.: NTIS, 1976.

U.S. Nuclear Regulatory Commission. *Reactor Safety Study: An Assessment of Accident Risks in U.S. Commercial Nuclear Power Plants,* WASH-1400. Washington, D.C: NRC, 1975.

Webb, Richard E. *The Accident Hazards of Nuclear Power Plants.* Amherst: University of Massachusetts Press, 1976.

Willrich, Mason, and Taylor, Theodore B. *Nuclear Theft: Risks and Safeguards.* Cambridge: Ballinger, 1974.

Winter-Berger, Robert. *The Washington Pay-Off.* New York: Dell, 1972.

Glossary

abnormal occurrence. An unscheduled incident or event which the Nuclear Regulatory Commission determines is significant from the standpoint of public health and safety, and which must be reported to NRC. Also referred to as a "reportable occurrence."

actinide. One of a series of heavy, radioactive metallic elements of increasing atomic number beginning with actinium (89) or thorium (90) and ending with atomic number 103.

Advisory Committee on Reactor Safeguards (ACRS). The group of reputable engineers, industry leaders, and nuclear engineering scholars appointed by the Nuclear Regulatory Commission to review and recommend action on nuclear plant license applications. The ACRS also reviews and approves the safety of reactor designs and systems.

architect-engineer (A-E). A corporation hired by an electric power company to construct the buildings and prepare the site for installation of the reactor system and turbine-generator. Frequently hired as architect-engineers are Bechtel, Ebasco, Stone & Webster, and Sargent & Lundy. Occasionally, the electric power company will act as its own A-E, as in the case of the Tennessee Valley Authority (TVA). Distinguished from "reactor systems manufacturer."

atom. The smallest unit of matter that remains unchanged after a chemical reaction. Most atoms are composed of fundamental particles, including but not limited to electrons, protons and neutrons. The nucleus of a typical atom includes protons (positive electric charge) and neutrons (neutral charge, each approximately the same mass as a proton). The nucleus is surrounded by a much less dense, outer domain consisting of electrons in orbit around the nucleus. Each electron has a mass much smaller than a proton's, and a negative charge equal in magnitude to a proton's. (See "neutron," "isotope" and "atomic particles.")

Atomic Energy Commission (AEC, USAEC). Until 1975, the AEC was the federal agency which controlled military use of atomic materials and which regulated and promoted civilian uses of atomic energy. In 1975, Congress split the AEC into two agencies: ERDA (Energy Research and Development Administration) and NRC (Nuclear Regulatory Commission). (See H. Peter Metzger, *The Atomic Establishment,* Simon and Schuster, 1972, for further information.)

Atomic Industrial Complex (AIC). The informal, interlocking directorate of electric power companies, federal agencies, and private corporations that promote and construct nuclear power plants, and that design, fabricate, and install plant systems and components. Also included are the industrial and professional engineering organizations (e.g., American Nuclear Society, Atomic Industrial Forum, Electric Power Research Institute) which favor electric generation by nuclear fission. The term is often used pejoratively by critics who believe that AIC organizations tend to follow a self-serving "party line" and to stonewall when criticized in public.

atomic particles. The nuclei of many radioactive substances spontaneously emit atomic particles when undergoing fission. An *alpha particle* is a helium nucleus (2 protons and 2 neutrons) which, although heavy and energetic, is able to penetrate only short distances into matter. Most alpha particles are stopped by a sheet of paper. A *beta particle* is an elementary particle emitted from a nucleus during radioactive decay, with a single electrical charge and a mass equal to $1/1837$ that of a proton. A negatively charged beta particle is identical to an electron. Beta particles penetrate matter much further than alphas; a few millimeters of aluminum will stop beta particles. More damaging than either alpha or beta particles is the *gamma ray,* which is characterized by far greater penetrating power and is considered a form of radiation rather than a particle although it, too, is emitted during fission of a radioactive substance.

availability. The percentage of time a plant is usable for generating electricity. In the following formula, "Unit Available Hours" refers to the time that the plant is operating normally during the Reporting Period. The formula for calculating availability factor is:

$$\frac{\text{Unit Available Hours}}{\text{Gross Hours in Reporting Period}} \times 100 = \frac{\text{Availability}}{\text{Factor}}$$

base-loaded. A *base-loaded* electric generating plant is set to generate its maximum output continuously because the owning corporation has determined that it produces electric power at a cost lower than other plants in the system. A base-loaded plant is distinguished from a *load-following* plant, which varies its power output according to the demands of the electric distribution system.

blowdown. A *controlled* reactor pressure vessel blowdown occurs when pressures and/or temperatures within the vessel exceed safe limits and the safety and/or relief valves open for a short time, discharging radioactive steam to dissipate in the drywell, suppression pool, and containment. An *uncontrolled* blowdown may occur when a pressure vessel steam or coolant line breaks, or when the vessel itself ruptures. Under these conditions, all the coolant may escape; enormous amounts of coolant and steam will accumulate in the drywell, suppression pool, and containment; and the reactor core may melt unless the decay heat is removed by deluging it with water.

boiling water reactor (BWR). A light water reactor system manufactured by General Electric in which water heated by the reactor core boils to form steam that is piped directly to a turbine. The BWR and pressurized water reactor (PWR) designs are the most common light water reactor (LWR) models.

breeder reactor (liquid metal fast breeder reactor—LMFBR). A nuclear reactor which uses molten sodium as a coolant and which produces more nuclear fuel than it consumes. Under bombardment with fast neutrons, plutonium-239 undergoes fission and uranium-238 is converted to fissionable plutonium-239. The conversion of uranium-238 occurs at a faster rate than the fissioning of plutonium-239, yielding a net fuel gain.

capacity factor. The amount of electricity a plant actually produced compared with the amount it would have produced if it had been operating at full power for the same period of time. The formula for calculating capacity factor is:

$$\frac{\text{Electricity Generated (MWe—hrs)}}{\text{Plant Rated Capacity (MWe)} \times \text{Hours on Line}} = \text{Capacity Factor}$$

carcinogen. A substance capable of producing cancer. Carcinogens identified in this book include plutonium, strontium-90, iodine-131, radon-222, and a number of radioisotopes.

cladding, fuel. In light-water reactors, the long, hollow tubing or rods in which reactor fuel pellets are hermetically sealed. Fuel cladding is manufactured from zircaloy metal, stainless steel or Magnox, substances which resist corrosion and withstand high temperatures. Fuel densification inside the cladding occurs when the fuel pellets shrink, leaving empty spaces which may cause the fuel rods to collapse or perforate. Cladding also refers to the layer of stainless steel which is applied by special welding techniques to the internal surfaces of the primary cooling system which are in contact with the liquid coolant.

component. Any machine, pump, valve, or constituent part of a system.

containment. The structure designed to confine radioactive gas, steam, and fluids after accidental release. A light-water reactor features three separate containments: (1) the reactor pressure vessel and piping, which contains the fuel and radioactive coolant; (2) the primary steel containment, which seals the volume containing the drywell and reactor pressure vessel; and (3) the concrete shield building, which encloses (1) and (2).

control rod. A rod containing stainless steel tubes, filled with boron carbide powder, which is inserted into the reactor core by hydraulic drives. The boron carbide absorbs neutrons emitted during fission. When control rods are withdrawn, less boron carbide is available for absorption, and the fission reaction *increases.* When control rods are inserted, neutrons are absorbed, and the rate of fission *decreases.* BWR control rods are inserted from the *bottom* of the reactor pressure vessel; PWR control rods are inserted from the *top.* In an emergency, control rods are inserted rapidly to terminate reactor operation (scram).

core. The complete assembly of fuel elements inside the reactor pressure vessel. Also present in the vessel on the periphery of the core are the recirculation systems, feedwater spargers, steam dryers, and separators that control the flow of steam and water inside the vessel.

critical. Refers to the fissioning process taking place among fuel elements in the reactor core. A *critical* reaction is one that is self-sustaining; the total number of neutrons in one "generation" of the chain reaction is the same as the number of neutrons in the next "generation" of the chain. The neutron density is neither increasing nor decreasing. A *supercritical* reaction is one where the increasing accumulation of neutrons results in an increase in power. A *subcritical* reaction is one where not enough neutrons are created to sustain a self-propagating chain reaction.

decay. The process by which a radioactive substance emits radiation in the form of alpha, beta, or gamma particles and, over a period of time, undergoes transformation from one nuclide into a different nuclide, or into a different energy state of the same nuclide. The process results in a decrease, with time, of the number of radioactive atoms in a sample. (See "half-life.")

decay heat. The heat in the reactor core generated, not by fission, but by radioactive decay of fission products. These products accumulate in nuclear fuel as it fissions; consequently, the longer a fuel pellet remains in the core, the greater the proportion of fission products and the greater the quantity of decay heat generated per pellet. Unlike atomic fission, decay heat cannot be "turned off" by inserting control rods, but must be controlled instead by circulating great quantities of coolant. A reactor with all control rods inserted but suffering a loss of coolant will rapidly overheat because of decay heat, and will melt down unless deluged with water from the emergency core cooling system.

derating. Restriction of power output of a reactor, usually by the NRC, to maintain its margin of safety. Derating occasionally follows the discovery of an unsafe condition that is not severe enough to warrant shutdown.

design basis accident (DBA). A hypothetical accident that a reactor system is designed to withstand without release of fissile material but which may still be severe enough to cause major damage. The DBA is used especially as the basis for designing reactor containment and engineered safeguards systems, the main purpose of which is to mitigate accident consequences. The DBA for light-water reactors is a loss-of-coolant accident caused by rupture of the coolant intake pipe. Note that DBA is not the *worst imaginable* accident. (See "maximum credible accident.")

design basis earthquake (DBE). A hypothetical earthquake that a reactor system is designed to withstand without release of fissile materials. (See "design basis accident.")

diversion. Theft, especially of nuclear material.

drywell. The open space between a boiling water reactor pressure vessel and the reinforced concrete drywell wall. If a steam or coolant line breaks, or if the reactor vessel ruptures, the drywell contains the escaping effluent.

emergency core cooling system (ECCS). A safety system designed to instantaneously flood the core of a light-water nuclear reactor with large amounts of water to prevent meltdown resulting from rapid heat accumulation. The ECCS is automatically activated if the reactor core water level drops below a certain point or if a loss-of-coolant accident (LOCA) occurs.

Energy Research and Development Administration (ERDA). The federal agency responsible for national energy programs including the promotion and development of nuclear energy. In 1975, the Atomic Energy Commission was split into two separate branches: ERDA and the Nuclear Regulatory Commission (NRC).

enriched uranium. Refined uranium in which the percentage of uranium-235 has been artificially increased. A sample of natural uranium normally includes 0.711 percent uranium-235. Enriched uranium fuel used in a nuclear reactor is approximately 97 percent uranium-238 and 3 percent uranium-235 *fast breeder.* See "breeder reactor."

fissile. See "fissionable material."

fission. The process by which a nucleus splits into fragments (fission products) and emits both heat and free neutrons. Fission may occur spontaneously or following the absorption of a neutron.

fissionable material. In reactor terminology, the radioactive material that is useful as fuel. *fissile* material is fissionable by neutrons that may be slow *or* fast (e.g. uranium-235). The meaning of *fissionable* material has been extended to include material that can be fissioned only by fast neutrons (e.g. uranium-238).

fission products. The fragments remaining after a nucleus fission. The fragments are nuclides, usually of lighter elements, which recoil from the split nucleus at high speeds for short distances. The kinetic energy is quickly transmitted to surrounding atoms and molecules by successive collisions, thereby raising their temperature. Most fission products are radioactive and decay, according to their characteristic half-lives, to stable nuclides. (See "decay heat.")

flux. A measure of neutron population and speed in a reactor. Also, the product of the speed and the number of neutrons per unit volume having that speed.

flux monitor. A device used to measure neutron flux in a reactor.

forced outage. Unplanned or unscheduled removal of a unit from service due to mechanical or safety problems resulting in loss of electric production. Distinguished from *scheduled outage.* The forced outage rate is calculated as follows:

$$\frac{\text{Forced Outage Hours}}{\text{Hours Generator on Line} + \text{Forced Outage Hours}} \times 100 = \frac{\text{Forced Outage Rate}}{}$$

fossil fuel. Coal, oil, or gas fuel. A *fossil-fueled plant* is one which uses one of these fuels to generate heat, as distinguished from a *nuclear plant,* which uses uranium fuel.

fuel assembly, fuel bundle. An assembly consisting of as many as 63 fuel rods held in precise geometric relation to one another. See Fig. 8 on p. 82. A *fuel bundle* is the collection of fuel rods. The fuel bundle is then sheathed in a protective metal capsule, called a "fuel channel." The composite bundle and channel together comprise a *fuel assembly.*

fuel rod. In light-water reactors, a zircaloy tube filled with uranium fuel pellets. Up to 63 fuel rods are loaded into a fuel assembly and up to 548 fuel assemblies are loaded into the reactor core. When the core is loaded and the control rods are withdrawn, over 100 tons of fuel create enough heat to convert large quantities of coolant to steam.

half-life. The length of time required for half of the atoms in any amount of radioactive substance to decay. Once the half-life duration of a given amount of radioactive material has elapsed, 50 percent of the volume of the material will have been transformed into one or more simpler, more stable isotopes or elements.

in-core instrumentation. Instruments attached to individual fuel elements to measure temperature and neutron flux in the reactor core.

ion. An atom or molecule which has lost or gained electrons and which has either a net positive charge (more protons than electrons) or a net negative charge (more electrons than protons). Examples of ions include: an alpha particle, which is a helium atom minus two electrons; a proton, which is a hydrogen atom minus its sole electron.

ionizing radiation. Alpha, beta and gamma radiation emitted by radionuclides. These radiations are energetic enough to dislodge planetary electrons from some of the atoms or molecules they encounter while traveling through matter. The "debris" left in the path of such ionizing radiation consists of *ion pairs*—positive and negative fragments of disrupted molecules—or freed planetary electrons and the positively charged residual atom. Living cells suffering ionizing radiation damage may later become cancerous. If the radiation damage occurs to chromosomes within the cell nucleus, the cell may mutate. If the damage occurs to germinal reproductive cells (spermatogonia and ova), mutations may result and be carried forward into every cell of an unborn human being, and on to future generations.

irradiation. The process of bombarding a material with neutrons.

isotope. Chemically identical variation of an element wherein both the element and its isotope have the same number of protons in the nucleus of each atom but different numbers of neutrons. Consequently, their atomic weights are different.

Joint Committee on Atomic Energy (JCAE). A permanent congressional committee including both senators and congressmen which, until 1977, provided oversight authority for the USAEC and which reviewed and (usually) approved USAEC, NRC and ERDA recommendations. The JCAE initiated legislation simultaneously for both Congress and the Senate and was stripped of its legislative power on December 8, 1976, by the House Democratic Caucus. (See H. Peter Metzger, *The Atomic Establishment,* Simon and Schuster, 1972, for the best popular study of the JCAE.)

kilo-. A prefix that multiplies a base unit by one thousand. A kilogram is one thousand grams.

light water reactor (LWR). Principally, boiling water reactors (BWR) and pressurized water reactors (PWR), which are both cooled by demineralized, deionized water and "moderated" by it (i.e., the coolant moderates the fission process by slowing down neutrons).

liquid metal fast breeder reactor (LMFBR). See "breeder reactor."

loss-of-coolant accident (LOCA). Rupture of the reactor pressure vessel or break in any one of the steam or coolant pipes penetrating the vessel at a point close enough to the penetration so that reactor coolant is lost. Likely results of a LOCA are overheating and meltdown due to decay heat, unless adequate cooling is immediately applied. Enormous quantities of coolant must be circulated rapidly and without interruption to carry off decay heat. Nearly all the coolant deluging the core from ECCS following a LOCA will be lost through the pipe or vessel break, in addition to the coolant that existed inside the reactor before the LOCA occurred.

maximum credible accident. The most serious reactor accident that can reasonably be imagined from any adverse combination of equipment malfunction, operating errors and other foreseeable causes. The *maximum hypothetical accident* is one in which the reactor pressure vessel ruptures, followed by rupture of the containment; this accident is one that the NRC does not use as a basis for safety analyses on the grounds that it is incredible.

mega-. A prefix that multiplies a base unit by one million. One megawatt is one million watts.

meltdown. The melting of some or all components in the reactor core that will occur if coolant is lost and emergency cooling systems fail to carry off heat emanating from fission or fission products. Several nuclear reactors have suffered expensive partial meltdowns. A full meltdown will result in the melting of the reactor fuel core through the bottom of the containment structure and into the earth, followed by radiation release into the atmosphere due to hydrogen or steam explosion.

micro-. A prefix that divides a basic unit by one million. One microgram is one-millionth of a gram.

milli-. A prefix that divides a basic unit by one thousand. One milligram is one thousandth of a gram.

moderator. A material of low atomic weight (light water, heavy water, graphite) used in the reactor core to slow down fast neutrons so that they will be more likely to collide with fissile atoms.

mutation. A change in hereditary material of a cell involving either a physical change in chromosome relations or a fundamental change in genes. (See "ionizing radiation.")

MWd. Megawatt-day. The thermal energy produced in twenty-four hours when operating at one-megawatt power level. 1 MWd = 24 hours × 1,000 kilowatts.

MWe. Megawatts of electrical energy.

MWt. Megawatts of thermal energy.

neutron. A neutral particle contained in the atomic nucleus. During fission, a neutron may be ejected at high energy. Neutron energy levels may be lowered by collision with atoms in the water or graphite moderator, and the neutron will be slowed. Such a neutron, if it collides with an atom of fissile material, may cause the fissile atom to split (fission) and release energy.

Nuclear Regulatory Commission (NRC). The federal agency responsible for licensing and overseeing all nuclear reactor systems in the United States.

nuclide. The nucleus of an isotope. The nucleus is positively charged, contains nearly all of the atom's mass, and measures about 10^{-5} times the diameter of the atom. (See "isotope.")

plutonium. A heavy artificial metal created by neutron irradiation of uranium-238. Plutonium is highly radioactive, extremely toxic chemically, and fissionable. Its half-life is 24,360 years.

posterity rights. A jurisprudential concept wherein unborn generations may be endowed by today's government and courts with certain rights and interests which are protected by court-appointed trustees. Posterity rights may include, but need not be limited to, the right to freedom from chemical and isotopic pollution generated by forebears, the right to adequate natural resources, wilderness and living space, and the right to an intact genetic heritage.

pressure vessel. The massive steel vessel which contains the reactor fuel core, moderator, reflector, thermal shield and control rods. The pressure vessel is designed to withstand high pressure and impact and is manufactured of stainless steel of the highest quality.

pressurized water reactor (PWR). A light-water reactor manufactured by Westinghouse, Babcock & Wilcox, and Combustion Engineering, consisting of two circulation loops. In the primary loop, pressurized water heated by the reactor core is circulated through a heat exchanger and back into the core. In the heat exchanger, the heat is transferred from radioactive water in the primary loop to nonradioactive water in the secondary loop, where it becomes steam, which is piped directly to a turbine.

Price-Anderson Act. A 1957 act of Congress, renewed in 1966 and 1975, limiting the insurance liability of electric power companies and their contractors in the event of a nuclear plant accident.

prompt critical. This refers to LMFBR operation. Prompt neutrons are those released at the time of fission. *Prompt critical* is a condition where a reactor is critical due to prompt neutrons alone and where the reaction rate increases extremely rapidly. *Prompt criticality* is the threshold for an explosive nuclear runaway.

quality assurance (QA). A management system for controlling quality of safety-related systems and components, assuring traceability and personal accountability. It is implemented through detailed procedures which state the quality objectives and regulate the performance of all safety-related activities from design through plant maintenance and decommissioning.

rad (radiation absorbed dose). A measure of radiation absorbed by a target expressed as the amount of energy (in ergs) per gram of absorbing material. (See also "rem.")

radiation. Particles or waves (neutrons, alpha, or beta particles, or gamma rays) which are emitted by atomic nuclei and which may fission or alter other atoms with which they collide.

radioactivity. A property of unstable elements whereby radiation is spontaneously emitted. (See "radiation.")

radioisotope. A radioactive isotope whose nuclei are capable of spontaneously emitting radiation. (See "radiation" and "isotope.")

radwaste (radioactive waste). Refers to the radioactive end products of nuclear plants and fuel processing.

reactor system. The entire nuclear steam supply system including reactor vessel, fuel load, turbine-generator, all valves, piping, instrumentation, control room, and associated systems.

reactor system manufacturer. The corporation hired by an electric power company to manufacture and install the reactor system. Reactor system manufacturers include General Electric, Westinghouse, Babcock & Wilcox, and Combustion Engineering. Distinguished from "architect-engineer."

refueling. Replacement of reactor core fuel assemblies containing depleted fuel. The refueling schedule varies: one reactor model is refueled once every six months; during each refueling, approximately one-third of the total fuel load is replaced. Thus, this reactor replaces its fuel entirely once every eighteen months. Fuel must be replaced because it tends to lose its reactivity during the fissioning process and because fission products which absorb neutrons build up in the fuel.

rem (roentgen equivalent man). Amount of rads absorbed by a target multiplied by the relative biological effectiveness of the type of radiation. Rads from heavy particles cause greater biological effects than rads from gamma or beta radiation.

reprocessing. Nuclear fuel that has undergone fission for several months is removed and reprocessed to recover valuable uranium and plutonium that can be used again. The recovered fissile material is sent to an enrichment plant where it is added to newly-mined uranium. The end-product is enriched nuclear fuel which contains about 3 percent uranium-235 instead of 0.7 percent. No commercial fuel reprocessing facilities were operating in the U.S. as of February, 1977.

runaway. Uncontrolled chain reaction or power excursion. An extremely rapid rise (to extreme peak levels) and fall in reactor power, resulting in an explosive burst of energy before the excursion is terminated.

scheduled outage. Planned removal of a unit from service for refueling, inspection or maintenance resulting in loss of electric production. Distinguished from "forced outage."

scram. Rapid shutdown of the reactor to prevent or to mitigate the consequences of an accident, by the rapid insertion of control rods. A reactor may be scrammed manually or automatically.

sparger. A perforated submerged pipe in a pressure vessel used for introducing fluid into the vessel below the surface of the core coolant.

startup. The phase of plant development immediately prior to operation, including checkout of steam and reactor systems at low power levels. The gradual withdrawal of control rods before a reactor "goes critical."

steam explosion. A hypothetical accident following a meltdown, in which the melted fuel comes into contact with water and, through a dynamic self-mixing process, most of the heat of the melted material is transferred to the water in a very short time, resulting in a violent explosion.

suppression pool. The water pool below and surrounding the drywell into which steam escaping from a BWR pressure vessel is channeled. The pool reduces the escaping steam pressure by condensing it and by providing additional volume for expansion. It also provides an additional water source for the emergency core cooling system.

torus. The vessel containing the pressure suppression pool in the Mark I BWR.

transient. A rapid change in the reactor temperature or pressure, leading to voids or other disturbances. Following a transient, fuel rods in the vicinity may be deprived of sufficient coolant and may rupture or melt. A transient is considered an emergency condition often justifying immediate scram to prevent core damage.

transmutation. The conversion of one element or nuclide into another, either naturally or artificially.

trip. A rapid, unscheduled shutdown of a reactor, synonymous with *reactor scram.* Also used with reference to operating components, as in a *turbine trip* (turbine is shut down) or *tripping* of recirculation pumps (the pumps are turned off).

turbine-generator. The coupled system including a steam turbine and electric generator which converts steam pressure to electricity.

vendor. A company which supplies components, materials, services, or systems to the electric power company, the reactor systems manufacturer, or the architect-engineer. Otherwise synonymous with *reactor manufacturer.*

watt. The unit of power equal to 1 joule/second in the metric system for measuring mechanical units, otherwise referred to as "the mks system," (meter, kilogram, second). The unit of work is one newton-meter, referred to as one joule, which is the work done by a unit force (newton) while moving a mass a unit distance (meter) in the direction of the force.

Organizations
and Periodicals

Two directories will be especially helpful to readers who wish to contact any of the organizations listed below. Current addresses and telephone numbers of federal agencies are published in the *United States Government Manual,* Office of the Federal Register, General Services Administration. Information on private organizations may be found in the *Encyclopedia of Associations,* vol. 1., Gale Research Company, Detroit, Michigan. Both reference texts are published annually and are available in most public libraries. The following information was correct as of March, 1977.

American Association for the Advancement of Science, 1515 Massachusetts Avenue NW, Washington, DC 20005. Telephone:202–467–4350. The largest national organization representing all scientific disciplines. Publication: *Science,* weekly. Annual subscription: $28 (members); $60 (nonmembers). Publications list on request.

American Nuclear Society, 555 North Kensington Avenue, La Grange Park, IL 60525. Telephone: 312–352–6611. Nuclear engineering, scientific, and related disciplines. U.S. professional organization for nuclear industry employees. Publication: *Nuclear News,* monthly. Annual subscription: $55 ($27.50 to members, allocated from dues). Publications list on request.

Atomic Industrial Forum, 7101 Wisconsin Avenue, Washington, DC 20014. Telephone: 301–654–9260. Trade organization for the nuclear industry. Publications: *Nuclear Industry,* monthly; *AIF Nuclear INFO,* monthly. Publications list on request.

Center for Science in the Public Interest, 1757 S Street NW, Washington, DC 20009. Telephone: 202–332–4250. Citizen action for energy policy reform in the public interest. Publication: *People & Energy,* monthly. Annual subscription: $10 (individuals); $16 (firms, libraries, government agencies). Publications list on request.

Citizens for a Better Environment, Suite 2610, 59 East Van Buren, Chicago, IL 60605. Telephone: 312–939–1984. 536 West Wisconsin Ave., Milwaukee, WI 53203. Telephone: 414–271–7475. Air and water pollution control, nuclear safety and economics, energy conservation. Toxic substances, environmental health research and litigation. Annual membership: $15. Monthly publication: CBE Environmental Review. Others on request.

Committee for Nuclear Responsibility, Inc., Main P.O. Box 11207, San Francisco, CA 94101. John W. Gofman, M.D., chairman. Hazards of and alternatives to nuclear power. Publications: Flyers, essays, technical reports. Donations tax-deductible.

Common Cause, 2030 M Street NW, Washington, DC 20036. Telephone: 202–833–1200. National lobby dedicated to making the federal and state governments accountable to citizens. Annual membership: $15.

Conservation Foundation, 1717 Massachusetts Avenue NW, Washington, DC 20036. Telephone: 202–797–4300. Environmental quality, research, and education. Publication: monthly newsletter. Annual subscription: $10. Others on request.

Council on Economic Priorities, 84 Fifth Avenue, New York, NY 10011. Telephone: 212–691–8550; 250 Columbus Avenue, San Francisco, CA 94133. Telephone: 415–989–7506; 5443 South Kenwood, Chicago, IL 60615. Telephone: 312–288–2934. Comparative studies of corporate performance, economic impacts, nuclear industry. Corporate accountability research. Publication: newsletter, 10 issues per year. Tax-deductible annual subscription: $15; $7.50 for students, unemployed, and retired; tax-deductible. Others on request.

Council on Environmental Quality, 722 Jackson Place, Washington, DC 20006. Telephone: 202–382–1415. Executive Office of the President. Recommendations for national environmental policy. Annual report on environmental conditions and trends.

Critical Mass, P.O. Box 1538, Washington, DC 20013. A Ralph Nader organization. Citizen movement for safe energy and public education regarding nuclear hazards, utility rates, and economic and political problems of the nuclear industry. Publication: *Critical Mass*, monthly newspaper. Annual subscription: $7.50 (individuals); $37.50 (businesses and institutions).

Ecology Law Quarterly, School of Law, Boalt Hall, University of California, Berkeley, CA 94720. Annual subscription: $8.

Edison Electric Institute, 90 Park Avenue, New York, NY 10016. Telephone: 212–573–8741. The principal association of the nation's investor-owned electric utilities. Publications: weekly statistical reports on electric power output; bimonthly magazine.

Educational Foundation for Nuclear Science, Inc., 1020–24 East 58th Street, Chicago, IL 60637. Telephone: 312–363–5225. Publication: *The Bulletin of the Atomic Scientists*, ten issues per year. Internationally recognized for authoritative analyses of the impact of science and technology on public affairs. Annual subscription: $18.

Electric Power Research Institute, 3412 Hillview Avenue, Palo Alto, CA 94304. Telephone: 415–493–4800. Management and funding of research pertaining to electric generation and transmission. Publications list on request.

The Energy Daily, 1239 National Press Building, Washington, DC 20045. Telephone: 202–638–4260. Published Monday through Friday. Annual subscription: $475. Formerly *Weekly Energy Report*. Detailed, comprehensive coverage of all national energy news, federal energy legislation, nuclear industry events.

Energy Research and Development Administration (ERDA), 20 Massachusetts Avenue NW, Washington, DC 20545. Telephone: 202–976–4000. Federal agency responsible for national energy research and development. Publications available at ERDA Public Document Room, or write to Office of Public Affairs, ERDA headquarters.

Environmental Action Foundation, 724 Dupont Circle Building, Washington, DC 20036. Telephone: 202–659–9682. Research and public education pertaining to energy, environmental, solid-waste-management and transportation issues. Publication: *The Power Line*, monthly. Annual subscription: $15 (individuals and firms); $7.50 (citizen and public interest organizations).

Environmental Action, Inc., Suite 731, 1346 Connecticut Avenue NW, Washington, DC 20036. Telephone: 202–833–1845. Publication: *Environmental Action*, biweekly. Annual subscription: $15.

Environmental Defense Fund, 162 Old Town Road, East Setauket, NY 11733. Telephone: 516-751-5191.

Environmental Law. The Lewis and Clark Law School, Northwestern School of Law, 10015 S.W. Terwilliger Blvd., Portland, OR 97219. Annual subscription: $8.

Environmental Protection Agency (EPA), Waterside Mall, 401 M Street SW, Washington, DC 20460. Telephone: 202-655-4000. Federal agency responsible for pollution control, environmental policy enforcement, research and development, radiation standards. Publications available from EPA Office of Public Affairs.

Federal Energy Administration (FEA), New Post Office Building, 12th Street and Pennsylvania Avenue NW, Washington, DC 20461. Telephone: 202-566-2000. Federal agency for U.S. energy supply, priorities, policy. For publications, write FEA Office of Public Affairs.

Federal Power Commission (FPC), 825 North Capitol Street NE, Washington, DC 20426. Telephone: 202-655-4000. Federal agency responsible for regulating interstate activities of electric power and natural gas industries. Publications available from FPC Office of Public Information.

Federation of American Scientists (FAS), 307 Massachusetts Ave. SE, Washington, DC 20002. Telephone: 202-546-3300. Public interest lobby. Research, education and publication of scientific and social developments. Reviews of bioethical controversies. Publications: *FAS Public Interest Report* and *Professional Bulletin.* Annual subscription: $20.

Friends of the Earth (FOE), 124 Spear St., San Francisco, CA 94105. Telephone: 415-495-4770. Leading international conservation organization. Promotes debate on all major environmental issues. Biweekly environmental newspaper: *Not Man Apart.* Annual subscription and dues: $20 (regular); $10 (student and retired). Other publications on request.

Institute for Energy Analysis—Oak Ridge Associated Universities, P.O. Box 117, Oak Ridge, TENN 37830. Publications list in preparation.

National Intervenors, Inc., 1757 S Street NW, Washington, DC 20009. Telephone: 202-543-1642. Formerly Consolidated National Intervenors. Clearinghouse for environmental and citizen groups representing public interest opposition to nuclear power. Periodic publications.

National Technical Information Service (NTIS), U.S. Department of Commerce, Springfield, VA 22161. Distribution center for all federal scientific and technical publications. List on request.

Natural Resources Defense Council, (NRDC), 20 West 43rd Street, New York NY 10036; 917 15th Street NW, Washington, DC 20005; 2345 Yale Street, Palo Alto, CA 94306. Environmental protection, litigation in the public interest; extensive, scholarly publications on LMFBR, waste disposal, fuel cycle. Annual membership: $15, includes quarterly newsletter. Publications list on request.

Nuclear Regulatory Commission (NRC), Washington, DC 20555. Telephone: 301–492–7715; Office of Public Affairs. Federal agency responsible for regulation and licensing of commercial U. S. nuclear plants, facilities and materials. Inspect or copy publications at Main Public Document Room, 1717 H Street NW, Washington, DC 20555. For local document rooms, write Public Document Room, USNRC, Washington, DC 20555.

Nuclear Safety Information Center (NSIC), Oak Ridge National Laboratory, Post Office Box Y, Oak Ridge, TENN 37830. Telephone: 615–483–8611, ext. 3–7253. National clearinghouse for nuclear safety information. Computer file of document abstracts. Technical inquiry and computer search service. Publications: *Nuclear Safety*, six issues per year, $16.70, through the U. S. Government Printing Office.

Nucleonics Week, McGraw-Hill, Inc., 1221 Avenue of the Americas, New York, NY 10020. Telephone: 212–997–3194. Weekly in-depth summary of nuclear industry news, relevant federal legislation. Annual subscription: $440.

Public Citizen, P.O. Box 19404, Washington, DC 20036. Telephone: 202–293–9142. A Ralph Nader organization. Clearinghouse for citizen advocates. Groups include: Tax Reform, Litigation, Health Research, Citizen Action, Congress Watch and Visitors' Center. Supported by public donations. Publications: newspaper, annual report. Others on request.

Public Interest Research Group (PIRG), 1346 Connecticut Avenue NW, P.O. Box 19312, Washington, DC 20036. A Ralph Nader organization for public policy research, including nuclear power and alternatives. List of publications on request.

Resources for the Future, 1755 Massachusetts Avenue NW, Washington, DC 20036. Research, education and publications of conservation, quality of the environment. Publications: Annual report. Others on request.

Scientific American, 415 Madison Avenue, New York, NY 10017. Frequent scholarly, readable articles on nuclear technology, weapons proliferation. Annual subscription: $18.

Scientists' Institute for Public Information, 560 Trinity Avenue, St. Louis, MO 63130. Publication: *Environment,* ten issues per year. Thorough coverage of all environmental issues. Annual subscription: $12.75 (individuals); $17.50 (institutions).

Sierra Club, 530 Bush Street, San Francisco, CA 94108. Telephone: 415–981–8634. Leading national conservation organization. Publishes scientific and general studies on environmental issues. Publications: *Sierra Club Bulletin, Weekly National News Report.* List of books on request.

Superintendent of Documents, U.S. Government Printing Office, Washington, DC 20402 Telephone: 202–783–3238. Congressional and agency publications. For technical and scientific reports, see NTIS.

Task Force Against Nuclear Pollution, Box 1817, Washington, DC 20013. Supports restrictions on nuclear industry and assists congressional representatives. Encourages solar energy research and legislation. Donations requested. Publications: quarterly Progress Reports.

Union of Concerned Scientists, 1208 Massachusetts Avenue, Cambridge, MA 02138. Telephone: 617–547–5552. Coalition of scientists, engineers, and other professionals concerned with impact of technology. Energy policy, nuclear safety, radioactive-waste disposal, nuclear weapons proliferation, environmental hazards. One of the most effective nuclear industry intervenors. Publications list on request.

Worldwatch Institute, 1776 Massachusetts Avenue NW, Washington, DC 20036. Telephone: 202–452–1999. Interdisciplinary research of global issues: energy, nuclear fission, weapons proliferation, environmental problems, the changing role of women, economic and population studies. Publications list on request.

Zero Population Growth (ZPG), 1346 Connecticut Avenue NW, Washington, DC 20036. Nonprofit political organization committed to sustaining the quality of life by reducing per capita consumption of resources and energy, and by stabilizing population. Membership: $15.

Notes

Congressional committee publications may be ordered from the Superintendent of Documents, U.S. Government Printing Office, Washington, D.C. 20402. Technical and scientific reports published by federal agencies are available from the National Technical Information Service, U.S. Department of Commerce, Springfield, Virginia 22161. Public Document Rooms maintained by the Nuclear Regulatory Commission (NRC) and by the Energy Research and Development Administration (ERDA) are government libraries for all correspondence, reports, and other publications pertaining to the nuclear industry which are not available through municipal or university libraries. The location of local public document rooms throughout the U.S. can be obtained by writing to the NRC and ERDA. In the list of Organizations and Periodicals which follows the Notes, addresses and identifications of publishing organizations are provided.

Abbreviations Used in the Notes

AAAS	American Association for the Advancement of Science		Resources, Land Use and Energy
AEC	U.S. Atomic Energy Commission	ERDA	U.S. Energy Research and Development Administration
APS	American Physical Society	FOE	Friends of the Earth
CAC	California Assembly Subcommittee on	FRC	Federal Radiation Council

GAO	U.S. General Accounting Office	RSS	USNRC Reactor Safety Study
IAEA	International Atomic Energy Agency	TVA	Tennessee Valley Authority
JCAE	Congressional Joint Committee on Atomic Energy	UCS	Union of Concerned Scientists
MIT	Massachusetts Institute of Technology	USAEC	U.S. Atomic Energy Commission
NRC	U.S. Nuclear Regulatory Commission	USGAO	U.S. Government Accounting Office
NRDC	Natural Resources Defense Council	USGPO	U.S. Government Printing Office
NTIS	National Technical Information Service	USNRC	U.S. Nuclear Regulatory Commission

Foreword

1. Lee Schipper and A. Lichtenberg, *Efficient Energy Use and Well-Being: The Swedish Example*, Report LBL-4430 (Berkeley, Calif.: Lawrence Berkeley Laboratory, Energy Resources Group, 1976).

Introduction

1. Amory Lovins, *Energy Strategy: The Road Not Taken?* draft, March, 1976, p. 1. Published in *Foreign Affairs*, 55 (October, 1976): 65–96.
2. National Council of Churches of Christ, *The Plutonium Economy: A Statement of Concern*, background report, New York, September, 1975, p. 1. See also Harold Feiveson et al., "The Plutonium Economy," *Bulletin of the Atomic Scientists*, 32 (December, 1976): 10–14 et seq.
3. CAC, *Reassessment of Nuclear Energy in California: A Policy Analysis of Proposition 15 and Its Alternatives*, Sacramento, May 10, 1976.
4. Ibid., p. 2.
5. Ibid., p. 95.
6. John Berger, *Nuclear Power: The Unviable Option* (Palo Alto, Calif.: Ramparts Press, 1976), chs. 10, 11, 12. See also Marc Ross and Robert Williams, "Assessing the Potential for Fuel Conservation," forthcoming in *Technology Review*, 1977, also in CAC, *Hearings*, December 2, 1975, pp. 220–258. Bethe acknowledges conservation's potential contribution but insists "that nuclear fission is the only major nonfossil power source the U.S. can rely on" for the indefinite future; see Hans Bethe, "The Necessity of Fission Power," *Scientific American*, January, 1976, pp. 21–31.
7. John Gofman, *The Cancer Hazard from Inhaled Plutonium*, May, 1975, and *Estimated Production of Human Lung Cancers by Plutonium from Worldwide Fallout* (Yachats, Ore.: Committee for Nuclear Responsibility, July, 1975).
8. Gofman, *Cancer Hazard* and *Estimated Production of Human Lung Cancers.* Although most reactor plutonium, which is several times more toxic than plutonium-239, remains unprocessed in spent-fuel pools throughout the U.S., enough has been manufactured during the fission process to inflict severe damage, even if only a small portion is released. As of 1976, approximately 75 million gallons of high-level radioactive waste and 51 million cubic feet of low-level waste had been produced. Although plutonium constitutes only a

small portion of these wastes, the aggregate of all radioactive waste material stored throughout the U.S. poses a serious health problem, according to Mason Willrich, former visiting MIT professor of nuclear engineering. See David Burnham, "Radioactive Waste at the Nation's Nine Storage Centers Is Called a Major Health Hazard," *The New York Times,* September 8, 1976.

9. Union of Concerned Scientists, *The Nuclear Power Issue: A Source Book of Basic Information* (Cambridge: UCS, 1974), ch. 6; CAC, *Hearings,* Daniel Ford, October 21, 1976, pp. 74–79, 82; CAC, *Hearings,* Robert Augustine, October 29, 1975, p. 67.; Anthony Ripley, "Critics Say AEC Suppressed Safety Test Data," *The New York Times,* October 31, 1972.

10. *NRDC vs USNRC and Vermont Yankee Corp.* (D.C. Cir., July 21, 1976) at 6, 8, 13, 17 et seq. See especially testimony by Robert Pollard reprinted here on pp. 000–000, and resignation letter by Robert Fluegge reprinted here on pp. 000–000. A thorough analysis of the government hearing process, nuclear power plant licensing, and the impediments facing intervenors is Steven Ebbin and Raphael Kasper, *Citizen Group Uses of Scientific and Technological Information in Nuclear Power Cases* (Washington, D.C.: George Washington University, Program of Policy Studies in Science and Technology, NSF Grant GI-33432, 1973), pp. 5–12. Cf. CAC, *Hearings,* Bernard Rusche, October 22, 1975, pp. 32–54, for USNRC version of the licensing, hearing, and review process.

11. *NRDC vs USNRC* (July 21, 1976), p. 18 note. Cross-examination restrictions and USAEC hearings procedures may have resulted in inadequate ventilation of the issues. Ebbin and Kasper, *Citizen Group Uses of Scientific and Technological Information,* conclude that substantive due process was denied intervenors and citizen groups during USAEC hearings. See also CAC, *Hearings,* Daniel Ford, October 21, 1975, p. 77; Common Cause, *Stacking the Deck: A Case Study of Procedural Abuses by the Joint Committee on Atomic Energy,* Washington, D.C., 1976.

12. "NRC Proposes Amending Regulations," *Not Man Apart,* December, 1976, p. 9.

13. Amory Lovins and John Price, *Non-Nuclear Futures: The Case for an Ethical Energy Strategy* (San Francisco: FOE and Cambridge: Ballinger, 1975), pp. xviii–xix. Cf. Lovins et al., *Soft Energy Paths: Toward a Durable Peace* (San Francisco: FOE, 1977).

The Incident at Browns Ferry

1. USNRC press release, April 3, 1975. Also USNRC news release, August 5, 1975. After details of the fire were publicized, NRC's policy of maintaining a low-key, benign attitude toward safety problems appears to have been applied. See USNRC, *Report to the Congress on Abnormal Occurrences, January-June 1975,* NUREG 75/090 (Washington, D.C.: USNRC, October, 1975), and statements by Robert Pollard, *Nuclear News,* March, 1976, p. 47.

2. *Final Report of the Preliminary Investigating Committee: Fire at Browns Ferry Nuclear Plant on March 22, 1975,* TVA report, May 7, 1975.

3. USNRC, Office of Inspection and Enforcement, Region II, *Regulatory Investigation Report.* Signatory: Charles E. Murphy, chief, Facilities Construction Branch, July 25, 1975.

4. H. Peter Metzger, *The Atomic Establishment* (New York: Simon & Schuster, 1972), chs. 3–5. Richard S. Lewis, *The Nuclear Power Rebellion* (New York: Viking, 1972), chs. 4, 5. John G. Fuller, *We Almost Lost Detroit* (New York:

Reader's Digest Press/Crowell, 1975), chs. 3, 4, 9–11. David Burnham, "AEC Files Show Effort to Conceal Safety Perils," *The New York Times,* November 10, 1974. Don Cullimore, "Radioactive Wastes Buried by the Bureaucracy," *Environmental Action,* August 28, 1976, charges that top federal officials are suppressing concerns of government scientists over the environmental hazards posed by radioactive wastes.

5. Paraphrase of a debate by deposition conducted by *The Catalyst: The Publication of the Stanford Forum* between General Electric Company and Joel Primack, assistant professor of physics at the University of California. See *The Catalyst,* 2 (Spring, 1976): 19–20.

6. Denis Hayes, *Nuclear Power: The Fifth Horseman* (Washington, D.C.: World-watch Institute, 1976), p. 65, which cites TVA Public Information Office release of April 9, 1976 as source for these figures.

7. Material based on testimony of Henry W. Kendall in connection with a hearing on the Price-Anderson Act (Cambridge: UCS, released September 28, 1976). On page 15, Kendall points out that "later in the afternoon, with the accident still in course, the *last 4 remaining valves were lost.* If the course of the fire had been slightly different, and [if] all of these valves had been lost early on, melting might well have occurred. There would have been *no other way* of relieving the pressure" (emphasis Kendall's).

8. Thomas Whiteside, "A Countervailing Force," *The New Yorker,* October 15, 1973, p. 47.

9. See the sources mentioned in note 4. See also testimony by Robert Pollard, pp. 315–319; resignation letter by Robert Fluegge, pp. 184–185; and "Negligence and Cover-up," UCS, *Nuclear Power Issue,* ch. 6.

10. Letter dated February 6, 1976, from Robert D. Pollard to William A. Anders, with attached report.

11. Conversation with David Comey, October 5, 1976, based on correspondence with Bernard Rusche, USNRC, Office of Reactor Regulation, and confirmed by Robert D. Pollard. See also Daniel Ford et al., *Browns Ferry: The Regulatory Failure* (Cambridge: UCS, June, 1976).

The complete report of the federal investigation of the Browns Ferry fire may be obtained from the Superintendent of Documents, USGPO, Washington, D.C. 20402: JCAE, *Hearings,* 94th Cong., 1st Sess., "Browns Ferry Nuclear Plant Fire," September 16, 1975. Other reports are USNRC, *Report to the Congress on Abnormal Occurrences: January-June 1975;* and USNRC, *Recommendations Related to the Browns Ferry Fire,* NUREG-0050 (Washington, D.C.: USNRC, February, 1976).

Malignant Giant

1. See "The Incident at Browns Ferry," note 4.

2. Thomas Scortia and Frank Robinson, *The Prometheus Crisis* (New York: Doubleday, 1975), p. 11.

The first federal investigation of Karen's death ended with the publication by the USGAO of *Federal Investigation into Certain Health, Safety, Quality Control and Criminal Allegations at Kerr-McGee Corporation,* May 10, 1975. Results of the second investigation are included in the House Small Business Committee, Subcommittee on Energy and the Environment, *Hearings,* 95th Cong., April 26–27 and May 20–21, 1976. A third investigation was scheduled for 1976–1977 by the House Subcommittee on Energy and the Environment until its chairman, Representative John Dingell (D-Mich.) was removed by the

House Democratic Caucus on December 6, 1976. In *Rolling Stone*, January 13, 1977, pp. 30–39, Howard Kohn alleges that Karen may have been murdered because the documents in the manila folder she was carrying the night she died included evidence of a plutonium smuggling ring within Kerr-McGee, although she may not have been aware of either the ring or the real nature of the documents.

Incidents reported in "Malignant Giant" were published in *The New York Times* on November 19, 1974, December 24, 1974, January 8, 1975. and January 22, 1975. *Nuclear News* carried the industry's version of the story in these issues: January, 1975 (pp. 38–40); February, 1975 (pp. 39–40); April, 1975 (pp. 62–71); June, 1975 (p. 60); November, 1975 (pp. 116–118); and June, 1976 (pp. 58–61).

For studies of the plutonium economy, see: Robert Gillette, "Plutonium I: Questions of Health in New Industry," *Science*, September 20, 1974, pp. 1027–1032; "Plutonium II: Watching and Waiting for Adverse Effects," *Science*, September 27, 1974, pp. 1140–1143; "On Inhaling Plutonium: One Man's Long Story," *Science*, September 20, 1974, pp. 1028–1029; Arthur Tamplin and Thomas Cochran, *Radiation Standards for Hot Particles*, (Washington, D.C.: NRDC, 1974); "Plutonium and the 'Hot Particle Problem,'" *Science*, March 1, 1974, pp. 834–835; J. G. Speth et al., "Plutonium Recycle: The Fateful Step," *Bulletin of the Atomic Scientists*, 30 (November, 1974): 15–22; Harold Feiveson et al., "The Plutonium Economy, *Bulletin of the Atomic Scientists*, 32 (December, 1976): 10–14 et seq. Environmental contamination is reported in *The Plutonium Situation at Rocky Flats*, Environmental Action of Colorado, November, 1975. See also S. E. Poet and Edward Martell, "Plutonium-239 and Americium-241 Contamination in the Denver Area," *Health Physics*, 23 (October, 1972): 537–548. R. W. Ayres, "Policing Plutonium: The Civil Liberties Fallout," *Harvard Civil Rights-Civil Liberties Review*, 10 (Spring, 1975): 367–443. For industry position, see Cyril Comar, "Plutonium Facts and Inferences," *EPRI Journal* (1900).

We Almost Lost Detroit

1. *Power Reactor Development Co. vs Int'l. Union of Elec. Workers*, 81 S. Ct. 1529 (1961).
2. Ibid., p. 1539.
3. Richard Webb, *The Accident Hazards of Nuclear Power Plants* (Amherst: University of Massachusetts Press, 1976), ch. 10. Confirmed by USGAO, *The Liquid Metal Fast Breeder Reactor: Promises and Uncertainties*, B164105 (Washington, D.C., July 31, 1976), p. 65.
4. See note 2.
5. Richard Danzig, "Six Legal Puzzles," *The Stanford Magazine*, Stanford University, Fall/Winter, 1975, pp. 42–43.
6. Ibid.
7. U.S. Senate, Committee on Government Operations, Subcommittee on Reorganization, Research and International Organizations, *Hearings*, 93rd Cong., 2nd Sess., "To Establish a Department of Energy and Natural Resources, Energy Research and Development Administration, and a Nuclear Safety and Licensing Commission," March 12, 1974, pp. 217–226.
8. See note 1.
9. *NRDC vs USNRC*, July 21, 1976.

10. *Calvert Cliffs' Coordinating Comm. vs USAEC,* 419 F. 2nd 1109 (D.C. Cir. 1971).
11. Terry Lash et al., *Citizens' Guide: The National Debate on the Handling of Radioactive Wastes from Nuclear Power Plants* (Palo Alto, Calif.:NRDC, 1975), p. 1.
12. "Udall Hearings Explore Nuclear Power," *Public Interest Report* 28 (Washington, D.C.: Federation of American Scientists, May-June, 1975).
13. Allen Hammond, "Complications Indicated for the Breeder," *Science,* August 30, 1974, p. 768.; "Morton and Zarb Join in Suggesting a Slowdown on Nuclear Breeder Reactors and Call for More Research," *The New York Times,* June 10, 1975.
14. Berger, *Nuclear Power,* pp. 149–150.
15. Thomas Cochran, *The Liquid Metal Fast Breeder Reactor: An Environmental and Economic Critique* (Baltimore: The Johns Hopkins Press/Resources for the Future, 1974); Amory Lovins, "The Case Against the Fast Breeder Reactor," *Bulletin of the Atomic Scientists,* 29 (March, 1973): 29–35; Sheldon Novick, "Nuclear Breeders," *Environment,* 16 (July-August, 1974); Irvin Bupp and Jean-Claude Derian, "The Breeder Reactor in the U.S.," *Technology Review,* 76 (July-August, 1974): 26–36.
16. Les Amis de la Terre, *L'Escroquerie Nucleaire,* Lutter/Stock No. 2 (Paris: Editions Stock, 1975). See also Berger, *Nuclear Power,* p. 139.
17. "Anders Reveals Doubts on Breeder Reactor," *Baltimore Sun,* March 18, 1976. As the Ford administration ended in 1976, signs pointed increasingly toward cancellation of the Clinch River demonstration breeder. Ford's decision in November, 1976, to delay spent fuel reprocessing (*Energy Daily,* December 7, 1976, p. 3) was followed by incoming President Carter's announcement that $200 million originally allocated to the breeder would be shunted to development of less dangerous energy sources.
18. CAC, *Reassessment,* pp. 60–61. See also John Holdren, "Uranium Availability and the Breeder Decision," *Energy Systems and Policy,* 1 (1975): 205–232; and Brian Chow, *The Liquid Metal Fast Breeder Reactor: An Economic Analysis,* American Enterprise Institute for Public Policy Research, December, 1975.

Primary sources for this article were R. L. Scott, Jr., "Fuel-Melting Incident at the Fermi Reactor on Oct. 5, 1966," *Nuclear Safety,* 12 (March–April, 1971): 122–134; Atomic Power Development Associates, *Report on the Fuel Melting Incident in the Enrico Fermi Atomic Power Plant,* USAEC Report APDA-233, December 15, 1968; JCAE, *Hearings,* January 30 and 31, and February 5 and 6, 1968, pp. 215–221; Atomic Power Development Associates, *Monthly Reports* 1–3, August–October, 1966. P. V. Evans, ed., *Fast Breeder Reactors* (New York: Pergamon, 1967) is a thorough review of the technology.

What You Don't Know Will Kill You

1. J. Martin Brown, "Health, Safety and Social Issues of Nuclear Power and the Nuclear Initiative," in William C. Reynolds, ed., *The California Nuclear Initiative* (Stanford, Calif.: Stanford University, Institute of Energy Studies, 1976), p. 139. See also John Gofman and Arthur Tamplin, *Poisoned Power* (Emmaus, Penn.: Rodale Press, 1971).
2. Paraphrase of Medical Research Council, *The Hazards to Man of Nuclear and Allied Radiations* (London: Her Majesty's Stationary Office, 1956), pp. 11–12.
3. See notes 1 and 2.

4. Gould Andrews, "Radiation Accidents," in John M. Fowler, ed., *Fallout* (New York: Basic Books, 1960), p. 112. See also Charles F. Zimmerman, ed., "Accidents in the Nuclear Industry: An Information File," unpublished compilation, Cornell University, 1973).

5. Karl Morgan and J. Turner, *Principles of Radiation Protection* (New York: Wiley, 1967). See also National Academy of Sciences/National Research Council, Report of the Advisory Committee on the Biological Effects of Ionizing Radiations (BEIR), *The Effects on Populations of Exposure to Low Levels of Ionizing Radiation* (Washington, D.C., 1972).

6. National Academy of Sciences/National Research Council, *Report of the Committee on Pathologic Effects of Atomic Radiation*, publication no. 452 (Washington, D.C., 1956), p. IV-23.

7. Ibid., p. IV-22.

8. Ibid., pp. I-5, I-6, I-10. See also B. Watson, ed., *Delayed Effects of Whole Body Radiation* (Baltimore: The Johns Hopkins Press, 1960), pp. 38–47.

9. Medical Research Council, *Hazards to Man*, pp. 20–22.

10. Compare D. W. Robinson and James Boley, "Irradiation Fibromatosis and Fibrosarcoma," *American Surgeon*, 22 (January, 1956): 33–41, for several similar cases reported in the U.S.

11. G. R. Merriam and E. F. Focht, "A Clinical and Experimental Study of the Effect of Single and Divided Dose Radiation on Cataract Production," *Transactions of the American Opthalmological Society*, 60 (1962): 35–52.

12. Medical Research Council, *Hazards to Man*, pp. 18–20.

13. National Academy of Sciences/National Research Council, *The Biological Effects of Radiation*, Summary Reports (Washington, D.C., 1956), p. 37.

14. NAS/NRC, *Pathologic Effects of Atomic Radiation*, pp. V-5–V-26.

15. Arell Schurgin and Thomas Hollocher, "Radiation-Induced Lung Cancers Among Uranium Miners," in UCS, *The Nuclear Fuel Cycle* (Cambridge: MIT Press, 1975), ch. 2.

16. N. P. Knowlton, Jr., *The Value of Blood Counts on Individuals Exposed to Ionizing Radiation* (Los Alamos, N.M.: Los Alamos Scientific Laboratory, undated). See also NAS/NRC, *Pathologic Effects of Atomic Radiation*, Bibliography, pp. IB-2–IB-42 for other Ingram and Knowlton research reports. See also Medical Research Council, *Hazards to Man*, pp. A-6 to A-8.

17. W. C. Hueper, "Recent Developments in Environmental Cancer," *American Medical Association Archives of Pathology*, 58 (October, 1954): 360–399; (November, 1954): 475–523; (December, 1954): 645–682.

18. I. Michael Lerner and William Libby, *Heredity, Evolution and Society*, 2nd ed. (San Francisco: Freeman, 1976). See especially ch. 16, "Mutation."

19. H. J. Muller, "Artificial Transmutation of the Gene," *Science*, 66 (1927): 84–87; and "Nature of Genetic Effects Produced by Radiation," *Radiation Biology*, 1 (1954): 351–473.

20. James Crow, "Radiation and Future Generations," in J. Crow, ed., *Fallout*, p. 103.

21. Ibid.

22. Richard Curtis and Elizabeth Hogan, *Perils of the Peaceful Atom* (New York: Ballantine, 1969), pp. 41–42.

23. Rachel Carson, *Silent Spring* (New York: Fawcett, 1964). See also C. Tyler Miller, *Living in the Environment: Concepts, Problems and Alternatives* (Belmont, Calif.: Wadsworth, 1975).

24. Whiteside, "A Countervailing Force," pp. 47–48.

25. Ibid.
26. See notes to "Poisoned Power."
27. Gofman and Tamplin, *Poisoned Power*, ch. 11. See also Edward Martell, National Center for Atmospheric Research, *Neglected Aspects of the Health Effects of Low Levels of Ionizing Radiation on Man* (February, 1975); Martell, *Unresolved Health Effects of Internal Alpha Emitters* (1976); and Martell, *Basic Considerations in the Assessment of Cancer Risks and Standards for Internal Alpha Emitters* (1975).
28. Scientific American, *Arms Control* (San Francisco: Freeman, 1973), p. 131.
29. Stockholm International Peace Research Institute, *Nuclear Proliferation Problems* (Cambridge: MIT Press, 1974). See also U. S. Senate, Committee on Government Operations, *Peaceful Nuclear Exports and Weapons Proliferation. A Compendium.* (Washington, D.C.: USGPO, 1975).

How a Nuclear Reactor Works

1. Charles Zimmerman, *Accidents in the Nuclear Industry,* (Cornell University, unpublished information file, 1975); Webb, *Accident Hazards of Nuclear Power Plants;* Fuller, *We Almost Lost Detroit.*
2. Berger, *Nuclear Power*, p. 94. See also *Palo Alto Times*, December 8, 1976, p. 5, where Leo Tumerman, emeritus professor of physical chemistry, Weizmann Institute of Science, Rehovoth, Israel, reports observing in 1961 the results of a major nuclear catastrophe in the Ural Mountains, enroute to Beloyarsk, U.S.S.R. His account is confirmed by former Soviet scientist Zhores Medvedev, who alleged that a 1958 explosion in Kyshtim killed hundreds and led to a massive evacuation.
3. *Nuclear News*, April, 1975, pp. 35–36 and 82–88; also July, 1975, p. 21. Compare Levenson et al., "Economic Perspectives of the LMFBR," *Nuclear News*, April, 1976, pp. 54–59, with Comey, "The Uneconomics of Nuclear Power," note 32, and Cochran, *The Liquid Metal Fast Breeder Reactor.*
4. National Council of Churches, *Plutonium Economy;* John Adams, "Should We Commit Ourselves to a Plutonium Economy?" *Catalyst*, 4 (1977): 10–13.
5. A. E. Fitzgerald, *The High Priests of Waste* (New York: Norton, 1970) analyzes several U.S. weapons systems, their cost overruns, and performance deficiencies with special attention to the Lockheed C-5A transport. Other instructive examples not included in Fitzgerald's study are the F-111 aircraft, the Hound Dog missile, and the B-70 bomber.
6. See "Shakeup at General Electric" on General Electric Mark III containment concept.
7. Edward Teller, "How Shall Nuclear Technology be Applied?" in Mark Mills et al., eds., *Modern Nuclear Technology* (New York: McGraw-Hill, 1960), p. 306.
8. USNRC, *RSS* (Washington, D.C.: 1975).
9. Compare Garrett Harding, "Living with the Faustian Bargain," with Alvin Weinberg, "The Many Dimensions of Scientific Responsibility," both in *Bulletin of the Atomic Scientists*, 32 (November, 1976): 21–29. For a more technical discussion, compare CAC, *Hearings*, Joel Primack, October 28, 1975, pp. 136 et seq., with CAC, *Hearings*, A. P. Bray, October 21, 1975, pp. 95–109, 204–217. The burden of proof that reactors will not work as designed still lies with nuclear opponents. For a clear presentation of "defense in depth," see CAC, *Hearings*, Russell Ball, October 29, 1975, pp. 97–104.
10. Alvin Weinberg, "Social Institutions and Nuclear Energy," *Science*, July 7,

1972, p. 33. See also note 9. Compare Alvin Weinberg and Phillip Hammond, "Limits to the Use of Energy," *American Scientist*, 58 (1970): 412.

The Nuclear Fuel Cycle Simplified

1. See Berger, *Nuclear Power*, pp. 129, 160. W. K. Davis reports, with different assumptions, in *Converter Reactor Alternatives*, Atomic Industrial Forum (1975) that 3,800 to 5,400 tons are required per reactor.
2. ERDA, *Statistical Data of the Uranium Industry*, Grand Junction, Colo., Report no. GJO-100(76), January 1, 1976. For reports of possible U_3O_8 shortage, see note 24 in "Uneconomics of Nuclear Power." The amount of U_3O_8 ore that can be profitably mined depends on its final production cost: at $15 per ton, 600,000 tons of U_3O_8 ore reserves may be mined; at $30 per ton, 780,000 are available. ERDA estimates include an additional 3 million tons as "potential resources." Berger points out that most of these resources involve low-grade uranium that may be recoverable only at prices considerably higher than $30 per ton. See Berger, *Nuclear Power*, pp. 123–138.
3. There has been a significant downward trend. Compare Daniel Ford, "Nuclear Power: Some Basic Economic Issues," testimony prepared for the National Nuclear Energy Debate, *Hearings* before the House Committee on Interior and Insular Affairs, Subcommittee on Energy and the Environment, mimeographed reprint (Cambridge: UCS, April 28, 1975), which cites 1,000 reactors as the goal for 2000 A.D. that was developed in 1965, with USAEC Office of Planning and Analysis, *Nuclear Plant Growth 1974–2000*, WASH-1139 (Washington, D.C.: 1974). The nuclear industry is now skeptical that 600 reactors will be built; see CAC, *Hearings*, H. Arnold, October 22, 1975, p. 22.
4. "Nuexco Report Analyzes Price Increases," *Nuclear News*, December, 1976, pp. 46–51. The January, 1977, immediate delivery price per pound, U_3O_8, was $41.00, according to James Vaughn, Nuclear Exchange Corporation, Menlo Park, California (telephone conversation, January 18, 1977).
5. CAC, *Reassessment*, p. 26.
6. Thomas Hollocher and James MacKenzie, "Radiation Hazards Associated with Uranium Mill Operations," in *Nuclear Fuel Cycle*, p. 45. See also Schurgin and Hollocher, "Radiation-Induced Lung Cancers Among Uranium Miners," in *Nuclear Fuel Cycle*, ch. 2, for other consequences of exposure.
7. Robert Pohl, *Nuclear Energy: The Health Effects of Thorium-230* (Ithaca, N.Y.: Cornell University Press, 1975). Comey estimates long-run effects following short-run processing activity: the likely consequences of U.S. uranium processing through 2000 A.D. will be 5,741,500 lung cancer deaths over the next 800,000 years. See David Comey, "The Legacy of Uranium Tailings, *Bulletin of the Atomic Scientists*, 31 (September, 1975): 43–45.
8. *Wall Street Journal*, September 23, 1976, p. 25; *Nucleonics Week*, September 23, 1976, p. 1–2.
9. Ibid.
10. NRDC, *Radioactive Wastes: The AEC's Non-Solution*, reprint of *NRDC Newsletter*, Winter, 1974–1975. See also Robert Gillette, *Science*, August 30, 1974, pp. 770–771.
11. "White Elephant?" *Wall Street Journal*, February 17, 1976, p. 1. See also "Nuclear Waste Disposal and Transportation," in *CAC, Reassessment*, Staff Background Papers, section 3, 1976.
12. "President's New Policy for Plutonium May Benefit a Delayed Nuclear Facility," *Wall Street Journal*, October 5, 1976, p. 2.

13. Gadi Kaplan, "Bugs in the Nuclear Fuel Cycle," *IEEE Spectrum*, September, 1975. See also CAC, *Reassessment*, pp. 23–24.
14. Nuclear Services Corporation, Campbell, Calif., in 1975 announced a system of shielded compartments for spent-fuel pools that will allow increased storage capacity. The spent-fuel rack designed by NSC allows the close-proximity storage of up to thirty fuel assemblies and allegedly doubles a pool's storage capacity. If a national waste reprocessing system never develops, the "temporary" storage racks may become permanent repositories that present hazards and technical problems anticipated today only in a general way. The January, 1977, issue of *Not Man Apart* quoted a resident of the San Francisco area who, though technically untrained, perceived that the nuclear industry is having trouble grasping the most obvious alternative: "Everybody talks about what to do with about this stuff. They act like it's dog shit. Should they perfume it? Put sawdust on it? Hide it somewhere else? Nobody talks about maybe we should get rid of the dog." Interview with Conrad Golich originally published in the *Pacific Sun*, week of November 5–11, 1976. NRC suggested that spent fuel pools may become permanent repositories, in testimony by Victor Gilinsky, NRC Commissioner, before the California Energy Resources, Conservation and Development Commission, January 31, 1977. Gilinsky assured the Commission that all technical questions pertaining to spent fuel were under study and implied that neither private industry nor the federal government was likely to do anything until the incoming Carter administration's policy is clear.

The Rise and Fall of Nuclear Power

1. Ford, "Nuclear Power," p. 1.
2. Ibid. This view is confirmed by Edwin Triner, former acting director of the USAEC Office of Program Analysis, in a December 2, 1974 memorandum to USAEC Commissioner L. Manning Muntzing, "Improving the Reliability of Nuclear Power Plants—A Point of View," reprinted in CAC, *Hearings*, October 29, 1975, pp. 106–111. Shortly after his memorandum was leaked, Triner was transferred to a less prominent job.
3. Sheldon Novick, *The Electric War* (San Francisco: Sierra Club, 1976); and Irvin Bupp et al., "The Economics of Nuclear Power," in *Technology Review*, 77 (February, 1975): 14–25.
4. CAC, *Reassessment*, p. 18. See also Ford, "Nuclear Power."
5. Novick, *Electric War*, ch. 11; UCS, *Nuclear Power Issue*, ch. 6; *NRDC vs. USNRC* (July 21, 1976) at 17, 31. Metzger, *Atomic Establishment*, includes many examples.
6. Ford, "Nuclear Power," pp. 3–4; compare CAC, *Hearings*, Daniel Ford, October 21, 1976, p. 70. See also "Nuclear Survey: Cancellations and Delays," *Electrical World*, October 15, 1975, pp. 35–44. W. H. Arnold of Westinghouse claims that nine new reactors were ordered in 1975; see CAC, *Hearings*, Arnold, October 22, 1975, p. 56.
7. Ford, "Nuclear Power," pp. 3–4. See also Bupp et al., "Economics," and "Atomic Lemons," *Wall Street Journal*, May 3, 1973. A thorough study by an industry periodical is "Why Atomic Power Dims Today," *Business Week*, November 17, 1975, pp. 96–105.
8. UCS, *Nuclear Power Issue*, chs. 3, 7; Carl Hocevar, "Nuclear Power Safety," *Professional Engineer*, 45 (May, 1975): 25–29; unresolved safety questions are itemized in CAC, *Hearings*, letter from USNRC Advisory Committee on Reactor Safeguards, October 22, 1975, pp. 122–133; and discussed in CAC, *Hearings*, "Nuclear Reactor Safety," October 22, 1975, p. 152.

9. Ford, "Nuclear Power," pp. 2–3.
10. USNRC, *Operating Units Status Reports* NUREG–0020–6 (June, 1976), p. 1–1.
11. UCS, *Nuclear Power Issue*, ch. 6; "Allegations of Nuclear Employee Intimidation Rise Again," *Nucleonics Week*, April 17, 1975, pp. 2–3. See also note 2. Louis H. Roddis, Jr. pointed out nuclear plant capacity factor deficiencies in his November, 1972, speech before the Atomic Industrial Forum, "Maintaining the Output—Or—Can We Get There from Here?" while president of Consolidated Edison of New York. Compare this speech—reviewed in "Fulfilling Nuclear Expectations: A 'Hard Luck' Utility Makes Plea for Greater Maintainability," *Nuclear Industry*, November-December, 1972, pp. 20–22—with what appears to be an apologetic, revisionist parting statement by him before the American Nuclear Society, "Maintaining the Output—Volume 2," March 12, 1974, just before he "resigned" as Con Ed president. Fifteen months after his November, 1972, AIF speech, Roddis confirmed his earlier criticisms of nuclear plant performance and implied that the nuclear industry was gilding the lilly by issuing misleading and unjustifiably favorable data. See "Utilities Dispute Atom Plant Data," *The New York Times*, February 3, 1974. Five weeks later, Roddis presented the second speech, which expresses considerable faith in the industry he had just finished criticizing. Soon after, he left Con Ed.
12. Novick, *Electric War;* Lewis, *Nuclear Power Rebellion;* Berger, *Nuclear Power*, ch. 13.
13. Berger, *Nuclear Power*, pp. 178–183.
14. *NRDC vs. USNRC* (July 21, 1976).
15. "No On [Proposition] 15," Committee/Californians Against the Nuclear Shutdown, *California Energy Bulletin* (Glendale, Calif., Spring, 1976), pp. 1–8.
16. *NRDC vs USNRC* (July 21, 1976), pp. 24–35.
17. Ibid., pp. 16–23, 34, 38–39.
18. Ibid., p. 34, note 53; see also pp. 31–32, note 50, where substantive arguments appear to have been deflected by condescension. See also Novick, *Electric War*, ch. 11, again involving Kendall. According to Ford and Kendall, intervenors are often treated by agencies and hearing board members in a patronizing, hostile manner. The meaning of this should be clear to anyone experienced in rendering or cross-examining testimony.
19. Ebbin and Kasper, *Citizen Group Uses of Scientific and Technological Information*, p. 4, argue that "there is a common set of interests" shared by AEC hearing and licensing boards, the agencies staff, the utilities, and the reactor manufacturers that "effectively make them allies." Ebbin and Kasper demonstrate that citizen groups are denied due process and found that "despite lip service paid to citizen participation . . . that agency arrogance, expert elitism, stacked-deck proceedings and the consigning of citizens to helplessness before the steamroller of big government is more the rule than the exception."
20. Ford, "Nuclear Power," p. 4.
21. Ibid., p. 5; and U.S. Bureau of Mines, *Demonstrated Coal Reserve Base of the U.S. on January 1, 1974* (Washington, D.C.: Mineral Industry Surveys, 1974). See also D. A. Brobst and W. P. Pratt, eds., *U.S. Mineral Resources*, professional paper 820 (Washington, D.C.: U.S. Geological Survey, 1973), pp. 133–142.

22. "Nuclear Power in the U.S.: Chaos Reigns Supreme as 1975 Opens," *Nucleonics Week*, January 16, 1975; "Atomic Lemons," *Wall Street Journal*, May 3, 1973; *Nuclear News*, editorial, March, 1976, p. 29.
23. Ford, "Nuclear Power," pp. 6–7.

The Uneconomics of Nuclear Power

1. David Burnham, "Hope for Cheap Power from Atom Fading," *The New York Times*, November 16, 1975; Jim Harding, "The Deflation of Rancho Seco II," *Not Man Apart*, August, 1975. Harding's report contributed significantly to the cancellation of plans to build the second unit of the Rancho Seco plant; see William Walbridge, general manager, Sacramento Municipal Utility District, *Report on Future Generation*, January 8, 1976. Also *Nuclear News*, December, 1974, pp. 60–63, and September, 1975, p. 66.
2. Bupp et al., "Economics"; also Berger, *Nuclear Power*, p. 151.
3. The plant standardization effort is limited to the SNUPPS (Standardized Nuclear Unit Power Plant System) group in Maryland. The Westinghouse effort to develop standardized floating plants has yet to produce one and has suffered several cancellations after tens of millions of dollars invested.
4. Burnham, "Hope for Cheap Power;" Bupp et al., "Economics."
5. Atomic Industrial Forum, *Causes of Nuclear Plant Delays, 1974;* Ford, "Nuclear Power"; Bupp et al., "Economics." Also "Quicker Startup of Nuclear Plants Sought," *Chemical and Engineering News*, 51 (November 26, 1973): 7–8.
6. Interviews of utilities and labor leaders, 1974–1976, by Comey. Also Bupp et al., "Economics."
7. Leonard Reichle of Ebasco Services, Inc., claims that the cost of a nuclear reactor in late 1975 was $1,135 per kilowatt in "Is Nuclear Too Costly?" *The New York Times*, October 5, 1975, section 3, p. 1. Bupp et al., "Economics;" report $135 per kilowatt for 1965. Extrapolation of estimated costs graph in USAEC, *Power Plant Capital Costs WASH-1345*, October, 1974, yields per kilowatt plant cost of less than $100 for 1964.
8. Leonard Reichle, "The Economics of Nuclear Power," *The Catalyst*, 2 (Spring, 1976): 9.
9. David Comey, "Will Idle Capacity Kill Nuclear Power?" *Bulletin of the Atomic Scientists*, 30 (November, 1974): 23–28; "Atomic Lemons," *Wall Street Journal*. See also "Maintaining the Output," summarized in "Fulfilling Nuclear Expectations."
10. Nine Mile Point Nuclear Station *Final Environmental Statement*, Docket 50–220, January, 1974, p. 10–1. The capacity factor expected by the applicant was 85 percent. Eugene Cramer of Southern California Edison Company asserts that it is not the absolute capacity factor that determines plant economics but instead "the relative capacity you assign to alternatives." (*Energy Daily*, December 1, 1976, p. 3.) This may be true, but it dodges the main issue: Why do Final Environmental Statements consistently overestimate nuclear capacity factors? How can nuclear fission be a profitable way to manufacture electricity if plant capacities remain in the range of 54–59 percent? Do not these data significantly undermine the assumptions of past Statements?
11. Donald McCormick, "PWR Radiation Buildup—The Course Ahead," *Transactions of the American Nuclear Society*, 22 (November, 1975): 120. See also U.S. Senate, Committee on Government Operations, *Hearings*, March 12, 1974, pp. 217–226.

12. Roddis, "Maintaining the Output": "In the seven-month effort (to repair a leak in the makeup line to the Indian Point No. 1 reactor coolant system), in order not to exceed radiation exposure regulations, 700 men were required. This included 50 health physicists, 250 men from operations, 350 men from maintenance, and 50 contractors' men. The job required, at one time or another, the use of every welder in the Con Edison organization who was qualified in a certain welding technique. A similar repair effort, if made in a conventional plant, would have required about two weeks and probably would not have involved more than 25 men."

13. Ibid. Comey's and Roddis's data are confirmed by Charles Komanoff in *Power Plant Performance* (New York: Council on Economic Priorities, 1976), by far the most thorough study of capacity factors to date and the first to extensively review and compare nuclear and coal plant performance. Komanoff determined that commercial U.S. nuclear plants operated at a 59.3 percent capacity factor through 1975 and that capacity factors have declined with increasing unit capacity. Komanoff also stated that "new 1150-Mw PWR's and BWR's will be 16 percent and 31 percent more expensive, respectively, than 600-Mw supercritical coal units using Eastern low-sulfur coal, without scrubbers. If scrubbers rather than low-sulfur coal are used, 1150-BWR's will be only 12 percent more expensive. PWR's will then equal the Eastern coal unit cost. This is based on projections from the CEP capacity factor trend equations and standard cost assumptions for other parameters." *Power Plant Performance* is reviewed in *Nuclear News*, January, 1977, pp. 35–36. As expected, the industry protested strenuously, citing improper use of data and an inadequate data base (ibid.). Unquestionably, the number of units in the 1000-MWe range is not large, but a study that may be useful in determining whether any more plants should be built would have only historical value if it postponed performance comparisons until most large plants were built and operating. If it is conceded that Komanoff's data base for large reactors is inadequate, then the industry should reciprocate by granting that too few large reactors have been operating for too short a time to authoritatively claim that these systems are safe.

14. Peter Margen and Sören Lindhe, "The Capacity Factors of Nuclear Power Plants," *Bulletin of the Atomic Scientists*, 31 (October, 1975): 38–40.

15. ERDA's 1975 estimate is that we will need 720 gigawatts (720 thousand million watts) of nuclear plants on-line by the year 2000. At an average capacity factor of 80 percent, these plants would produce 576 gigawatt-years annually. But if they only average 42.7 percent capacity factor, then 1,349 gigawatts worth of nuclear plants would be needed to generate those 576 gigawatt-years of electricity each year. If construction costs of future nuclear plants are conservatively estimated at $1,000 per kilowatt, this difference represents a $629 billion cost overrun that would most likely result in increased electricity prices and a substantial drain on investment capital.

16. "Steam Station Cost Survey," *Electrical World*, November 15, 1975, p. 45.

17. See example and critique of Pacific Gas and Electric advertisement in Berger, *Nuclear Power*, pp. 185–188. Also Commonwealth Edison news release, January 20, 1975; Consolidated Edison of N.Y. press release, "Con Ed Earnings Summary," January 30, 1976. Also "Did Con Ed Generate False Data?" *New York Post*, August 25, 1975, and Charles Komanoff, *Responding to Con Edison: An Analysis of the 1974 Costs of Indian Point and Alternatives* (New York: Council on Economic Priorities, 1975). Compare "Utilities Dispute Atom Plant Data," *The New York Times*, February 3, 1974, p. 37, for a report

by the former Con Ed president that other utilities were publishing misleading plant performance data.

18. Koshkonong Nuclear Plant, Units 1 and 2, *Draft Environmental Statement*, Dockets 50–502 and 50–503, August, 1976, p. 9–4, Table 9.1.

19. Westinghouse based its earlier reactor marketing strategy on this assumption; see note 22 and "Westinghouse Eyes Loss on Uranium Contracts," *Energy Finance Week*, July 23, 1975. For uranium reserves generally, see M. A. Lieberman, "U.S. Uranium Resources—An Analysis of Historical Data," *Science*, April 30, 1976.

20. Conversation with James Vaughn, Nuclear Exchange Corporation, Menlo Park, California, January 18, 1977.

21. Mitchell-Hutchins, Inc., *The Uranium Stocks: Nuclear Industry Kaleidoscope Coming Together*, New York, January, 1976, p. 1.

22. *Nuclear News* coverage of the Westinghouse uranium suits was especially thorough. See October, 1975, pp. 22 and 50–52; November, 1975, pp. 25 and 80–82; December, 1975, p. 50; January, 1976, p. 65; February, 1976, pp. 46–47; April, 1976, p. 22. Westinghouse argued that it should be excused from the earlier contracts because today's uranium prices make the fulfillment of the contracts "commercially impracticable." Fifteen million pounds of Westinghouse-owned uranium were divided among the plaintiffs satisfactorily (*Nuclear News*, March, 1976, p. 59), and the questions of where the remaining 65 million pounds would come from and who will pay the difference between promised and present prices were referred to a committee. Westinghouse may yet persuade the court to invalidate the earlier contracts, but to do so it may have to prove the existence of the uranium cartel; this may weaken industry arguments which allege that nuclear power is a cheap source of energy and further undermine utility enthusiasm for buying reactors.

23. Mitchell-Hutchins, *Uranium Stocks*, pp. 39–42. See also note 19.

24. Hans Adler, *Geological Aspects of Foreign and Domestic Uranium Deposits*, ERDA, Division of Nuclear Fuel Cycle and Production (Washington, D.C.: October, 1975); Berger, *Nuclear Power*, pp. 123–138; Mitchell-Hutchins, *Uranium Stocks*.

25. Adler, *Geological Aspects of Foreign and Domestic Uranium Deposits*.

26. The uranium cartel is discussed in detail in "The Atomic Industrial Complex."

27. Mitchell-Hutchins, *Uranium Stocks*, p. 16.

28. Ibid., p. 9. Inadequate enrichment capacity is reported in CAC, *Reassessment*, pp. 26–27.

29. Mitchell-Hutchins, *Uranium Stocks*, p. 9. Daniel Ford quotes Edison Electric Institute report that fuel reprocessing may not be profitable. See CAC, *Hearings*, Ford, October 21, 1975, p. 71.

30. "Getty Oil Subsidiary Says It Won't Reopen," *Wall Street Journal*, September 23, 1976.

31. Fred Kramer, *PWR Fuel Performance: The Westinghouse View*, American Nuclear Society, April 19, 1975; also F. D. Judge et al., *General Electric Fuel Performance Update*, American Power Conference, 1975.

32. "Anders Reveals Doubts on Breeder Reactor," *Baltimore Sun*, March 18, 1976; "Morton and Zarb," *The New York Times*, June 10, 1975, p. 25.

33. JCAE, *Hearings* on draft version of Environmental Impact Statement regarding the liquid metal fast breeder reactor, 1974; JCAE, *Hearings on LMFBR program* 1975; ERDA, *Final Environmental Statement, LMFBR Program*,

ERDA-1535, 3 vols., December, 1975; NRDC press release, December 19, 1975, regarding unresolved plutonium recycling questions.

34. Conversations between David Comey and USNRC officials, January-October, 1976.

35. Don Kuhn, ERDA, Division of Nuclear Fuel Cycle and Production, private communication, 1975.

36. *Nuclear Energy and Nuclear Security* (New York: Council for Economic Development, September, 1976).

37. "President's New Policy for Plutonium," *Wall Street Journal*, October 5, 1976, p. 2.

38. "Federal Financing for Energy?" *Washington Post*, September 26, 1975.

The Silent Bomb: Radioactive Wastes

1. IAEA, *Impacts of Nuclear Releases into the Aquatic Environment*, Proceedings of the 1975 Otaniemi Symposium, STI/PUB 406 (New York: Unipub, 1975). See also note 23, "What You Don't Know Will Kill You"; IAEA, *Transuranium Nuclides in the Environment*, Proceedings of the 1975 San Francisco Symposium, STI/PUB 410 (New York: Unipub, 1975); *NRDC vs USNRC* (July 21, 1976), note 10 at pp. 7–8. Leaching of strontium-90 is reported by Henry Kendall in material based on testimony in connection with a hearing on the Price-Anderson Act (Cambridge: UCS, released September 28, 1976). Radwaste migration from burial sites cited in "GAO Reports New Nuclear Garbage Problem," *Science and Government Reports*, 6 (February 1, 1976): 8. Also *Business Week*, February 2, 1976, at 17; and "Nuclear Waste Found Leaking at Site Upstate," *The New York Times*, February 10, 1977, p. 27. Dispersion is more rapid following radwaste ocean dumping. Extensive ocean dumping by European nations is reported in JCAE, *Hearings*, November 19, 1975. See also Victor Noshkin, "Ecological Aspects of Plutonium Dissemination in Aquatic Environments," *Health Physics*, 22 (June, 1972): 537–549.

2. Lash et al., *Citizens' Guide*. See also Arthur Kubo and David Rose, "Disposal of Nuclear Wastes," *Science*, December 21, 1973, pp. 1205–1211.

3. Hannes Alfvén, "Energy and Environment," *Bulletin of the Atomic Scientists*, 28 (May, 1972): 5–7. See also U.S. Environmental Protection Agency, Office of Radiation Programs, *Environmental Dose Commitment: An Application to the Nuclear Power Industry*, Washington, D.C., 1974.

4. Lovins and Price, *Non-Nuclear Futures*, p. 13.

5. Deborah Shapley, "Plutonium Reactor Proliferation Threatens a Nuclear Black Market," *Science*, April 9, 1971, pp. 143–146, who quotes Delmar Crowson, then head of AEC's Office of Safeguards and Materials Management. Crowson reports that audits for plutonium-239 loss average within +0.18 to 0.51 percent, with 0.2 percent "not unusual." Terry Lash points out that this notation may be misleading by one magnitude and that the loss rate approximates *2 percent* (conversation with Lash on January 5, 1977). The health implications of this release rate are detailed by Edward Martell, *Actinides in the Environment and Their Uptake by Man*, NCAR-TN/STR-110, National Center for Atmospheric Research, May, 1975.

6. Shapley, "Plutonium Reactor Proliferation," p. 144.

7. Carson, *Silent Spring*. See also Melton Davis, "Under the Poison Cloud," *The New York Times Magazine*, October 10, 1976, pp. 20–38.

8. For Hanford radioactive spills, see CAC, *Reassessment*, p. 69. Past nuclear

weapons test explosions are cited in Scientific American, *Arms Control.* Recent plutonium contamination in Colorado following two major fires at the Rocky Flats Nuclear Weapons Plant is reported by Carl Johnson et al., "Plutonium Hazard in Respirable Dust on the Surface of Soil," *Science,* August 6, 1976, pp. 488–490. A more complete report is Poet and Martell, "Plutonium-239 and Americium-241 Contamination in the Denver Area." Gofman reports considerable plutonium in the atmosphere in *Estimated Production of Human Lung Cancers by Plutonium from Worldwide Fallout,* 1975.

9. Terry Lash and Richard Cotton, "Radioactive Wastes," unpublished, undated paper submitted to Friends of the Earth, p. 3: "[The] slow rate of decay [of radioactive wastes], coupled with rapidly increasing production, could lead to an irreversible buildup of radioactive elements in the environment." The issue of whether or not this point may be reached soon is subordinate to concern about the immediate genetic and physiological damage caused by released wastes. IAEA, *Biological and Environmental Effects of Low-Level Radiation,* Proceedings of the 1975 IAEA Symposium in Chicago, STI/PUB 409 (New York: Unipub, 1976).

10. Virginia Brodine, *Radioactive Contamination* (New York: Harcourt Brace Jovanovich, 1975), pp. 24–38, 166–173; Metzger, *Atomic Establishment,* pp. 99–102; "Strontium-90 Fallout," *Nuclear Information,* 5 (March-April, 1963).

11. Gofman and Tamplin, *Poisoned Power;* B. Schlien, "An Evaluation of Internal Radiation Exposure Based on Dose Commitments from Radionuclides in Milk, Food and Air," *Health Physics,* 18 (1970): 267–276; Philip Hatfield, "Nuclear Fuel Reprocessing: Radiological Impact of West Valley Plant," in *Nuclear Fuel Cycle,* ch. 7.

12. Whiteside, "A Countervailing Force," pp. 47–48.

13. Gofman and Tamplin, *Poisoned Power,* ch. 11; see especially Metzger, *Atomic Establishment,* pp. 83–91, and E. F. Schumacher, *Small Is Beautiful* (New York: Harper & Row, 1973), pp. 126–127.

14. Neville Grant, "Pollution Struggle in Japan," *Environment,* 18 (January-February, 1976): 5, 36–40. The Japanese court refused to accept the standards for releases set by the government and held that the defendant polluted at its own risk; see p. 39: "The defendant's plant discharged acetaldehyde waste water with negligence at all times, and even though the quality and content of the waste water of the defendant's plant satisfied statutory limitations and administrative standards, and even if the treatment methods it employed were superior to those taken at the work yards of other companies in the same industry, these are not enough to upset the said assumption . . . the defendant cannot escape from the liability of negligence." See also W. Eugene Smith and Eileen Smith, *Minamata: Words and Photographs* (New York: Holt, 1975).

15. Brown, "Health, Safety and Social Issues," in *California Nuclear Initiative,* p. 141. Also see note 8, and Schurgin and Hollocher, "Radiation-Induced Lung Cancers Among Uranium Miners, in *Nuclear Fuel Cycle.*

16. Note 15 and W. C. Hueper, "Environmental and Occupational Cancer," *Public Health Reports,* supplement 209, 1948.

17. JCAE, *Hearings* on radiation safety and regulation, June, 1961, and on environmental effects of producing electric power, 1969 and 1970.

18. "NRDC Sues to Stop Plutonium Recycle," *Not Man Apart,* mid-February, 1976, p. 9, where an USNRC spokesman is reported to have asked, "How do these environmentalists think we can assess the real hazards of plutonium use without using it? Interim licensing will allow valuable experience in the fission-

ing, handling, and processing of this critical natural [sic] resource." In *NRDC vs USNRC* (July 21, 1976) at 43, the court implied that the defendants were experimenting: "It has been commonplace among proponents of nuclear power to lament public ignorance. The public—the 'guinea pigs' who will bear the consequences of either resolution of the nuclear controversy—is apprehensive. But public concern will not be quieted by proceedings like [the litigated USAEC licensing hearings]."

19. Edward Martell, "Tobacco Radioactivity and Cancer in Smokers," *American Scientist*, 63 (July-August, 1975): 404–412.

20. John Gofman, address at Nuclear Energy Forum, San Luis Obispo, California, October, 1975. Until a proven, safe system for permanent storage is demonstrated and tested, it cannot be said that a *disposal* system exists. A partitioning and transmuting *concept* exists (see Lash et al., *Citizens' Guide*, p. 25), but this is no assurance that it will work, that all side effects are known and minimized, or that it will prove economical. Also, the error of assuming technical infallibility should be avoided here; see Lovins and Price, *Non-Nuclear Futures*, p. 13; Hardin, "Living with the Faustian Bargain," *Bulletin of the Atomic Scientists*, 32 (November, 1976): 25–29.

21. Miller, *Living in the Environment*, pp. E81–E84.

22. Gofman and Tamplin, *Poisoned Power*, ch. 3 et seq for critics' view. See also IAEA, *Biological and Environmental Effects of Low-Level Radiation;* Lerner and Libby, *Heredity, Evolution and Society*, 2nd ed., ch. 16 et seq; John Drake, *The Molecular Basis of Mutation* (San Francisco: Holden-Day, 1970), chs. 13–16.

23. Lash et al., *Citizens' Guide*, frontispiece and p. 1. The government's failure to develop and test a system is detailed in *NRDC vs USNRC* (July 21, 1976). Proposed disposal concepts and their weaknesses are reviewed in Thomas Hollocher, "Storage and Disposal of High-Level Radioactive Wastes, *Nuclear Fuel Cycle*, ch. 8, especially pp. 247–266.

24. CAC Staff Background Paper, *Nuclear Waste Disposal and Transporation*, November 3, 1975, p. 1. See also House Committee on Government Operations, Subcommittee on Conservation, Energy and Natural Resources, *Hearings* on radioactive waste, February, 1976, during which Henry Eschwege, USGAO, testified that the volume of low-level wastes expected to be produced by the year 2000 will, according to the Environmental Protection Agency, equal one billion cubic feet, or enough to cover a four-lane highway, coast-to-coast, to a depth of one foot (*Hearings*, February 23, 1976).

25. Berger, *Nuclear Power*, p. 186.

26. Berger, *Nuclear Power*, p. 99; and Lash et al., *Citizens' Guide*, p. 22.

27. Berger, *Nuclear Power*, p. 75. See also the notes for "Poisoned Power."

28. USNRC, *Generic Environmental Impact Statement on the Use of Mixed Oxide Fuels* (GESMO), Washington, D.C., 1975.

29. Shapley, "Plutonium Reactor Proliferation," p. 143.

30. Lash et al., *Citizens' Guide*, p. 1; and Hollocher, "Storage and Disposal" *Nuclear Fuel Cycle*, p. 219.

31. CAC Staff Background Paper, *Nuclear Waste Disposal*, p. 1. Lash et al. cite 500,000 gallons in *Citizens' Guide*, p. 30. Debate summarized in CAC, *Reassessment*, pp. 69–71. The cited data was provided by Lash in conversation on February 20, 1977.

32. Lash et al., *Citizens' Guide*, p. 33. See also Berger, *Nuclear Power*, pp. 109–110.

33. United Press International news release, January 13, 1976. An excellent summary of transportation accident risk is CAC, *Reassessment,* pp. 65–71.
34. "Giant Sponges Stalk West Coast," *Not Man Apart,* mid-October, 1976.
35. Ibid.
36. Metzger, *Atomic Establishment,* pp. 154–157ff. Also F. Pittman, director of USAEC Waste Management and Transporation Division, testimony, Subcommittee of the House Subcommittee on Appropriations, 93rd Cong. 1st sess., *Hearings,* 1973; and Kubo and Rose, "Disposal of Nuclear Wastes."
37. "Key A-Burial Site Rejected," *Denver Post,* October 13, 1975. See also Richard Lewis, "The Radioactive Salt Mine," *Bulletin of the Atomic Scientists,* 27 (June, 1971): 27–34. Kansas and New Mexico site problems are summarized by Boffey in "Radioactive Waste Site Search Gets into Deep Water," *Science,* October 24, 1975, p. 361.
38. Lash and Cotton, "Radioactive Wastes," p. 9.
39. Ibid., pp. 9–12.
40. *NRDC vs USNRC* (July 21, 1976).
41. *Nuclear News,* September, 1976, pp. 29–31.
42. *NRDC vs USNRC* (July 21, 1976), pp. 4, 8–9.
43. Ibid., p. 8.
44. Ibid., note 50 at p. 31.
45. Ibid., pp. 23–39, esp. p. 35.
46. Lovins and Price, *Non-Nuclear Futures,* p. 12.

Theft and Terrorism

1. David Krieger, "Terrorists and Nuclear Technology," *Bulletin of the Atomic Scientists,* 31 (June, 1975): 28–34.
2. Telephone conversation with L. Douglas DeNike, December 9, 1976.
3. John Holdren, "Hazards of the Nuclear Fuel Cycle, *Bulletin of the Atomic Scientists,* 30 (October, 1974): 14–23.
4. A Princeton student designed what was alleged to be a workable nuclear weapon. Representatives of at least two countries urged him to sell the design after FBA and university representatives instructed him to show the design to no one. "Nation Beats Path to Door of Princeton Senior for His Atom Bomb Design," *The New York Times,* February 10, 1977, p. 37.
5. U.S. Senate, Committee on Government Operations, *Hearings,* March 12, 1974, pp. 107–120.
6. CAC, *Hearings,* Theodore Taylor, November 19, 1975.
7. Fuller, *We Almost Lost Detroit,* p. 221.
8. CAC, *Hearings,* L. Douglas DeNike testimony, "Terrorist Attacks on California Nuclear Power Facilities: A Threat Analysis," November 19, 1975.
9. Report by Henry Eschwege, USGAO, to Dixy Lee Ray, USAEC, B-164105, October 16, 1974.
10. "Bombs Hit Nuclear Site in France," *Washington Post,* May 4, 1975; "A New Bombing of a French Nuclear Plant," *Nuclear News,* September, 1975, p. 134; "Le Programme Nucleaire Dans Les Choux!" *La Gueule Ouverte,* a French radical weekly, one of the June, 1975, issues.
11. USNRC, *RSS,* WASH-1400. Appendix VI.
12. Mason Willrich and Theodore Taylor, *Nuclear Theft: Risks and Safeguards* (Cambridge: Ballinger, 1974), p. 13. See also ibid., p. 25, Table 2–2, bearing in mind that reactor-grade plutonium is five to six times more radiotoxic than plutonium-239. The distribution of one-and-one-half ounces of plutonium-239

represents a value of one microcurie per square meter of the material. See also debate on plutonium safeguards in CAC, *Reassessment,* pp. 72–82.
13. DeNike, "Terrorist Attacks."
14. Ayres, "Policing Plutonium"; MITRE Corporation, *The Threat to Licensed Nuclear Facilities,* MTR-7022, September, 1975. H. C. Greisman et al., *Analysis of the Terrorist Threat to the Commercial Nuclear Industry,* BDM/75–176–TR, BDM Corporation, September, 1975. See also William Ophuls, "Return of Leviathan," *Bulletin of the Atomic Scientists,* 29 (March, 1973):50–52; and L. Douglas DeNike, "Radioactive Malevolence," *Bulletin of the Atomic Scientists,* 30 (February, 1974): 16–20.

Reactor Safety

1. Between refueling operations, a reactor operating at full power accumulates in the order of 1.7×10^{10} curies of radioisotopes. See "Report to the American Physical Society by the Study Group on Light Water Reactor Safety," *Reviews of Modern Physics,* 47, supplement 1 (Summer, 1975): S–23 (reprinted in CAC, *Hearings,* October 29, 1975, appendix, pp. 124–248).
2. John Holdren, "Hazards of the Nuclear Fuel Cycle," p. 14.
3. It is understood now that serious problems can occur in a reactor without a LOCA so that there is probably too much attention given to the LOCA and too little to other abnormal reactor conditions. But because of the historical significance of the LOCA and the central role it has played in designing reactors to cope with it, this chapter examines the problem in some detail. For special study of the LOCA, see CAC, *Hearings,* A. Stathoplos, October 22, 1975, pp. 96–104. For a general treatment of USNRC licensing and safety reviews with special attention to seismic and ECCS design, see CAC, *Hearings,* David Okrent, October 29, 1975, pp. 77–97. An excellent short introduction to safety problems is Joel Primack and Frank Von Hippel, "Nuclear Reactor Safety," *Bulletin of the Atomic Scientists,* 30 (October, 1974): 5–12.
4. CAC, *Hearings,* testimony by Carl Hocevar, October, 21, 1975, pp. 49–55. See also Hocevar, *Nuclear Reactor Licensing: A Critique of the Computer Safety Prediction Methods* (Cambridge: UCS, 1975). Hocevar's treatment of the ECCS controversy and inadequacies in present mathematical approaches to design verification is definitive in "Nuclear Power Safety: An Engineering Overview," *Professional Engineer,* 45 (May, 1975): 25–29. For confirmation of Hocevar on PWR designs, see CAC, *Hearings,* Keith Miller, "Memorandum [to USNRC]:Recommendations Relating to Licensing of Commercial Nuclear Power Plants in the USA," October 29, 1975, pp. 301–313. Deficiencies in computer codes, cross-checking, and further confirmation of Hocevar is Fred Finlayson, *Assessment of ECCS Effectiveness for Light-Water Nuclear Power Reactors* (Pasadena: California Institute of Technology, 1975).
5. CAC, *Hearings,* Hans Bethe, October 22, 1975, p. 16. NRC justifies parallel development of nuclear power with ongoing research on the grounds that the industry has reached a level at which reactor safety is acceptable in terms of accident consequences; see CAC *Hearings,* Bernard Rusche, October 22, 1975. pp. 32 et seq. The political issue of "how safe is safe enough" has never been openly debated in Congress, however.
6. CAC, *Hearings,* Russell Ball, October 29, 1975, p. 102. But see Daniel Ford and Henry Kendall, "Nuclear Misinformation," *Environment,* 17 (July-August, 1975).
7. APS Report, *Reviews of Modern Physics,* p. S-41

8. CAC, *Hearings*, Daniel Ford, October 21, 1975, p. 77.
9. Investor's Responsibility Research Center, *The Nuclear Power Alternative*, Washington, D.C., 1975. Quotation by Carl Hocevar, p. 68.
10. CAC, *Hearings*, David Okrent, October 29, 1975, pp. 79–87.
11. CAC, *Hearings*, Donald Reardon, October 21, 1975, pp. 118–123.
12. CAC, *Hearings*, Harold Lewis, October 29, 1975, p. 43.
13. APS Report, *Reviews of Modern Physics*, p. S-74.
14. CAC, *Hearings*, Saul Levine, November 19, 1975, pp. 28–38.
15. CAC, *Hearings*, Milton Kamins, "A Reliability Review of the Reactor Safety Study," October 28, 1975, p. 94.
16. Several witnesses commented that in California, with its earthquake record, the public should be as concerned with the catastrophic rupture of a dam as they appear to be about catastrophic reactor accidents. See CAC, *Hearings*, Edward Teller, October 21, 1975, p. 6; CAC, *Hearings*, Richard Wilson, October 28, 1975, pp. 5–7.
17. USNRC, *RSS*, WASH-1400, Executive Summary, p. 9.
18. Ibid., p. 107.
19. Ibid., p. 111. The normal incidence for the same population would be 17,000 cancer deaths per years, 8,000 thyroid nodules per year, and 8,000 genetic disorders per year.
20. Ibid., p. 107.
21. CAC, *Hearings*, Kamins, October 28, 1975, pp. 95, 170–197. Compare with USNRC Advisory Committee on Reactor Safeguards, Minutes of the Public Meeting of the Working Group on the Reactor Safety Study (Washington, D.C., January 4, 1977). Also APS Report, *Reviews of Modern Physics;* Kendall and Moglewer, *Preliminary Review of the AEC Reactor Safety Study.* But see David Bodansky and Fred Schmidt, "Safety Aspects of Nuclear Energy," in Arthur Murphy, ed., *The Nuclear Power Controversy* (Englewood Cliffs, N.J.: Prentice-Hall, 1976), especially pp. 28–43, and compare Berger, *Nuclear Power*, ch. 4. For critique of RSS cancer estimates, see Brown, "Health, Safety and Social Issues," in W. Reynolds, ed., *California Nuclear Initiative*, pp. 160–170, 196–199.
22. Ibid., 176.
23. CAC, *Hearings*, Kamins, October 28, 1975, pp. 179–180.
24. Ibid., p. 188.
25. Ibid., p. 177.
26. CAC, *Hearings*, Joel Primack, October 28, 1975, p. 47.
27. The APS Study Group did *not* attempt to critique the RSS since it was not available in time for their study, although they did comment on the general methodology.
28. APS Report, *Review of Modern Physics*, p. S-5.
29. CAC, *Hearings*, letter from A. E. Green, November 19, 1975, pp. 173–174.
30. CAC, *Hearings*, Okrent, October 29, 1975, p. 92.
31. CAC, *Hearings*, quoted by Dan Anderson, December 10, 1975, p. 52.
32. The criticisms are reprinted along with a response in the *RSS* itself.
33. But see CAC, *Hearings*, L. Douglas DeNike, November 19, 1975, pp. 118–123.
34. *Sacramento Bee*, March 19, 1976, p. A-13.
35. *Nucleonics Week*, May 8, 1975, p. 6.
36. CAC, *Hearings*, Kamins, October 28, 1975, p. 93.
37. USNRC, *RSS*, WASH-1400, Appendix VI, pp. 11–4 to 11–6.

38. CAC, *Hearings*, Primack, October 28, 1975, pp. 142, 159.
39. Kevin Shea, "An Explosive Reactor Possibility, *Environment*, 18 (January-February, 1976): 6–11.
40. The Atomic Industrial Forum has labeled this the attitude of the "what if" critics.
41. William Casto, "Operating Experiences: Dresden-2 Incident of June 5, 1970," *Nuclear Safety*, 12 (September-October, 1971): 538–546.
42. The AEC report of this incident dryly adds: "The coupling of a contaminated water system with a potable water system is considered poor practice in general" (USAEC *Operating Experiences: Reactor Safety*, ROE [Reactor Operating Experience] no. 69–10, Washington, D.C., 1969).
43. USAEC, *Operating Experiences: Reactor Safety*, ROE. no. 69–8.
44. USAEC, *Operating Experiences: Reactor Safety*, ROE. no. 71–72.
45. USAEC, *Operating Experiences: Reactor Safety*, ROE. no. 71–72.
46. USNRC, *Report to the Congress on Abnormal Occurrences*, January-June, 1975, p. A-15.
47. William Casto, "Safety-Related Occurences Reported in April-July, 1971," *Nuclear Safety*, 12 (November-December, 1971): 619.
48. William Casto, "Selected Safety-Related Occurences Reported in November and December, 1973," *Nuclear Safety*, 15 (March-April, 1974): 210–211.
49. *Nuclear Power Alternative*, p. 62.
50. USAEC, *Abnormal Occurrences Report*, p. A-19.
51. William Casto, "Selected Safety-Related Occurrences Reported in September and October, 1975," *Nuclear Safety*, 17 (January-February, 1976): 107–108.
52. USAEC, *Abnormal Occurrences Report*, p. A-7.
53. Ibid., p. A-4.
54. The NRC blamed inadequate quality assurance by TVA as a contributing cause of the Browns Ferry fire (*Nucleonics Week*, March 4, 1976, p. 4). Concurrently, a nuclear industry engineer cited sloppy quality assurance on critical safety items (*Nucleonics Week*, March 11, 1976, p. 4).
55. Any one or a combination of these causes may have led to a representative incident where the discharge from a safety valve impinged on the lifting levers of two other safety valves, cocking them open (USAEC, *Operating Experiences: Reactor Safety*, ROE. no. 71–72).
56. CAC, *Hearings*, Daniel Ford, October 21, 1975, p. 65; and Carl Hocevar, October 21, 1975, pp. 47–54.
57. Ralph Nader testified that, in his opinion, the *RSS* was the centerpiece in the phony public relations effort by the federal government (CAC, *Hearings*, December 2, 1975, p. 11).
58. CAC, *Hearings*, Bernard Rusche, October 22, 1975, pp. 47–51.
59. APS Report, *Reviews of Modern Physics*, p. S-6.
60. CAC, *Hearings*, Primack, October 28, 1975, p. 143.
61. Dr. Teller believes that the alleged safety advantages of the CANDU are the result of the briefer safety analysis done by the Canadians. He notes that U.S. reactor safety analyses are four feet high on the average; the Canadian reports are four inches high. Subjected to the same level of scrutiny, the CANDU reactor may be found to have safety complications similar to the U.S. light water reactors (CAC, *Hearings*, October 21, 1975, p. 15). A General Electric witness also commented that the CANDU reactor is not subject to the same safety criteria as U.S. reactors today (CAC, *Hearings*, A. Philip Bray, October 21, 1975, p. 109). However, ERDA is now considering licensing CANDU

reactors in this country because they conserve uranium and do not need enriched fuel (*Nucleonics Week*, August 7, 1975, p. 3).

62. See note 65. See also note 11, "Rise and Fall of Nuclear Power;" note 13, "Uneconomics of Nuclear Power"; Lovins and Price, *Non-Nuclear Futures*, p. 48.

63. The federal government audits the design and implementation of each electric company's quality assurance program, but does not itself inspect manufacture, installation, or operation of nuclear plant components. Responsibility for component and system quality assurance is given to the power company, which usually assigns it, in turn, to the reactor manufacturer and architect-engineer during the design, installation, and construction phases. During none of these phases is reliability contractually imposed. See U.S. Senate, Committee on Government Operations, *Hearings*, March 12, 1974; USAEC Triner memo cited in note 2, "Rise and Fall of Nuclear Power;" Lewis, *Nuclear Power Rebellion*, p. 131.

64. See page 109 in "Uneconomics of Nuclear Power." Data was derived from *Operating Units Status Reports*, NUREG-0020, issued monthly by NTIS and published by USNRC. See also *A Summary of Abnormal Occurrences Reported to the AEC during 1973* (Washington, D.C.:USAEC, 1974); and "Atomic Lemons," *Wall Street Journal*, May 3, 1973. Comprehensive summary of reactor reliability debate is CAC, *Reassessment*, pp. 61–65.

65. Norman Rasmussen, "Reactor Safety: Real Probabilities," *Combustion*, 45 (June, 1974): 10. See also "Public Safety and Nuclear Power," testimony by Henry W. Kendall prepared for the National Nuclear Energy Debate, *Hearings* before the House Committee on Interior and Insular Affairs, Subcommittee on Energy and the Environment, mimeographed reprint (Cambridge: UCS, April 29, 1975). Both sides of the debate suffer from the fact that nuclear plants have been on-line for only a small portion of their total operating lives. Consequently, the data base is inadequate both for those who point to industry's perfect safety record so far, and for those who predict that plants will continue to demonstrate unsatisfactorily low reliability. Far more persuasive are arguments that identify infinite accident possibilities due to human fallibility and that point to overreliance on nuclear. That a spectrum of failures due to human error was provided for in the RSS does not mean that allowance was made for *all possible* human and common-mode failures. For example, at no point in the draft RSS were the probability and consequences of a fire resulting from a workman's candle identified. If the Browns Ferry fire had *not* occurred to lend credence to this chain of events, the RSS staff would have refused to consider this and other "extremely unlikely" accidents. Their position would have been, justifiably, that all situations simply could not be provided for and that to consider the possibility of a fire started by a candle would open the door for consideration of an infinite number of variables. It is precisely on this point that nuclear opponents rest their case: that there are, indeed, an *infinite* number of ways that a nuclear accident can occur, and that the RSS represents an attempt to reassure Congress and the public with a study that considers only a *limited* number of variables.

The argument against reliance on nuclear power points to the economic consequences of one major accident or terrorist attempt which would bring home vividly to all citizens and the Congress both the vulnerability of this technology and its capacity for damage. A reasonable man might foresee that either citizens, Congress or technicians would demand that all plants be derated, shut down or expensively retrofitted after an accident involving only one plant. This would severely disrupt electric power generation on a national scale if a significant portion of America's electricity depended on nuclear fission. Coal- and oil-fired plants are far less vulnerable to generic shutdowns, for an accident involving a fossil-fueled plant is much less likely to lead to catastrophic damage capable of outraging the public. See CAC *Hearings*, Ralph Nader, December 2, 1975, pp. 2–3; Novick, *Electric War*, p. 313.

Insurance and Nuclear Power Risks

1. USAEC, *Theoretical Possibilities and Consequences of Major Accidents in Large Nuclear Power Plants*, WASH-740 (Washington, D.C., March, 1957); Clifford Beck et al. materials pertaining to revision of WASH-740, including meeting minutes, memoranda, and letters, 1965–1966, USNRC Public Document Room Docket No. 144–18; USNRC, *RSS*, WASH-1400, Appendix VI. See also Webb, *Accident Hazards*.
2. Renewal of Price-Anderson Act by Congress, 1975. See JCAE, *Hearings*, 93rd Cong., "Possible Modification or Extension of the Price-Anderson Insurance and Indemnity Act," H.R. 15323 (January 31 and March 27 and 28, 1974; See also JCAE, 94th Cong., 1st Sess., *Hearings* on H.R 8631: "To Amend and Extend the Price-Anderson Act," September 23 and 24, 1975, Washington, D.C.: USGPO, 1975.
3. Laurie Rockett et al., *Issues of Financial Protection in Nuclear Activities*, Legislative Drafting Research Fund, Columbia University, 1973. Compare with Harold Green, "Nuclear Power: Risk, Liability and Indemnity," *Michigan Law Review*, 71 (January, 1973): 479–510. The nuclear industry is explicitly unwilling to assume full liability; see CAC, *Hearings*, H. Arnold, October 22, 1975, pp. 54–86, where Arnold dodges the question several times and finally concedes this point when pressed by Assemblyman Charles Warren.
4. JCAE, *Selected Materials on Atomic Energy Indemnity and Insurance Legislation*, Joint Committee Print 93–2 (March, 1974). Also Congressional Research Service Report (1972) on liability limits.
5. Green, "Nuclear Power," pp. 509–510.
6. USAEC, *The Safety of Nuclear Power Reactors and Related Facilities*, WASH-1250, Washington, D.C., 1973, p. 8–5, quoting H. P. Green, "Safety Determinations in Nuclear Power Licensing: A Critical View," *Notre Dame Lawyer*, 43 (June, 1968). Excellent summary of debate in CAC, *Reassessment*, pp. 82–93.
7. USAEC, WASH-1250, p. 8–5.
8. CAC, *Reassessment*, pp. 82–83.
9. CAC, *Hearings*, November 20, 1975, p. 57.
10. S. David Freeman et al., *A Time to Choose: America's Energy Future* (Cambridge: Ballinger, 1974), pp. 222–223.

Shakeup at General Electric

1. L. C. Koke et al., "BWR/6: Optimizing the Design of BWR Power Plants," paper presented at the American Power Conference, April 18–20, 1972.
2. General Electric—Nuclear Energy Division, "Mark III: Helping Contain Nuclear Plant Construction Costs," p. 3 of brochure describing BWR/6 and Mark III containment concept. Cutaway rendering in the brochure shows *three solid concrete platforms* above the suppression pool and surrounding the drywell. Unidentified equipment appears installed on the platforms.

3. General Electric cutaway rendering, "Mark III Containment," GEZ-4385A, apparently a later design, shows *metal gratings substituted for the concrete platforms* referred to in note 2. See also note 5, *PEPCO memorandum.*

4. Conversation with Dale Bridenbaugh, October 10, 1976, regarding share of cost overruns due to design changes. For data on construction cost coverruns, see Bupp et al., "Economics."

5. Paul Dragoumis, memorandum to W. Reid Thompson and Ellis T. Cox, "Report of the Nuclear Engineering and Construction Group for the Period January 21–25, 1974" (February 1, 1974). This memorandum and attachments are the primary sources for material in "Shakeup at General Electric" pertaining to Mark III containment problems. Other sources confirming certain details of the Dragoumis memo were Dale Bridenbaugh, Gregory Minor, and Richard Hubbard, MHB Technical Associates, Palo Alto, California.

6. Dragoumis, February 1, 1974, memorandum.

7. "The Silent Bomb: Another Problem for General Electric?" *Palo Alto Times,* February 25, 1976. Although General Electric and its customers were, as Dragoumis claims in the *Times* article, sincerely trying to develop a concept, the events alleged in Dragoumis's memorandum and General Electric's admitted changes reveal that not all aspects of the Mark II design had been thoroughly studied prior to marketing.

8. See note 5.

9. See note 7.

10. Dragoumis memorandum, February 1, 1974, pp. 3–4.

11. *Nuclear News,* June, 1975, p. 38. The *Nuclear News* article is misleading in its reference to "new data" and to "previously unidentified hydrodynamic forces" in the second paragraph. According to the Dragoumis memo, GE was aware of the suppression pool swell problem as early as June of 1973, and PEPCO's nuclear engineering group rebuked GE at that time for not having investigated this problem more thoroughly. If NRC's request for information resulted from the data referred to by Dragoumis, then at the time of the *Nuclear News* article, some of these data were at least two years old. The hydrodynamic forces were "unidentified" previous to December of 1973, not spring of 1975, as the article implies. If the NRC was *not* aware of this problem until shortly before the spring, 1975, announcement, then several agencies were remiss in their responsibility to report it sooner. If AEC *was* aware of this problem earlier, why did it delay in initiating the investigation (consider that the problem raised very serious questions of containment-system integrity regarding Mark I, II, and III models)? It is known that the Dragoumis memo was leaked during the second week of February, 1974, and that several copies of it were circulating around Washington, D.C., during the spring and summer of that year. The suspicion arises that Dragoumis' memo may well have played an important part in forcing the federal government to review the problem publicly. Whether it did or not is a subordinate question to whether the nuclear industry blamed Dragoumis for the events that led to the leaking of his memo.

12. See testimony of the former General Electric engineers in Appendix A.

13. Ibid., and conversations with Bridenbaugh, Minor, and Hubbard. Other confirmation is note 8, "Incident at Browns Ferry."

14. Conversations with Bridenbaugh, Minor, and Hubbard, October 20, 1976.

15. *Nuclear News,* January, 1977, pp. 18, 105. *San Francisco Chronicle,* December 14, 1976, p. 5.

16. CAC, *Reassessment*, p. 39, indicates that the LOFT tests will not begin until 1977. No other full-scale, operational tests of ECCS have been accomplished, principally because they would entail synthesizing conditions that might lead to a core meltdown. See also Daniel Ford and Henry Kendall, *An Assessment of the Emergency Core Cooling Systems Rulemaking Hearings* (Cambridge: UCS and San Francisco: FOE, 1974). For the nuclear industry's view of the need for and usefulness of full-scale safety tests, see CAC *Hearings*, W. H. Arnold, October 22, 1975, pp. 60–63.

17. Robert Gillette, "Nuclear Fuel Reprocessing: GE's Balky Plant Poses Shortage," *Science*, August 30, 1974, pp. 710–711, which identifies General Electric errors that led to the Morris plant failure as "overconfident engineering" and "failure to test the new (reprocessing concept) . . . fully in intermediate stages." Gillette suggests that it may be that "competitive pressures are forcing engineers to scale up the size and increase the efficiency of new technologies much too rapidly" (ibid.). This is a crucial point. General Electric's apparent haste in constructing the Morris plant may also be seen in the way it handled the Mark III design concept. According to Bertram Wolfe, a GE manager, the company underestimated the difficulty of going from a laboratory environment to a remotely operated production plant (ibid.). As with the Mark III, GE appears to have been in a rush to demonstrate industry leadership in an important new technology. Competitive pressures from without and top-management marketing orientation from within may have precipitated what Gillette refers to as "overconfident engineering and a failure to test the new (Morris) process fully in intermediate stages" (ibid.).

18. Lash et al., *Citizens' Guide*, p. 30.

A Moment of Truth

1. Jim Harding, "GE Engineers Resign, Will Work to Pass Nuclear Initiative," *Not Man Apart*, March, 1976. The material pertaining to the three former General Electric engineers in this chapter is an edited version of Harding's article.

2. Conversation with Dale Bridenbaugh, October 15, 1976.

3. JCAE, *Hearings*, 94th Cong. 2nd Sess., "Investigation of Charges Relating to Nuclear Reactor Safety," February 18–24 and March 2–4, 1976, 2 vols. (Washington, D.C.: USGPO, 1976).

4. "On Being Our Own Best Critic," *Nuclear News*, March, 1976, p. 29.

5. Ibid., pp. 46–47.

6. Ibid., p. 47.

7. Ibid., p. 46. See also "Nothing New, says NRC of 'GE Three' Testimony," *Nuclear News*, April, 1976, pp. 35ff.

8. JCAE, *Hearings*, February 18, 1976. See testimony in Appendix A, pp. 000–000, and also CAC, *Hearings*, Daniel Ford, October 21, 1975, pp. 79–80.

9. "Pressure Tactic Suspected in Nuclear Resignations," *Palo Alto Times*, February 12, 1976.

10. Ibid., February 14 and 20, 1976. All major California newspapers carried the story: "Nuclear Shocker," February 3, 1976, and "GE Colleagues Assail Nuclear Rebels' Story," February 12, 1976, *San Francisco Chronicle*.

11. Lovins and Price, *Non-Nuclear Futures*, p. 12.

12. Save, perhaps, manned space flight. In *The New Yorker* (November 11, 18, 1972) and in *Thirteen: The Flight that Failed* (New York: Dial, 1973), H. S. F. Cooper, Jr., gives an account of the impact of human fallibility on Apollo 13.

13. Ebbin and Kasper, *Citizen Group Uses of Scientific and Technological Information*, pp. 5–8. See also Common Cause, *Stacking the Deck: A Case Study of Procedural Abuses by the Joint Committee on Atomic Energy*, Washington, D.C., 1976.
14. "The Rising Tide of Citizen Action," in Berger, *Nuclear Power*, ch. 13. Also, generally, Lewis, *The Nuclear-Power Rebellion*.
15. U.S. Senate, Committee on Government Operations, *Hearings*, March 12, 1974, pp. 217–226.

The Atomic Industrial Complex

1. Berger, *Nuclear Power*, pp. 339–356.
2. Fritz Heimann, "How Can We Get the Nuclear Job Done?" in Arthur Murphy, ed., *The Nuclear Power Controversy* (Englewood Cliffs, N.J.: Prentice-Hall, 1976), p. 98. Subsequently, Heimann suggests that the problem of public acceptance may be solved when "political leaders whom the public trusts endorse the conclusion that the benefits of nuclear power clearly outweigh the risks" (ibid.). No mention is made of the moral problems associated with imposing the radwaste custodial and risk burdens on future generations. Not surprisingly, the nuclear industry rarely initiates discussion of these broader issues (see Alvin Weinberg, "The Moral Imperatives of Nuclear Energy," *Nuclear News*, December, 1971) and appears to dismiss them when raised (Novick, *Electric War*, p. 212, and CAC, *Hearings*, H. Arnold, October 22, 1975, p 77).

Nevertheless, a number of prominent scientists, including Albert Einstein, have declared that a national commitment to nuclear technology is subject to approval by the people. If Heimann is correct that the public is easily scared, then public approval of nuclear power may be secured only by the exercise of political leadership he proposes and may be as contrived as voter rejection of state initiatives following massive public relations campaigns to secure this result, funded by the nuclear industry. Perhaps the most significant fact is that all recent initiatives and referendums by which the issues raised by nuclear power have been presented to the public have been organized initially by citizen groups, never by the nuclear industry. For rigorous treatment of nuclear decision-making and public policy problems, see Lovins and Price, *Non-Nuclear Futures*, ch. 8, especially pp. 47ff; Brown, "Health, Safety and Social Issues," in *California Nuclear Initiative*, pp. 127–136; and Weinberg, "Social Institutions and Nuclear Energy," *Science*, July 7, 1972, pp. 27–34; compare with Weinberg, *Bulletin of the Atomic Scientists*, 22 (1966):9–13. The industry position is presented in Robert Campana and Sidney Langer, eds., *Nuclear Power and the Environment: Questions and Answers* (Hinsdale, Ill.: American Nuclear Society, 1975), ch. 13. A much broader approach is William Lowrance, *Of Acceptable Risk: Science and the Determination of Safety* (Los Altos, Calif.: Kaufmann, 1976). Jacques Ellul, *The Technological Society* (New York: Vintage, 1964) is a thorough, classical study.
3. The *San Francisco Examiner* article by John Hall was based on attendance at the Senate hearings and on the following documents: U.S. Senate, Select Committee on Presidential Campaign Activities, *Hearings*, 93rd Cong., 1st Sess., "Presidential Campaign Activities of 1972," phase III, book 13, testimony of Claude Wild, Jr., November 14, 1973, pp. 5459–5481; *Executive Session Hearings*, book 25, testimony of Carl Arnold, January 28, 1974, pp. 12019–12090 (Washington, D.C.: USGPO, 1974); *Securities and Exchange*

Commission vs Gulf Oil Corp. and Claude C. Wild, Jr., civil action no. 75–0324, U.S. District Court for the District of Columbia; *Report of the Special Review Committee of the Board of Directors of Gulf Oil Corporation*, John J. McCloy, chairman (Pittsburgh: Gulf Oil Corporation, December 30, 1975). See also John Brooks, "Annals of Business: Funds Gray and Black," *The New Yorker*, August 9, 1976, pp. 28–44; and Wyndham Robertson, "The Directors Woke Up Too Late at Gulf," *Fortune*, June, 1976, pp. 121–124, 206–210.

4. A December, 1976, study by Common Cause concluded that twelve members of the JCAE during the 94th Congress "represented six states which received over 50 percent of all ERDA funding in the fiscal years 1976 and 1977—nearly $5.5 billion." See Common Cause, *Stacking the Deck: A Case Study of Procedural Abuses by the Joint Committee on Atomic Energy* (Washington, D.C., 1976) for other charges, including manipulation of hearings and witnesses, and avoidance of proliferation issues. The House Democratic Caucus stripped the JCAE of its power to write legislation on December 8, 1976. Nuclear supporters in Congress were unable to sidetrack this and other tactics designed to abolish the JCAE's long-standing stranglehold on nuclear legislation. One pro-nuclear congressman, Rep. John Anderson (R-Ill.) hoped to salvage some of the JCAE's machinery by establishing a single Energy Committee (*Energy Daily*, December 13, 1976, pp. 1–2). Another view of Anderson's proposal is that the AIC hopes to reestablish central control over Congress: an Energy Committee would become a *de facto* JCAE under a new name once the flow of "political contributions" from the AIC to Energy Committee members resumed.

5. While the system for laundering Gulf funds was being set up in 1960, W. K. Whiteford, chief executive officer and chairman of the board of Gulf Oil, met with two Gulf vice-presidents to discuss the operation that ultimately channeled millions of dollars through Wild to scores of Washington officials. "Whiteford told them [the vice-presidents] that he had talked to top management of some other major oil companies and learned that all of them had set up arrangements similar to that which Whiteford planned." Quote from *Report of the Special Review Committee of the Board of Directors of Gulf Oil Corporation*, p. 35.

6. Berger, *Nuclear Power*, pp. 165–170, for industry-government-banking interlocks. Also *Critical Mass*, March 1976, for a series of articles on this topic.

7. John Berger, "Common Cause Study Reveals Conflicts of Interest," *Not Man Apart*, mid-September, 1976. p. 14. Berger's article constitutes the portion of "The Atomic Industrial Complex" which pertains to the Common Cause report. California Assemblyman Lawrence Kapiloff alleges that NRC is well staffed with industry appointees; see CAC, *Hearings*, Kapiloff, October 22, 1975, p. 78.

8. John Berger, "Uranium Pricefixing: FOE Reveals Secret Uranium Cartel," *Not Man Apart*, October, 1976, p. 8. Again, Berger's article is reprinted in this chapter with minor editing.

9. "Westinghouse Sues 29 Uranium Suppliers," *Nuclear News*, November, 1976, p. 29.

10. List of defendants published in *Nuclear News*, December, 1976, p. 41.

11. See note 8.

12. Berger, *Nuclear Power*, ch. 8.

13. Whiteside, "A Countervailing Force," pp. 47–48. See also Ralph Nader and Mark Green, *Corporate Power in America* (New York: Grossman, 1973).
14. Whiteside, "A Countervailing Force," p. 48.
15. Berger, *Nuclear Power*, ch. 8.
16. Mark Sharefkin, *The Fast Breeder Reactor Decision: An Analysis of Limits and the Limits of Analysis*, Report to the Joint Economic Committee (Washington, D.C.: Resources for the Future, 1976). According to Sharefkin, the long-term storage of radioactive wastes entails substantial costs that "will be transferred inequitably and incompensably onto future generations." Because present generations, who will benefit from nuclear power, have not developed a system for compensating later, "unluckier" generations, "the cost-benefit criterion loses both its rigorous basis and its aura of fairness. This is a dilemma that cannot be resolved by cost-benefit analysis." See also James Gardner, "Discrimination Against Future Generations: Equal Protection and the Concept of Time," article in preparation; and Van Rensselaer Potter, *Bioethics: Bridge to the Future* (Englewood Cliffs, N.J.: Prentice-Hall, 1971). For a rigorous study of the problem of gaining legal standing for natural objects and, by analogy, future generations, see Christopher Stone, *Should Trees Have Standing? Toward Legal Rights for Natural Objects* (Los Altos, Calif.: Kaufmann, 1974).
17. Vance Packard, *The Hidden Persuaders* (New York: McKay, 1958). See also Daniel Boorstin, "Tomorrow: The Republic of Technology," *Time*, January 17, 1977, pp. 36–38.

Dawning of the Nuclear Age

In addition to the references cited below, sources for this article were issues of the *San Francisco Chronicle*, *Nucleonics Week*, *Environment* (then *Nuclear Information*), *Nuclear News*, and *The New York Times* between 1959 and 1964. Additional sources were materials filed with the federal government by PG&E; see USAEC Docket 50–205, "Preliminary Hazards Analysis."

1. The official history of the USAEC from 1939 to 1952 is published by the Pennsylvania State University Press in two volumes, with a third in preparation: Richard Hewlett and Oscar Anderson, Jr., *The New World, 1939–1946* (1962); and Richard Hewlett and Francis Duncan, *The Atomic Shield, 1947–1952* (1969). Both are available through NTIS. For the period 1953 to 1976, see Metzger, *Atomic Establishment;* Novick, *Electric War;* and C. Allardice and E. Trapnell, *The Atomic Energy Commission* (New York: Praeger, 1974).
2. USAEC, *Theoretical Possibilities and Consequences of Major Accidents in Large Nuclear Power Plants*, WASH-740 (Washington, D.C.: USGPO, March, 1957). This study was updated by the Brookhaven National Laboratory in 1965; estimated consequences of a large reactor major accident were severe enough to persuade USAEC officials to suppress the study until 1973. See CAC, *Hearings*, Daniel Ford, October 21, 1975, pp. 74–77 and note, 1, "Insurance and Nuclear Power Risks." Also see Henry Kendall and Sidney Moglewer, *Preliminary Review of the AEC Reactor Safety Study* (Cambridge: UCS and San Francisco: Sierra Club, 1974), pp. 24, 101–109, and appendix C, for detailed review of the WASH-740 update.
3. JCAE, 84th Cong., 2nd Sess., *Hearings* on government indemnity (Washington, D.C.: USGPO, 1956).

4. For a complete report on PG&E operation and plants, see Charles Komanoff et al., *The Price of Power: Electric Utilities and the Environment* (Cambridge: MIT Press, 1974), pp. K-1 to K-31. *Sierra Club Bulletin* issues since 1960 carry the account of conflicts between PG&E and California environmental organizations.

5. *Northern California Association to Preserve Bodega Head vs Public Utilities Commission*, 61 Cal. 2d 126 (1964).

6. Lindsay Mattison and Richard Daly, "Bodega: The Reactor, The Site, The Hazard," *Nuclear Information*, 6 (now *Environment*) (April, 1964):1–12. See also editorial, *Nuclear Information*, 7 (November, 1964).

7. California Public Utilities Commission, *Hearings*, Bodega Bay Nuclear Plant, 1962 et seq.

8. Ibid.

9. Ibid.

10. Pierre Saint-Amand, *Geologic and Seismologic Study of Bodega Head*, Northern California Association to Preserve Bodega Head and Harbor (1963). See also Julius Schlocker and Manuel Bonilla, U.S. Department of the Interior, Geological Survey, *Engineering Geology of the Proposed Nuclear Power Plant Site on Bodega Head, Sonoma County, California*, TEI-884 (December, 1963).

11. USAEC, Division of Reactor Licensing, *Report on Siting, Bodega Head Nuclear Plant* (October, 1964).

12. "PG&E Decision—No A-Plant at Bodega Bay," *San Francisco Chronicle*, October 31, 1964, pp. 1, 10.

How Safe Is Diablo Canyon?

1. Clarence Hall, Jr., "San Simeon—Hosgri Fault System, Coastal California: Economic and Environmental Implications," *Science*, December 26, 1975.

2. David Perlman, "New Delay on Atom Plant Opening—Quake Hazard," *San Francisco Chronicle*, January 16, 1976, p. 5.

3. See notes for "Dawning of the Nuclear Age," especially note 5.

4. H. F. Perla, "Power Plant Siting Concepts in California," *Nuclear News*, October, 1973, pp. 47–51. Compare S. T. Algermissen, *Seismic Risk Studies in the U.S.*, U.S. Department of Commerce, Environmental Science Services Administration, Coast and Geodedic Survey (January 14, 1969).

5. R. H. Morris, *Humboldt Bay Power Plant Site—Status of Geological Review*, draft report, U.S. Geological Survey (February 16, 1973).

6. Statement of Frederick A. Slautterback, March 8, 1976, with three attachments.

7. "Complaints by a QA Man about Some Nuclear Plant Electrical Cable," *Nucleonics Week*, July 1, 1976, p. 11.

8. Correspondence and telephone conversations with Slautterback, March through August, 1976.

9. Ibid. See also Ford et al., *Browns Ferry: The Regulatory Failure*, which identified TVA and federal government confusion over cable separation criteria as an important contributing cause of the Browns Ferry fire. Quality assurance deficiencies at the plant (see "Browns Ferry: Significant QA Deficiencies Blamed for Plant Fire," *Nuclear News*, April, 1976, pp. 46–51) revealed during investigation of the fire, when viewed together with evidence of performance shortfall (see note 13, "Uneconomics of Nuclear Power," and Table 2, p.109), and Slautterback's allegations, confirms testimony by the three former General

Electric engineers that NRC has not regulated effectively (see Appendix A). For a thorough review of nuclear industry QA problems, see note 7, "We Almost Lost Detroit."

10. Memorandum by Dale Bridenbaugh, "Earthquake Hazards at the Diablo Canyon Nuclear Generating Station," MHB Technical Associates, Palo Alto, Calif., February 16, 1977.

Isotopes and Hogwash

Primary sources for this article were reports on the Monticello controversy published between 1967 and 1971 in *Nucleonics Week, Nuclear News, St. Paul Pioneer Press,* and *Minneapolis Tribune,* in addition to the references cited in the notes that follow.

1. National Council on Radiation Protection and Measurements, *Basic Radiation Protection Criteria,* NCRP Report no. 39 (Washington, D.C., 1971); Federal Radiation Council, *Background Material for the Development of Radiation Protection Standards—Report No. 1* (Washington, D.C.: U.S. Department of Commerce, 1960); Federal Radiation Council *Staff Reports No. 1* (May, 1960), *No. 2* (September, 1961). See also JCAE, *Hearings* on radiation standards, June, 1962.

2. JCAE, *Hearings,* June 22, 1971. But see Metzger, *Atomic Establishment,* chs. 3–5; Gofman and Tamplin, *Poisoned Power.*

3. USAEC, *The Safety of Nuclear Power Reactors,* WASH-1250, pp. 4–49 and 8–15 to 8–20. See also note 2.

4. JCAE, 91st Cong., *Hearings* on environmental effects of producing electric power, 1st sess. (1969) and 2nd sess. (1970). See also John Gofman and Arthur Tamplin, "Radiation: The Invisible Casualties," *Environment,* 12 (April, 1970):12–19, 49; and notes, "Poisoned Power."

5. Komanoff et al., *Price of Power,* pp. I-1 to I-18.

6. Dean Abrahamson, "Ecological Hazards from Nuclear Power Plants," in M. T. Farvar and J. Milton, eds., *The Careless Technology* (New York: Natural History Press, 1972). See also Abrahamson, "A Nuclear Moratorium," *Environment,* 15 (June, 1973): 27–28.

7. Ernest C. Tsivoglou, *Radioactive Pollution Control in Minnesota,* final report to Minnesota Control Agency, March, 1969.

8. Correspondence from Robert H. Engels to Dr. Howard A. Anderson, chairman, Minnesota Pollution Control Agency, April 2, 1971.

9. *Minnesota vs Northern States Power Co.,* 447 F. 2nd 1143 (8th Cir. 1971) *aff'd per curiam* 405 U.S. 1035 (1972). The U.S. Supreme Court held that the Atomic Energy Act of 1954 preempted state laws regulating effluents from nuclear plants. Gofman, "Nuclear Power and Ecocide: An Adversary View of New Technology," *Bulletin of the Atomic Scientists,* 27 (September, 1971): 28–32.

10. Ibid. Gofman, "Time for a Moratorium," *Congressional Record,* 119 (March 15, 1973).

11. JCAE, *Hearings* on the environmental effects of producing electric power, 1969–1970. See also Harry Foreman, ed., *Nuclear Power and the Public* (Minneapolis: University of Minnesota Press, 1970) for papers presented at Minnesota symposium following the Monticello case during which the major issues were aired. Gofman, "The Cancer and Leukemia Consequences of Medical X-Rays," *Osteopathic Annals,* 3 (November, 1975): 24–42.

12. Proposed amendment to Code of Federal Regulations, 10 CFR 50, "Licensing of Production and Utilization Facilities—Light-Water-Cooled Nuclear Power Reactors," *Federal Register*, 36 (June 9, 1971): 11113. See also JCAE, *Hearings*, June 22, 1971.

Poisoned Power

1. Ralph and Mildred Buchsbaum, *Basic Ecology* (Pittsburgh: Boxwood Press, 1957), p. 20.

For a thorough review of the 1969–1970 Gofman-Tamplin controversy, see Lewis, *Nuclear-Power Rebellion.* Compare Metzger, *Atomic Establishment,* pp. 276–278, note 17. Rebutted by USAEC in *Safety of Nuclear Power Reactors,* WASH-1250, pp. 8–15 to 8–20, with extensive citations. These publications carry the controversy through 1973. There are no indications that it will be resolved. The following bibliography of Gofman's and Tamplin's publications may be helpful.

- Gofman and Tamplin, "Low Dose Radiation, Chromosomes and Cancer," paper presented at the Nuclear Science Symposium, Institute of Electrical and Electronics Engineers, San Francisco (October 29, 1969).
- Gofman and Tamplin testimony, JCAE, *Hearings* on environmental effects of producing electric Power, 91st Cong., 1st Sess. (October-November, 1969), and 2nd Sess. (January-February, 1970).
- Gofman and Tamplin, *Population Control through Nuclear Pollution* (Chicago: Nelson-Hall, 1970).
- Gofman and Tamplin, "Radiation: The Invisible Casualties," *Environment,* 12 (April, 1970):12–19, 49.
- Gofman and Tamplin, testimony, State of Pennsylvania, Senate Committee to Investigate Construction of Atomic Generating Plants in Pennsylvania, *Hearings* (August 20, 1970).
- Gofman and Tamplin, "Epidemiologic Studies of Carcinogenesis by Ionizing Radiation," *Proceedings* of the Sixth Berkeley Symposium on Mathematical Statistics and Probability (Berkeley: University of California Press, 1971).
- Gofman and Tamplin, "Radiation as an Environmental Hazard," Symposium on Fundamental Cancer Research, Houston, 1971.
- Gofman and Tamplin, "The Question of Safe Radiation Thresholds for Alpha-Emitting Bone Seekers in Man," *Health Physics,* 21 (July, 1971): 47–51.
- Gofman, *The Cancer Hazard from Inhaled Plutonium* and *Estimated Production of Human Lung Cancers by Plutonium from Worldwide Fallout* (Yachats, Ore.: Committee for Nuclear Responsibility, 1975).
- Gofman, "Radiation Doses and Effects in a Nuclear Power Economy," *Congressional Record,* 122 (May 6, 1976).

Clean Energy Alternatives

1. Statement before a hearing panel on energy policy, Washington, D.C., co-chaired by Senator Mike Gravel and Representative Hamilton Fish, November 17, 1975. Dr. Kistiakowsky is emeritus professor of chemistry, Harvard University. See also George Kistiakowsky, "Nuclear Power: How Much is Too Much?" in Arthur Murphy, ed., *The Nuclear Power Controversy,* ch. 6.
2. Marc Ross and Robert Williams, "Assessing the Potential for Fuel Conservation," forthcoming in *Technology Review,* 1977; also in CAC, *Hearings,* December 2, 1975, pp. 220–258. See also Freeman, *A Time to Choose,* and David Goldstein and Arthur Rosenfeld, "Conservation and Peak Power: Cost

and Demand," CAC, *Hearings*, pp. 188–207. Excellent rebuttal of industry skepticism re conservation is Lester Lees, "Energy Conservation: Will it Work?" *Engineering and Science*, April-May, 1975, pp. 3 et seq.

3. Freeman, *A Time to Choose*, p. 6.

4. Lee Schipper, *Explaining Energy: A Manual of Non-Style for the Energy Outsider Who Wants In!* Report LBL-4458 (Berkeley, Calif.: Lawrence Berkeley Laboratory, Energy and Environment Division, 1975).

5. Lee Schipper, *Energy Conservation: Its Nature, Hidden Benefits and Hidden Barriers* (Berkeley: University of California, Energy Resources Group, 1975). This report also appears in *Energy Communications* (1976) and in *Annual Review of Energy*, 1 (Palo Alto, Calif., 1976).

6. "Energy: The Real Debate, Management's View," *Orange Disk* (Pittsburgh: Gulf Oil Corporation, July-August, 1975). See also "Gas Spokesman Sees Cut in Oil Use Slowing Economy," *The New York Times*, December 11, 1974.

7. "Energy Growth . . . And Waste," editorial, *The New York Times*, December 11, 1974.

8. "Double-Duty Steam Can Save Electricity, Study Finds," *The New York Times*, October 10, 1975. See also "Waste Industrial Steam Could Generate Electricity to Spare," *Not Man Apart*, November, 1975.

9. Charles Berg, "Conservation in Industry," in Philip Abelson, ed., *Energy Use, Conservation and Supply* (Washington, D.C.: AAAS, 1974), p. 30.

10. American Institute of Architects, *Energy and the Built Environment* (Washington, D.C., 1974), and *A Nation of Energy-Efficient Buildings by 1990* (Washington, D.C., 1974).

11. "Transmission Lines: Three Ways to Carry Electricity," in Allen Hammond et al., eds., *Energy and the Future* (Washington, D.C.: AAAS, 1973).

12. Philip Steadman, *Energy, Environment and Building* (Cambridge, Eng.: Cambridge University Press, 1975).

13. W. Seifert et al., *Energy and Development—A Case Study*, MIT Report No. 25 (Cambridge: MIT Press, 1973).

14. NSF/NASA Solar Energy Panel, *Solar Energy as a National Energy Resource*, NSF/RA/N–73–001, NSF Grant GI-32488 (Washington, D.C.: USGPO, 1973), pp. 1, 3.

15. Statement of Edwin Rothschild, Consumers Electric Power Corporation before Federal Energy Administration, Special Advisory Committee, hearings on consumer affairs, April 17, 1975.

16. U.S. Senate, Select Committee on Small Business, *Hearings*, "The Role of Small Business in Solar Energy Research, Development and Demonstration," interim report (Washington, D.C.: USGPO, October 7, 1975).

17. MITRE Corporation, *Solar Energy Research Program Alternaitves* (McLean, Va., December, 1973).

18. U.S. Senate, 94th Cong., *Hearings*, "Role of Small Business," testimony by Donald Craven, former FEA administrator, "The Question of Large Company Domination of the Solar Energy Industry," October, 1975. See also *People and Energy* (Washington, D.C.: Center for Science in the Public Interest, April, 1976), which reports that large oil, automobile, and aerospace companies are aggressively buying up patent rights on solar systems, and that large U.S. corporations receive the majority of government funds allocated to solar research and development. Also John Keyes, *The Solar Conspiracy* (Dobbs Ferry, N.Y.: Morgan & Morgan, 1975).

Index

After receiving degrees from Yale and Stanford Universities, Peter Faulkner served as a flying officer with the Strategic Air Command. Upon returning to civilian life, he worked with the Behavioral Research Laboratories as a consultant and with the Advanced Technology Group, Ampex Corporation, as a technical associate. As a systems application engineer for Nuclear Services Corporation, he gained firsthand experience at several nuclear plants for which he designed and developed quality assurance programs, preoperational test procedures, and management information systems. He was fired from this position in 1974, after presenting a paper to a subcommittee of the U.S. Senate Committee on Government Operations in which he alleged widespread mismanagement in the nuclear industry and recommended tighter quality assurance and management controls by the federal government. He is currently a nuclear consultant with Friends of the Earth. He has co-authored several technical papers and is a contributing author to the text, *Computers and the Problems of Society*, published in 1972. Mr. Faulkner lives in Palo Alto, California.

Paul R. Ehrlich is Professor of Biological Sciences at Stanford University. His research in theoretical and experimental population biology includes studies of coevolution and the dynamics and genetics of natural populations. Recent books by Professor Ehrlich are *The Population Bomb* and *How to Be a Survivor* (with R. L. Harriman). He coauthored *The End of Affluence* and *Population/Resources/Environment* with his wife, Anne H. Ehrlich, and has published numerous papers and articles since 1959 while engaged in teaching and research at Stanford, where he is director of graduate studies in biological sciences. He is honorary president and founder of Zero Population Growth.

Friends of the Earth (FOE) was founded in 1969 by David R. Brower, former executive director of the Sierra Club. FOE is a nonprofit membership organization, streamlined for litigation and legislative activity in the United States and abroad, aimed at protecting the environment and preserving remaining wilderness areas. In addition to its main office in San Francisco, FOE maintains local branches in sixty U. S. locations and international offices in twelve countries.